Hybrid Advanced Techniques for Forecasting in Energy Sector

Hybrid Advanced Techniques for Forecasting in Energy Sector

Special Issue Editor
Wei-Chiang Hong

MDPI • Basel • Beijing • Wuhan • Barcelona • Belgrade

MDPI

Special Issue Editor
Wei-Chiang Hong
Schoo of Computer Science and Technology,
Jiangsu Normal University
China

Editorial Office
MDPI
St. Alban-Anlage 66
Basel, Switzerland

This is a reprint of articles from the Special Issue published online in the open access journal *Energies* (ISSN 1996-1073) from 2012 to 2013 (available at: https://www.mdpi.com/journal/energies/special_issues/forecasting_energy)

For citation purposes, cite each article independently as indicated on the article page online and as indicated below:

LastName, A.A.; LastName, B.B.; LastName, C.C. Article Title. *Journal Name* **Year**, *Article Number*, Page Range.

ISBN 978-3-03897-290-7 (Pbk)
ISBN 978-3-03897-291-4 (PDF)

Contents

About the Special Issue Editor

Wei-Chiang Hong, Jiangsu Distinguished Professor, School of Computer Science and Technology, Jiangsu Normal University, China. His research interests mainly include computational intelligence (neural networks, evolutionary computation) and application of forecasting technology (ARIMA, support vector regression, and chaos theory). In May 2012, his paper was evaluated as a "Top Cited Article 2007–2011" by *Applied Mathematical Modelling* (Elsevier). In August 2014, he was elected to be awarded the "Outstanding Professor Award" by the Far Eastern Y. Z. Hsu Science and Technology Memorial Foundation (Taiwan). In November 2014, he was elected to be awarded the "Taiwan Inaugural Scopus Young Researcher Award—Computer Science", by Elsevier, in the Presidents' Forum of Southeast and South Asia and Taiwan Universities. He was awarded as one of the "Top 10 Best Reviewers" of Applied Energy in 2014, and as one of Applied Energy's "Best Reviewers" in 2016.

Preface to "Hybrid Advanced Techniques for Forecasting in Energy Sector"

Accurate forecasting performance in the energy sector is a primary factor in the modern restructured power market, accomplished by any novel advanced hybrid techniques. Particularly in the Big Data era, forecasting models are always based on a complex function combination, and energy data are always complicated by factors such as seasonality, cyclicity, fluctuation, dynamic nonlinearity, and so on. To comprehensively address this issue, it is insufficient to concentrate only on simply hybridizing evolutionary algorithms with each other, or on hybridizing evolutionary algorithms with chaotic mapping, quantum computing, recurrent and seasonal mechanisms, and fuzzy inference theory in order to determine suitable parameters for an existing model. It is necessary to also consider hybridizing or combining two or more existing models (e.g., neuro-fuzzy model, BPNN-fuzzy model, seasonal support vector regression–chaotic quantum particle swarm optimization (SSVR-CQPSO), etc.). These advanced novel hybrid techniques can provide more satisfactory energy forecasting performances.

This book contains articles from the Special Issue titled "Hybrid Advanced Techniques for Forecasting in the Energy Sector", which aimed to attract researchers with an interest in the research areas described above. As Fan et al. [1] indicate, the research direction of energy forecasting in recent years has concentrated on proposing hybrid or combined models, such as: (1) hybridizing or combining these artificial intelligence models with each other; (2) hybridizing or combining with traditional statistical tools; and (3) hybridizing or combining with superior evolutionary algorithms. Therefore, this Special Issue was interested in contributions to these recent developments (i.e., hybridizing or combining any advanced techniques in energy forecasting). The hybrid forecasting models should be with the superior capabilities over the traditional forecasting approaches, with the ability to overcome some embedded drawbacks, and with the very superiority to achieve significant improved forecasting accuracy.

The 14 articles collected in this compendium all display a broad range of cutting-edge topics in the hybrid advanced technologies. The preface author believes that the applications of hybrid technologies will play an important role in energy forecasting accuracy improvements, such as hybrid different evolutionary algorithms/models to overcome some critical shortcomings of single evolutionary algorithm/models or direct improvements of these shortcomings by innovative theoretical arrangements.

Based on these collected articles, an interesting emergent issue for future research is how to help researchers to employ the proper hybrid technology for different data sets. This is because the most important problem for any analytical model (e.g., classification, forecasting, etc.) is how to capture patterns in the data and apply the learned patterns or rules to achieve satisfactory performance (i.e., the key factor in success is determining how to suitably search for data patterns). However, each model has an excellent ability to capture a specific data pattern. For example, exponential smoothing and ARIMA models focus on strict increasing (or decreasing) time series data (i.e., linear patterns). They even have a seasonal modification mechanism to analyze seasonal (cyclic) change. Due to the use of an artificial learning function to adjust the training rules, artificial neural networks (ANNs) excel only if a historical data pattern has been learned. They lack a systematic explanation of how the accurate forecasting results are obtained. Support vector regression (SVR) can achieve superior performance only if there is a proper parameters determination for the search algorithms. Therefore, it is essential to construct an inference system to collect the characteristic rules to determine the data pattern category.

The next main problem in model development is assigning the appropriate approach to implement forecasting: For (1) ARIMA or exponential smoothing approaches, only their differential or seasonal parameters need to be adjusted. (2) In ANN or SVR models, the forthcoming problem is how to determine the best combination of parameters (e.g., number of hidden layers, units of each layer, learning rate—also called hyper-parameters) to achieve superior forecasting performance. Particularly, for the focus of this discussion, in order to determine the most proper parameter combination, a series of evolutionary algorithms should be employed to test their compatibility with the data pattern. Experimental findings demonstrated that those evolutionary algorithms also had merits and drawbacks. For example, genetic algorithm (GA) and immune algorithm (IA) performed excellently with regular trend data patterns (real numbers) [2,3], SA excelled with fluctuating or noisy data patterns (real numbers) [4], Tabu search algorithm (TA) performed well with regular cyclic data pattern (real numbers) [5], and ant colony optimization algorithm (ACO) did well in integer number searching [6].

As mentioned previously, it is possible to build an intelligent support system to improve the efficiency of hybrid evolutionary algorithms/models or to make improvements by innovative theoretical arrangements (chaotization and cloud theory) in all forecasting/prediction/classification applications. Firstly, the original data should be filtered by a data base with a well-defined characteristic data pattern rules set (e.g., linear, logarithmic, inverse, quadratic, cubic, compound, power, growth, exponential, etc.), in order to recognize the appropriate data pattern (fluctuating, regular, or noisy). The recognition decision rules should include two principles: (1) the change rate of two continuous data; and (2) the decreasing or increasing trend of the change rate (i.e., behavior of the approached curve). Secondly, adequate improvement tools should be selected (e.g., hybrid evolutionary algorithms, hybrid seasonal mechanism, chaotization of decision variables, cloud theory, and any combination of all tools). In order to avoid becoming trapped in local optima, improvement tools can be employed into these optimization problems to obtain an improved, satisfactory solution.

This discussion of the work by the author of this preface highlights work in an emerging area of hybrid advanced techniques that has come to the forefront over the past decade. The collected articles in this text span many cutting edge areas that are truly interdisciplinary in nature.

Wei-Chiang Hong
Guest Editor

Reference

1. Fan, G.F.; Peng, L.L.; Hong, W.C. Short term load forecasting based on phase space reconstruction algorithm and bi-square kernel regression model. *Appl. Energy* **2018**, *224*, 13–33.
2. Hong, W.C. Application of seasonal SVR with chaotic immune algorithm in traffic flow forecasting. *Neural Comput. Appl.* **2012**, *21*, 583–593.
3. Hong, W.C.; Dong, Y.; Zhang, W.Y.; Chen, L.Y.; Panigrahi, B.K. Cyclic electric load forecasting by seasonal SVR with chaotic genetic algorithm. *Int. J. Electr. Power Energy Syst.* **2013**, *44*, 604–614.
4. Geng, J.; Huang, M.L.; Li, M.W.; Hong, W.C. Hybridization of seasonal chaotic cloud simulated annealing algorithm in a SVR-based load forecasting model. *Neurocomputing* **2015**, *151*, 1362–1373.
5. Hong, W.C.; Pai, P.F.; Yang, S.L.; Theng, R. Highway traffic forecasting by support vector regression model with tabu search algorithms. In Proceedings the IEEE International Joint Conference on Neural Networks, Vancouver, BC, Canada, 16–21 July 2006, pp. 1617–21.
6. Hong, W.C.; Dong, Y.; Zheng, F.; Lai, C.Y. Forecasting urban traffic flow by SVR with continuous ACO. *Appl. Math. Modelling* **2011**, *35*, 1282–1291.

energies

MDPI

Article

Support Vector Regression Model Based on Empirical Mode Decomposition and Auto Regression for Electric Load Forecasting

Guo-Feng Fan [1], **Shan Qing** [1,*], **Hua Wang** [1], **Wei-Chiang Hong** [2] and **Hong-Juan Li** [1]

[1] Engineering Research Center of Metallurgical Energy Conservation and Emission Reduction, Ministry of Education, Kunming University of Science and Technology, Kunming 650093, China; guofengtongzhi@163.com (G.-F.F.); wanghua65@163.com (H.W.); fxzwlihongjuan@163.com (H.-J.L.)

[2] Department of Information Management, Oriental Institute of Technology/58 Sec. 2, Sichuan Rd., Panchiao, Taipei 220, Taiwan; samuelsonhong@gmail.com

* Author to whom correspondence should be addressed; yanls22@163.com; Tel.: +86-1388-855-2395; Fax: +86-0871-6515-3405.

Received: 28 November 2012; in revised form: 2 February 2013; Accepted: 25 March 2013; Published: 2 April 2013

Abstract: Electric load forecasting is an important issue for a power utility, associated with the management of daily operations such as energy transfer scheduling, unit commitment, and load dispatch. Inspired by strong non-linear learning capability of support vector regression (SVR), this paper presents a SVR model hybridized with the empirical mode decomposition (EMD) method and auto regression (AR) for electric load forecasting. The electric load data of the New South Wales (Australia) market are employed for comparing the forecasting performances of different forecasting models. The results confirm the validity of the idea that the proposed model can simultaneously provide forecasting with good accuracy and interpretability.

Keywords: electric load prediction; support vector regression; empirical mode decomposition auto regression

---

1. Introduction

Electric energy is an unstored resource, thus, electric load forecasting plays a vital role in the management of the daily operations of a power utility, such as energy transfer scheduling, unit commitment, and load dispatch. With the emergence of load management strategies, it is highly desirable to develop accurate, fast, simple, robust and interpretable load forecasting models for these electric utilities to achieve the purposes of higher reliability and better management [1].

In the past decades, researchers have proposed lots of methodologies to improve load forecasting accuracy. For example, Bianco *et al.* [2] proposed linear regression models for electricity consumption forecasting; Zhou *et al.* [3] applied a grey prediction model for energy consumption; Afshar and Bigdeli [4] proposed an improved singular spectral analysis method for short-term load forecasting (STLF) for the Iranian electricity market; and Kumar and Jain [5] applied three time series models—Grey-Markov model, Grey-Model with rolling mechanism, and singular spectrum analysis—to forecast the consumption of conventional energy in India. By employing artificial neural networks, references [6–9] proposed several useful short-term load forecasting models. By hybridizing the popular method and evolutionary algorithm, the authors of [10–13] demonstrated further performance improvements which could be made for energy forecasting. Though these methods can yield a significant proven forecasting accuracy improvement in some cases, they have usually focused on the improvement of the accuracy without paying special attention to the interpretability.

Recently, expert systems, mainly developed by means of linguistic fuzzy rule-based systems, allow us to deal with the system modeling with good interpretability [14]. However, these models have strong dependency on an expert and often cannot generate good accuracy. Therefore, combination models, based on the popular methods, expert systems and other techniques, are proposed to satisfy both high accurate level and interpretability.

Based on the advantages in statistical learning capacity to handle high dimensional data, the SVR (support vector regression) model, especially suitable for small sample size learning, has become a popular algorithm for many forecasting problems [15–17]. As a disadvantage of an SVR method, it is easily trapped into a local optimum during the nonlinear optimization process of the three parameters, in the meanwhile, its robustness and sparsity are also lacking satisfactory levels. On the other hand, empirical mode decomposition (EMD) and auto regression (AR), a fast, easy and reliable unsupervised clustering algorithm, has been successfully applied to many fields, such as communication, society, economy, engineering, and has achieved good effects [18–20]. Particularly, the EMD method can effectively extract the components of the basic mode from nonlinear or non-stationary time series [21], *i.e.*, the original complex time series can be transferred into a series of single and apparent components. It can effectively reduce the interactions among lots of singular values and improve the forecasting performance of a single kernel function. Thus, it is useful to employ suitable kernel functions for forecasting the medium-and-long-term tendencies of the time series.

In this paper, we present a new hybrid model with clear human-understandable knowledge on training data to achieve a satisfactory level of forecasting accuracy. The principal idea is hybridizing EMD with SVR and AR, namely creating the EMDSVRAR model, to receive better solutions. The proposed EMDSVRAR model has the capability of smoothing and reducing the noise (inherited from EMD), the capability of filtering dataset and improving forecasting performance (inherited from SVR), and the in capability of effectively forecasting the future tendencies of data (inherited from AR). The forecasting outputs of an unseen example by using the hybrid method are described in the following section.

To show the applicability and superiority of the proposed algorithm, half-hourly electric load data (48 data points per day) from New South Wales (Australia) with two different sample sizes are employed to compare the forecasting performances among the proposed model and other four alternative models, namely the PSO-BP model (BP neural network trained by a particle swarm optimization algorithm), SVR model, PSO-SVR model (optimal combination of SVR parameters determined by a PSO algorithm), and the AFCM model (an adaptive fuzzy combination model based on a self-organizing map and support vector regression). This study also suggests that researchers and practitioners should carefully consider the nature and intention in using these electric load data while neural networks, statistical methods, and other hybrid models are being determined to be the critical management tools in electricity markets. The experimental results indicate that this proposed EMDSVRAR model has the following advantages: (1) simultaneously satisfies the need for high levels of accuracy and interpretability; (2) the proposed model can tolerate more redundant information than the original SVR model, thus, it has better generalization ability.

The rest of this paper is organized as follows: in Section 2, the EMDSVRAR forecasting model is introduced and the main steps of the model are given. In Section 3, the data description and the research design are outlined. The numerical results and comparisons are presented and discussed in Section 4. A brief conclusion of this paper and the future research are provided in Section 5.

2. Support Vector Regression with Empirical Mode Decomposition

2.1. Empirical Mode Decomposition (EMD)

The EMD method is based on the simple assumption that any signal consists of different simple intrinsic modes of oscillations. Each linear or non-linear mode will have the same number of extreme and zero-crossings. There is only one extreme between successive zero-crossings. Each mode should be independent of the others. In this way, each signal could be decomposed into a number of intrinsic mode functions (IMFs), each of which should satisfy the following two definitions [22]:

a   In the whole data set, the number of extreme and the number of zero-crossings should either equal or differ to each other at most by one.

b   At any point, the mean value of the envelope defined by local maxima and the envelope defined by the local minima is zero.

An IMF represents a simple oscillatory mode compared with the simple harmonic function. With the definition, any signal $x(t)$ can be decomposed as following steps:

1   Identify all local extremes, and then connect all the local maxima by a cubic spline line as the upper envelope.

2   Repeat the procedure for the local minima to produce the lower envelope. The upper and lower envelopes should cover all the data among them.

3   The mean of upper and low envelope value is designated as m_1, and the difference between the signal $x(t)$ and m_1 is the first component, h_1, as shown in Equation (1):

$$h_1 = x(t) - m_1 \tag{1}$$

Generally speaking, h_1 will not necessarily meet the requirements of the IMF, because h_1 is not a standard IMF. It needs to be determined for k times until the mean envelope tends to zero. Then, the first intrinsic mode function c_1 is introduced, which stands for the most high-frequency component of the original data sequence. At this point, the data could be represented as Equation (2):

$$h_{1k} = h_{1(k-1)} - m_{1k} \tag{2}$$

where h_{1k} is the datum after k times siftings. $h_{1(k-1)}$ stands for the data after $k-1$ times sifting. Standard deviation (SD) is used to determine whether the results of each filter component meet the IMF or not. SD is defined as Equation (3):

$$SD = \sum_{k=1}^{T} \frac{\left| h_{1(k-1)}(t) - h_{1k}(t) \right|^2}{h_{1(k-1)}^2(t)} \tag{3}$$

where T is the length of the data.

The value of standard deviation SD is limited in the range of 0.2 to 0.3, which means when $0.2 < SD < 0.3$, the decomposition process can be finished. The consideration for this standard is that it should not only ensure $h_k(t)$ to meet the IMF requirements, but also control the decomposition times. Therefore, in this way, the IMF components could retain amplitude modulation information in the original signal.

4   When h_{1k} has met the basic requirements of SD, based on the condition of $c_1 = h_{1k}$, the signal $x(t)$ of the first IMF component c_1 can be obtained directly, and a new series r_1 could be achieved after deleting the high frequency components. This relationship could be expressed as Equation (4):

$$r_1 = x(t) - c_1 \tag{4}$$

The new sequence is treated as the original data and repeats the steps (1) to (3) processes. The second intrinsic mode function c_2 could be obtained.

5   Repeat previous steps (1) to (4) until the r_n can not be decomposed into the IMF. The sequence r_n is called the remainder of the original data $x(t)$. r_n is a monotonic sequence, it can indicate the

3

Energies **2013**, *6*, 1887–1901

overall trend of the raw data $x(t)$ or mean, and it is usually referred as the so-called trend items. It is of clear physical significance. The process is expressed as Equations (5) and (6):

$$r_1 = x(t) - c_1, r_2 = r_1 - c_2, \ldots, r_n = r_{n-1} - c_n \tag{5}$$

$$x(t) = \sum_{i=1}^{n} c_i + r_n \tag{6}$$

The original data can be expressed as the IMF component and remainder.

2.2. Support Vector Regression

The notions of SVMs for the case of regression are introduced briefly. Given a data set of N elements $\{(X_i, y_i), i = 1, 2, \cdots, N\}$, where X_i is the i-th element in n-dimensional space, *i.e.*, $X_i = [x_{1i}, \cdots, x_{ni}] \in \mathfrak{R}^n$, and $y_i \in \mathfrak{R}$ is the actual value corresponding to X_i. A non-linear mapping $(\cdot): \mathfrak{R}^n \to \mathfrak{R}^{n_h}$ is defined to map the training (input) data X_i into the so-called high dimensional feature space (which may have infinite dimensions), $\mathfrak{R}^{n_h}$ (Figure 1a,b). Then, in the high dimensional feature space, there theoretically exists a linear function, f, to formulate the non-linear relationship between input data and output data. Such a linear function, namely SVR function, is shown as Equation (7):

$$f(X) = W^T \varphi(X) + b \tag{7}$$

where $f(X)$ denotes the forecasting values; the coefficients $W(W \in \mathfrak{R}^{n_h})$ and b ($b \in \mathfrak{R}$) are adjustable. As mentioned above, the SVM method aims at minimizing the empirical risk, shown as Equation (8):

$$R_{emp}(f) = \frac{1}{N} \sum_{i=1}^{N} \Theta_{\varepsilon}(y_i, W^T \varphi(X_i) + b) \tag{8}$$

where $\Theta_{\varepsilon}(y, f(x))$ is the ε-insensitive loss function (indicated as a thick line in Figure 1c) and defined as Equation (9):

$$\Theta_{\varepsilon}(Y, f(X)) = \begin{cases} |f(X) - Y| - \varepsilon, & if |f(X) - Y| \geq \varepsilon \\ 0, & otherwise \end{cases} \tag{9}$$

Energies **2013**, *6*, 1887–1901

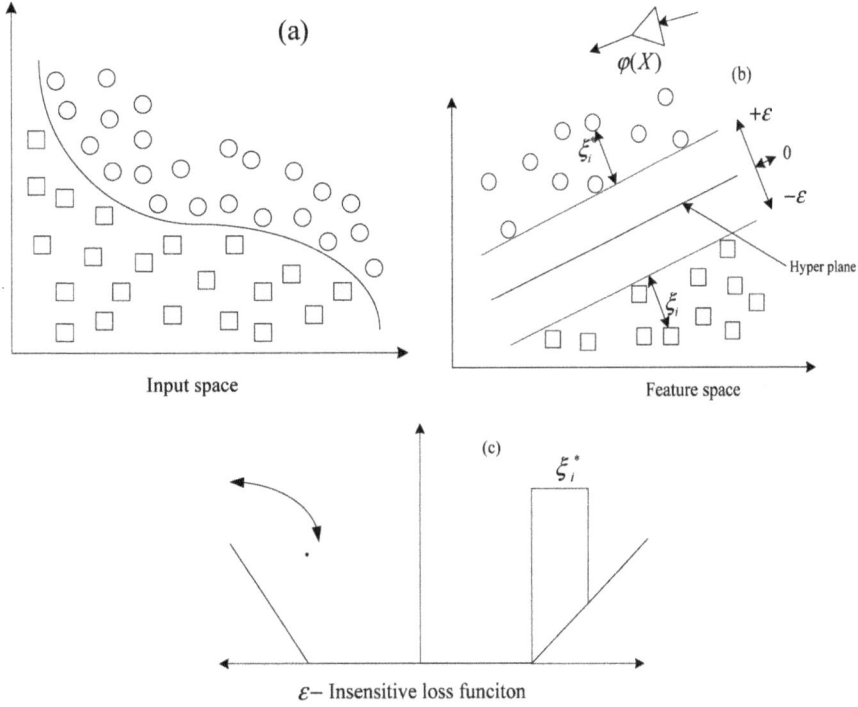

Figure 1. Transformation process illustration of a SVR model. (**a**) Input space; (**b**) Feature space; (**c**) ε-insensitive loss function.

In addition, $\Theta_\varepsilon(Y, f(X))$ is employed to find out an optimum hyperplane on the high dimensional feature space (Figure 1b) to maximize the distance separating the training data into two subsets. Thus, the SVR focuses on finding the optimum hyper plane and minimizing the training error between the training data and the ε-insensitive loss function. Then, the SVR minimizes the overall errors, shown as Equation (10):

$$\underset{W,b,\xi^*,\xi}{Min}\ R_\varepsilon(W,\xi^*,\xi) = \frac{1}{2}W^TW + C\sum_{i=1}^{N}(\xi_i^* + \xi_i) \tag{10}$$

with the constraints:

$$\begin{aligned} Y_i - W^T\varphi(X_i) - b &\leq \varepsilon + \xi_i^*, & i &= 1,2,...,N \\ -Y_i + W^T\varphi(X_i) + b &\leq \varepsilon + \xi_i, & i &= 1,2,...,N \\ \xi_i^* &\geq 0, & i &= 1,2,...,N \\ \xi_i &\geq 0, & i &= 1,2,...,N \end{aligned} \tag{11}$$

The first term of Equation (10), employing the concept of maximizing the distance of two separated training data, is used to regularize weight sizes to penalize large weights, and to maintain regression function flatness. The second term penalizes training errors of $f(x)$ and y by using the ε-insensitive loss function. C is the parameter to trade off these two terms. Training errors above ε are denoted as ξ_i^*, whereas training errors below $-\varepsilon$ are denoted as ξ_i (Figure 1b).

After the quadratic optimization problem with inequality constraints is solved, the parameter vector w in Equation (7) is obtained as Equation (12):

$$W = \sum_{i=1}^{N}(\beta_i^* - \beta_i)\varphi(X_i) \tag{12}$$

where ζ_i^*, ζ_i are obtained by solving a quadratic program and are the Lagrangian multipliers. Finally, the SVR regression function is obtained as Equation (13) in the dual space:

$$f(X) = \sum_{i=1}^{N} (\beta_i^* - \beta_i) K(X_i, X) + b \tag{13}$$

where $K(X_i, X)$ is called the kernel function, and the value of the kernel equals the inner product of two vectors, X_i and X_j, in the feature space $\varphi(X_i)$ and $\varphi(X_j)$, respectively; that is, $K(X_i, X_j) = \varphi(X_i)\varphi(X_j)$. Any function that meets Mercer's condition [23] can be used as the kernel function.

There are several types of kernel function. The most used kernel functions are the Gaussian radial basis functions (RBF) with a width of σ : $K(X_i, X_j) = \exp(-0.5\|X_i - X_j\|^2/\sigma^2)$ and the polynomial kernel with an order of d and constants a_1 and a_2: $K(X_i, X_j) = (a_1 X_i X_j + a_2)^d$. However, the Gaussian RBF kernel is not only easy to implement, but also capable of non-linearly mapping the training data into an infinite dimensional space, thus, it is suitable to deal with non-linear relationship problems. Therefore, the Gaussian RBF kernel function is specified in this study. The forecasting process of a SVR model is illustrated in Figure 2.

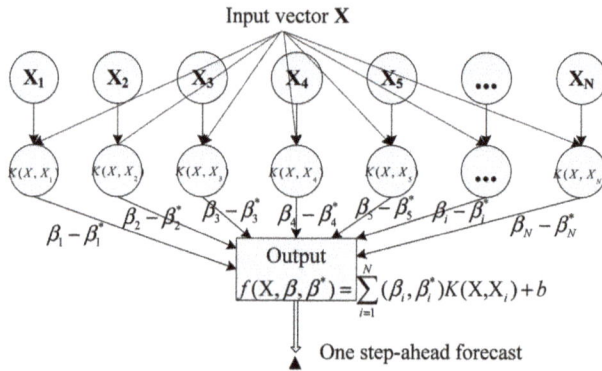

Figure 2. The forecasting process of a SVR model.

2.3. AR Model

Equation (14) expresses a *p*-step autoregressive model, referring as $AR(p)$ model [24]. Stationary time series $\{X_t\}$ that meet the model $AR(p)$ is called the $AR(p)$ sequence. That $a = (a_1, a_2, \cdots, a_p)^T$ is named as the regression coefficients of the $AR(p)$ model:

$$X_t = \sum_{j=1}^{p} a_j X_{t-j} + \varepsilon_t, t \in Z \tag{14}$$

3. Numerical Examples

In the first experiment, the proposed model is trained by electric load from New South Wales (Australia) from 2 May 2007 to 7 May 2007, and testing electric load is from 8 May 2007. The employed electric load data is on a half-hourly basis (*i.e.*, 48 data points per day). The data size contains only 7 days, to differ from the other example with more sample data, this example is so-called the small sample size data, and illustrated in Figure 3a.

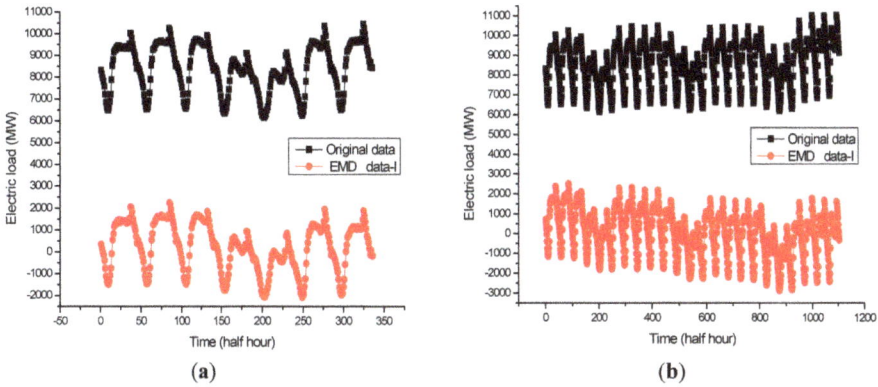

Figure 3. (a) Half-hourly electric load in New South Wales from 2 May 2007 to 8 May 2007; (b) Half-hourly electric load in New South Wales from 2 May 2007 to 24 May 2007.

Too large training sets should avoid overtraining during the learning process of the SVR model. Therefore, the second experiment with 23 days (1104 data points from 2 May 2007 to 24 May 2007) is modeled by using part of all the training samples as training set. This example is so-called the large sample size data, and illustrated in Figure 3b.

3.1. Results after EMD

After being decomposed by EMD, the data can be divided into eight groups, which are shown in Figure 4a–h and the last group (Figure 4h) is a trend term (remainders). The so-called high frequency item is obtained by adding the preceding seven groups. From Figure 3a,b, the trend of the high frequency item is the same as original data, and its the structure is more regular, *i.e.*, it is more stable. Then, the high frequency item (data-I) and the remainders (data-II) have good effect of regression by the SVR and AR, respectively, and will be described as follow.

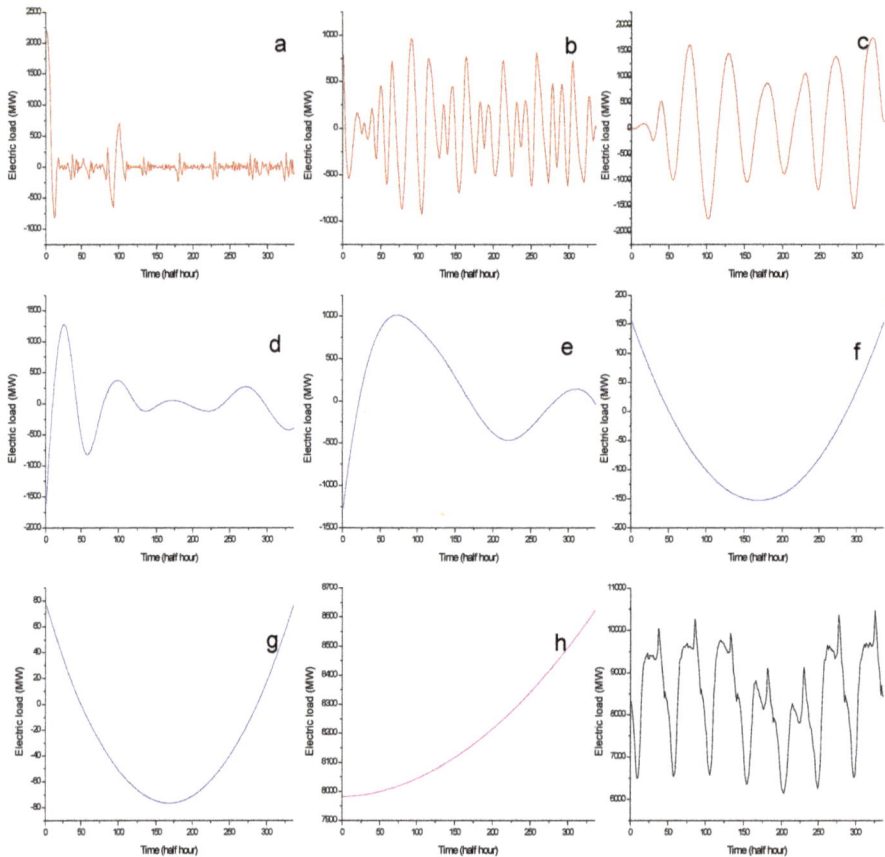

Figure 4. For ease of prevention, the graphs (**a**–**h**) show our section of plots at different IMFs for the small sample size.

3.2. Forecasting Using SVR for Data I (The High Frequency Item)

Firstly, for both small sample and large sample data, the high-frequency item is simultaneously employed for SVR modeling, and the better performances of the training and testing (forecasting) sets are shown in Figure 5a,b, respectively. The correlation coefficients of training effects are 0.9912 and 0.9901, respectively, of the forecast effects are 0.9875 and 0.9887, accordingly. This implies that the decomposition is helpful to improve the forecasting accuracy. The parameters of a SVR model for data I are shown in Table 1.

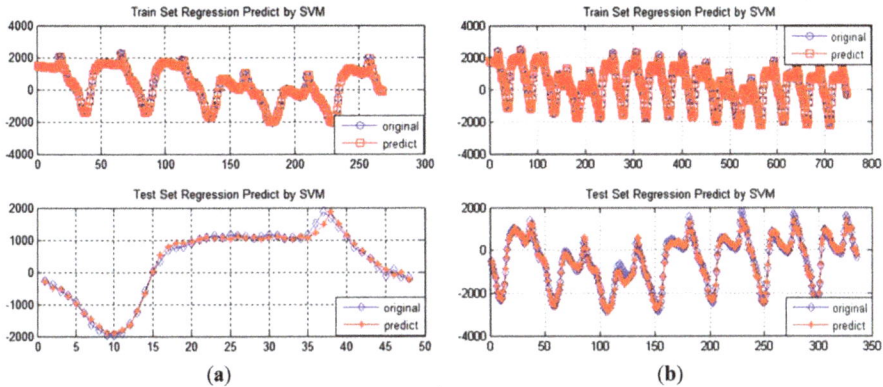

Figure 5. Comparison of the data-I and the forecasted electric load of train and test by the SVR model for the small sample and large sample data: (**a**) One-day ahead prediction of 8 May 2007 are performed by the model; (**b**) One-week ahead prediction from 18 May 2007 to 24 May 2007 are performed by the model.

Table 1. The SVR's parameters for data-I and data-II.

Sample size	m	σ	C	ε	Testing MAPE
The high frequency item (data-I)	20	0.1	100	0.0061	9.85
The remainders (data-II)	20	0.35	181	0.0034	5.1

3.3. Forecasting Using AR for Data II (The Remainders)

Then, according to the geometric decay of the correlation coefficient and partial correlation coefficients fourth-order truncation for data II (the remainders), it can be regarded as AR (4) model. The parameters of a SVR model for data II are shown in Table 1.

As shown in Figure 6a,b, the remainders, for both small sample and large sample data, almost are in a straight line. The good forecasting results are shown in Table 2, and the errors have reached the level of 10^{-7} for the small or large amount of data. It has demonstrated the superiority of the AR model.

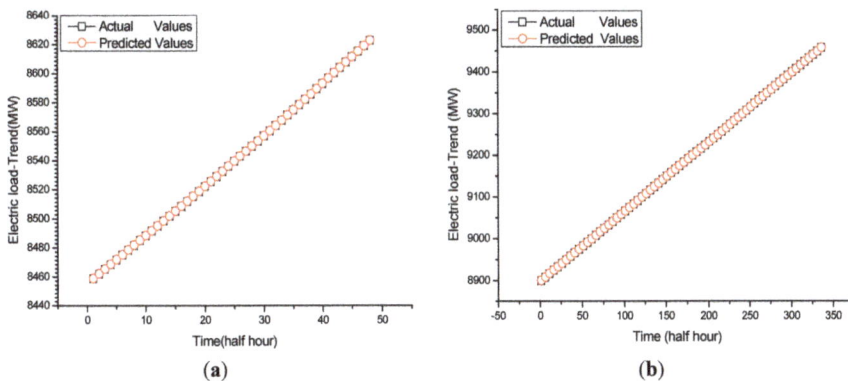

Figure 6. Comparison of the data-II and the forecasted electric load by the AR model for the two experiments: (**a**) One-day ahead prediction of 8 May 2007 performed by the model; (**b**) One-week ahead prediction from 18 May 2007 to 24 May 2007 performed by the model.

9

Table 2. Summary of results of the AR forecasting model for data-II.

Remainders	MAE	Eqution
The small sample size	6.5567×10^{-7}	$x_n = 8417.298 + 1.013245x_{n-1} + 0.490278x_{n-2} - 0.011731x_{n-3} - 0.491839x_{n-4}$
The large sample size	1.8454×10^{-7}	$x_n = 8546.869 + 1.000046x_{n-1} + 0.499957x_{n-2} - 5.18 \times 10^{-5}x_{n-3} - 0.499951x_{n-4}$

4. Result and Analysis

This section focuses on the efficiency of the proposed model with respect to computational accuracy and interpretability. To consider the small sample size modeling ability of the SVR model and conduct fair comparisons, we perform a real case experiment with relatively small sample size in the first experiment. The next experiment with 1104 datapoints is focused on illustrating the relationship between sample size and accuracy.

4.1. Parameter Settings of the Employed Forecasting Models

As mentioned by Taylor [25], and to be based on the same comparison condition with Wang *et al.* [26], some parameter settings of the employed forecasting models are set as followings. For the PSO-BP model, we use 90 percent of all training samples as the training set, and the rest as the evaluation set. The parameters used in the PSO-BP are as follows: (i) The first set related to BP neural network: input layer dimension *indim* = 2, hidden layer dimension *hiddennum* = 3, output layer dimension *outdim* = 1; (ii) The second set related to PSO: maximum iteration number *itmax* = 300, number of particles $N = 40$, length of particle $D = 3$, weight $c_1 = c_2 = 2$.

Because the PSO-SVR model embeds the construction and prediction algorithm of SVR in the fitness value iteration step of PSO, it will take a long time to train the PSO-SVR using the full training dataset. For the above reason, we draw a small part of all training samples as training set, and the rest as evaluation set. The parameters used in the PSO are as follows: For small sample size: maximum iteration number itmax = 50, number of particles $N = 20$, length of particle $D = 3$, weight $c_1 = c_2 = 2$. For large sample size: maximum iteration number *itmax* = 20, number of particles $N = 5$, length of particle $D = 3$, weight $c_1 = c_2 = 2$.

4.2. Forecasting Evaluation Methods

For the purpose of evaluating the forecasting capability, we examine the forecasting accuracy by calculating three different statistical metrics, the root mean square error (RMSE), the mean absolute error (MAE) and the mean absolute percentage error (MAPE). The definitions of RMSE, MAE and MAPE are expressed as Equations (15–17):

$$RMSE = \sqrt{\frac{\sum_{i=1}^{n} (P_i - A_i)^2}{n}} \tag{15}$$

$$MAE = \sqrt{\frac{\sum_{i=1}^{n} |P_i - A_i|}{n}} \tag{16}$$

$$MAPE = \sqrt{\frac{\sum_{i=1}^{n} \left| \frac{P_i - A_i}{A_i} \right|}{n}} * 100 \tag{17}$$

Where P_i and A_i are the i-th predicted and actual values respectively, and n is the total number of predictions.

4.3. Empirical Results and Analysis

For the first experiment, the forecasting results (the electric load on 8 May 2007) of the original SVR model, the PSO-SVR model and the proposed EMDSVRAR model are shown in Figure 7a. Notice that the forecasting curve of the proposed EMDSVRAR model fits better than other alternative models.

The second experiment shows the one-week-ahead forecasting for the large sample size data. The peak load values of testing set are bigger than that of training set shown in Figure 3b. The detailed forecasted results of this experiment are shown in Figure 7b. It indicates that the results obtained from the EMDSVRAR model fits the peak load values exceptionally well. In other words, the EMDSVRAR model has better generalization ability than the three comparison models.

The forecasting results from these models are summarized in Table 3. The proposed EMDSVRAR model is compared with four alternative models. It is found that our hybrid model outperforms all other alternatives in terms of all the evaluation criteria. One of the general observations is that the proposed model tends to fit closer to the actual value with a smaller forecasting error.

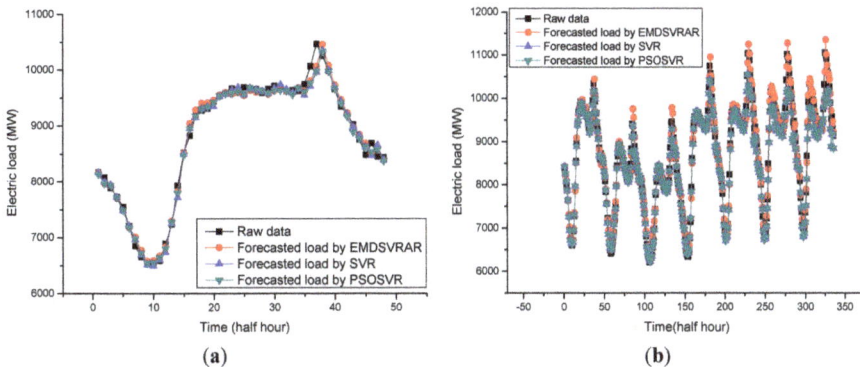

Figure 7. Comparison of the original data and the forecasted electric load by the EMDSVRAR Model, the SVR model and the PSO-SVR model for (**a**) the small sample size (One-day ahead prediction of 8 May 8, 2007 are performed by the models); (**b**) the large sample size (One-week ahead prediction from May 18, 2007 May 24, 2007 are performed by the models).

The proposed model shows the higher forecasting accuracy in terms of three different statistical metrics. In view of the model effectiveness and efficiency on the whole, we can conclude that the proposed model is quite competitive against four comparison models, the PSO-BP, SVR, PSO-SVR, and AFCM models. In other words, the hybrid model leads to better accuracy and statistical interpretation.

Table 3. Summary of results of the forecasting models.

Algorithm	MAPE	RMSE	MAE	Running Time(s)
For the first experiment (small sample size)				
Original SVR	11.6955	145.865	10.9181	180.4
PSO-SVR	11.4189	145.685	10.6739	165.2
PSO-BP	10.9094	142.261	10.1429	159.9
AFCM [24]	9.9524	125.323	9.2588	75.3
EMDSVRAR	9.8595	117.159	9.0967	80.7
For the second experiment (large sample size)				
Original SVR	12.8765	181.617	12.0528	116.8
PSO-SVR	13.503	271.429	13.0739	192.7
PSO-BP	12.2384	175.235	11.3555	163.1
AFCM [26]	11.1019	158.754	10.4385	160.4
EMDSVRAR	5.100	134.201	9.8215	162.0

Several observations can also be noticed from the results. Firstly, from the comparisons among these models, we point out that the proposed model outperforms other alternative models. Secondly, the EMDSVRAR model has better generalization ability for different input patterns as shown in the second experiment. Thirdly, from the comparison between the different sample sizes of these two experiments, we conclude that the hybrid model can tolerate more redundant information and construct the model for the larger sample size data set. Finally, since the proposed model generates good results with good accuracy and interpretability, it is robust and effective as shown in Table 3. Overall, the proposed model provides a very powerful tool to implement easily for forecasting electric load.

Furthermore, to verify the significance of the accuracy improvement of the EMDSVRAR model, the forecasting accuracy comparison among original SVR, PSO-SVR, PSO-BP, AFCM, and EMDSVRAR models is conducted by a statistical test, namely a Wilcoxon signed-rank test, at the 0.025 and 0.05 significance levels in one-tail-tests. The test results are shown in Table 4. Clearly, the proposed EMDSVRAR model has statistical significance (under a significant level 0.05) among the other alternative models, particularly comparing with original SVR, PSO-SVR, PSO-BP, and AFCM models.

Table 4. Wilcoxon signed-rank test.

Compared models	Wilcoxon signed-rank test	
	$\alpha = 0.025$; W = 4	$\alpha = 0.05$; W = 6
EMD-SVR-AR *vs.* original SVR	8	3 [a]
EMD-SVR-AR *vs.* PSO-SVR	6	2 [a]
EMD-SVR-AR *vs.* PSO-BP	6	2 [a]
EMD-SVR-AR *vs.* AFCM	6	2 [a]

[a] denotes that the EMDSVRAR model significantly outperforms other alternative models.

5. Conclusions

The proposed model achieves superiority and outperforms the original SVR model while forecasting based on the unbalanced data. In addition, the goal of the training model is not to learn an exact representation of the training set itself, but rather to set up a statistical model that generalizes better forecasting values for the new inputs. In practical applications of a SVR model, if the SVR model is overtrained to some sub-classes with overwhelming size, it memorizes the training data and gives poor generalization of other sub-classes with small size. The EMD term of the proposed

EMDSVRAR model has been employed in the present research, details of which have discussed in the above section.

The interest in applying the EMD forecast systems arises from the fact that those systems consider both accuracy and comprehensibility of the forecast result simultaneously. To this end, a combined model has been proposed and its effectiveness in forecasting the electric load data has been compared with three other alternative models. In this study, various data characteristics of electric load are identified where the proposed model performs better than the other algorithms in terms of its forecasting capability. Based on the obtained experimental results, we conclude that the proposed EMDSVRAR model algorithm can generate not only human-understandable rules, but also better forecasting accuracy levels. Our proposed model also outperforms other alternative models in terms of interpretability, forecasting accuracy and generalization ability, which are especially true for forecasting with unbalanced data and very complex systems.

Acknowledgments: This work was supported by National Natural Science Foundation of China under Contract (No. 51064015) to which the authors are greatly obliged, and National Science Council, Taiwan (NSC 100-2628-H-161-001-MY4; NSC 101–2410–H– 161-001).

References

1. Bernard, J.T.; Bolduc, D.; Yameogo, N.D.; Rahman, S. A pseudo-panel data model of household electricity demand. *Resour. Energy Econ.* **2010**, *33*, 315–325.
2. Bianco, V.; Manca, O.; Nardini, S. Electricity consumption forecasting in Italy using linear regression models. *Energy* **2009**, *34*, 1413–1421.
3. Zhou, P.; Ang, B.W.; Poh, K.L. A trigonometric grey prediction approach to forecasting electricity demand. *Energy* **2006**, *31*, 2839–2847.
4. Afshar, K.; Bigdeli, N. Data analysis and short term load forecasting in Iran electricity market using singular spectral analysis (SSA). *Energy* **2011**, *36*, 2620–2627.
5. Kumar, U.; Jain, V.K. Time series models (Grey-Markov, Grey Model with rolling mechanism and singular spectrum analysis) to forecast energy consumption in India. *Energy* **2010**, *35*, 1709–1716.
6. Topalli, A.K.; Erkmen, I. A hybrid learning for neural networks applied to short term load forecasting. *Neurocomputing* **2003**, *51*, 495–500.
7. Kandil, N.; Wamkeue, R.; Saad, M.; Georges, S. An efficient approach for short term load forecasting using artificial neural networks. *Int. J. Electr. Power Energy Syst.* **2006**, *28*, 525–530.
8. Beccali, M.; Cellura, M.; Brano, V.L.; Marvuglia, A. Forecasting daily urban electric load profiles using artificial neural networks. *Energy Convers. Manag.* **2004**, *45*, 2879–2900.
9. Topalli, A.K.; Cellura, M.; Erkmen, I.; Topalli, I. Intelligent short-term load forecasting in Turkey. *Electr. Power Energy Syst.* **2006**, *28*, 437–447.
10. Pai, P.F.; Hong, W.C. Forecasting regional electricity load based on recurrent support vector machines with genetic algorithms. *Electr. Power Syst. Res.* **2005**, *74*, 417–425.
11. Hong, W.C. Electric load forecasting by seasonal recurrent SVR (support vector regression) with chaotic artificial bee colony algorithm. *Energy* **2011**, *36*, 5568–5578.
12. Hong, W.C. Application of chaotic ant swarm optimization in electric load forecasting. *Energy Policy* **2010**, *38*, 5830–5839.
13. Pai, P.F.; Hong, W.C. Support vector machines with simulated annealing algorithms in electricity load forecasting. *Energy Convers. Manag.* **2005**, *46*, 2669–2688.
14. Yongli, Z.; Hogg, B.W.; Zhang, W.Q.; Gao, S.; Yang, Y.H. Hybrid expert system for aiding dispatchers on bulk power systems restoration. *Int. J. Electr. Power Energy Syst.* **1994**, *16*, 259–268.
15. Basak, D.; Pal, S.; Patranabis, D.C. Support vector regression. *Neural Inf. Process. Lett. Rev.* **2007**, *11*, 203–224.
16. Burges, C.J.C. A tutorial on support vector machines for pattern recognition. *Data Min. Knowl. Discov.* **1998**, *2*, 121–167.
17. Schölkopf, B.; Smola, A.; Williamson, R.C.; Bartlet, P.L. New support vector algorithms. *Neural Comput.* **2000**, *12*, 1207–1245.

Energies **2013**, *6*, 1887–1901

18. Meng, Q.; Peng, Y. A new local linear prediction model for chaotic time series. *Phys. Lett. A* **2007**, *370*, 465–470.

19. Xie, J.X.; Cheng, C.T. A new direct multi-step ahead prediction model based on EMD and chaos analysis. *J. Autom.* **2008**, *34*, 684–689.

20. Fan, G.; Qing, S.; Wang, H.; Shi, Z.; Hong, W.C.; Dai, L. Study on apparent kinetic prediction model of the smelting reduction based on the time series. *Math. Probl. Eng.* **2012**. [CrossRef]

21. Bhusana, P.; Chris, T. Improving prediction of exchange rates using differential EMD. *Expert Syst. Appl.* **2013**, *40*, 377–384.

22. Huang, N.E.; Shen, Z. A new view of nonliner water waves: The Hilbert spectrum. *Rev. Fluid Mech.* **1999**, *31*, 417–457.

23. Vapnik, V. *The Nature of Statistical Learning Theory*; Springer-Verlag: New York, NY, USA, 1995.

24. Gao, J. Asymptotic properties of some estimators for partly linear stationary autoregressive models. *Commun. Statist. Theory Meth.* **1995**, *24*, 2011–2026.

25. Taylor, J.W. Short-term load forecasting with exponentially weighted methods. *IEEE Trans. Power Syst.* **2012**, *27*, 458–464.

26. Che, J.; Wang, J.; Wang, G. An adaptive fuzzy combination model based on self-organizing map and support vector regression for electric load forecasting. *Energy* **2012**, *37*, 657–664.

Article

An Improved Quantum-Behaved Particle Swarm Optimization Method for Economic Dispatch Problems with Multiple Fuel Options and Valve-Points Effects

Qun Niu [1,*], Zhuo Zhou [1], Hong-Yun Zhang [1] and Jing Deng [2]

[1] Shanghai Key Laboratory of Power Automation Technology, School of Mechatronics Engineering and Automation, Shanghai University, Shanghai 200072, China; zhouzhuo1101@yahoo.com.cn (Z.Z.); zhanghongyun@shu.edu.cn (H.-Y.Z.)

[2] Intelligent Systems and Control Group, School of Electronics, Electrical Engineering and Computer Science, Queen's University of Belfast, Belfast BT9 5AH, UK; jdeng01@qub.ac.uk

* Author to whom correspondence should be addressed; comelycc@hotmail.com; Tel./Fax: +86-21-56334241.

Received: 2 July 2012; in revised form: 10 August 2012; Accepted: 22 August 2012; Published: 19 September 2012

Abstract: Quantum-behaved particle swarm optimization (QPSO) is an efficient and powerful population-based optimization technique, which is inspired by the conventional particle swarm optimization (PSO) and quantum mechanics theories. In this paper, an improved QPSO named SQPSO is proposed, which combines QPSO with a selective probability operator to solve the economic dispatch (ED) problems with valve-point effects and multiple fuel options. To show the performance of the proposed SQPSO, it is tested on five standard benchmark functions and two ED benchmark problems, including a 40-unit ED problem with valve-point effects and a 10-unit ED problem with multiple fuel options. The results are compared with differential evolution (DE), particle swarm optimization (PSO) and basic QPSO, as well as a number of other methods reported in the literature in terms of solution quality, convergence speed and robustness. The simulation results confirm that the proposed SQPSO is effective and reliable for both function optimization and ED problems.

Keywords: economic dispatch; quantum-behaved particle swarm optimization; valve-point effects; multiple fuel options

---

1. Introduction

Economic dispatch (ED) is considered to be one of the key functions in electric power system operation. The main objective of ED is to determine the optimal scheduling of power outputs for all generating units that minimizes the total fuel cost while satisfying all the equality and inequality constraints of units and system. Due to valve-point effects, prohibited operating zones and multiple fuel effects, the characteristics of power generating units are inherently highly nonlinear [1].

Multiple fuel options problem (coal, nature gas or oil) is one of the important kinds of ED problems and each part of the hybrid cost function implies some information about the fuel being burned or the operation cost of units. Taking valve-point effects and multiple fuel options into consideration, the ED problem can be represented as a non-smooth optimization problem, which causes difficulties in finding the global or near global optimization solution using conventional approaches.

Over the past two decades, many modern meta-heuristic methods have been applied to ED problems, such as genetic algorithm (GA) [2], particle swarm optimization (PSO) [3], differential evolution (DE) [4], ant colony optimization (ACO) [5] and simulated annealing (SA) [6]. Among these methods, PSO has recently attracted more attention due to its rapid convergence and algorithmic accuracy compared with other optimization methods.

Energies **2012**, *5*, 3655–3673

PSO is a population based optimization algorithm, which was introduced by Kennedy and Eberhart in 1995 [7]. PSO is motivated by the simulation of social behaviour of animals such as fish schooling and bird flocking. In the conventional PSO mechanism, a swarm of individuals (called particles) fly within the search space. Each particle represents a potential solution to the optimization problem. The position of a particle is influenced by the best position (*pbest*) found by itself (*i.e.*, its own experience) and the position of the best particle in the whole swarm (*gbest*) (*i.e.*, the experience of neighbouring particles).

Although PSO can converge quickly towards the optimal solution, it has difficulties in reaching a global optimum and suffers from premature convergence. Moreover, PSO has several control parameters. The convergence of the algorithm depends heavily on the value of its control parameters.

Taking advantage of both PSO mechanism and quantum mechanics, in 2004, a new version of PSO, quantum-behaved particle swarm optimization, named QPSO, was proposed by Sun, Xu and Feng [8], which is inspired by quantum mechanics and trajectory analysis of PSO. As a quantum system is an uncertain system that is different from classical stochastic system in which every particle can appear at any position with a certain probability, the swarm can search in the whole feasible region [9]. Besides, unlike PSO, there are no velocity vectors for particles in QPSO, and it has fewer parameters to be adjusted, which makes it easier to implement. In [10–12], convergence analysis and other varients of QPSO have been presented. As an efficient algorithm, QPSO has been applied to many optimization problems, such as system identification [13], non-linear programming problems [14], power system [15], *etc.* Although Coelho *etal.* proposed a quantum-inspired HQPSO using the harmonic oscillator potential well to solve economic dispatch problems [16], Sun and Lu applied QPSO to ED problems [15], and Chakraborty *et al.* presented a hybrid QPSO to solve the ED problems [17], to the best of our knowledge, it has not been used yet to solve ED problems with multiple fuel options.

In this paper, an improved QPSO namely SQPSO is proposed to solve ED problems with multiple fuel options and valve-points effects. In the proposed SQPSO, a new selective probability operator is introduced into the updating mechanism of QPSO, which can balance the global and local searching abilities and enhance the diversity of QPSO. In particular, based on the selective probability operator, *pbest* and *gbest* are used to generate the local attractor of QPSO, with user defined selective probability, to enhance the local search performance. This modification on the original QPSO together with a recombination operator will maintain the best information of the swarm and, in the same time, exchange information between individuals to increase the population diversity.

To show the performance of the proposed SQPSO, five popular benchmark functions and two ED problems with valve-point effects and multi-fuel options are tested. The results obtained by SQPSO are analyzed and compared with PSO, DE and QPSO, as well as some other optimization methods reported in recent literature. The remainder of this paper is organized as follows: Section 2 is the formulation of the ED problem and Section 3 presents the conventional PSO, QPSO and proposed SQPSO, respectively. Section 4 gives the experimental results. Finally, Section 5 concludes the paper.

2. Formulation of the ED Problem

The main objective of solving the ED problem is to minimize the total fuel cost of each thermal generating unit in electric power system while satisfying a variety of equality and inequality constraints. The total fuel cost function of ED problem is described as:

$$\min F_T = \sum_{i=1}^{n} F_i(P_i) \tag{1}$$

where F_T is the total generation cost, n is the total number of generating unit, P_i is the power of the ith generator and F_i is its corresponding fuel cost, which is defined by the following equation as:

$$F_i(P_i) = a_i + b_i P_i + c_i P_i^2 \tag{2}$$

where a_i, b_i and c_i are the cost coefficients and subject to:

$$\sum_{i=1}^{n} P_i = P_D, i = 1, 2, \ldots\ldots, n \qquad (3)$$

$$P_i^{min} < P_i < P_i^{max} \qquad (4)$$

where P_D is the total demand of the power system, P_i^{min} and P_i^{max} are the minimum and maximum output of the *i*th generation unit, respectively.

2.1. The ED Problem with Valve-Point Effects

A valve-point is the rippling effect added to the generation unit curve when each steam admission valve in a turbine starts to open [2]. This curve poses higher order non-linearity and discontinuity, which makes the problem of finding the optimum more difficult and increases the number of local minima in the fuel cost function. Considering the valve-point effects, sinusoidal functions are added to the quadratic cost function, which is defined by the following equation:

$$F_i(P_i) = a_i + b_i P_i + c_i P_i^2 + \left| e_i \sin(f_i(P_i^{min} - P_i)) \right| \qquad (5)$$

where e_i, f_i are the coefficients of generator i, reflecting the valve-point.

2.2. ED Problem with Multiple Fuels and Valve-Point Effects

To give a more accurate description of the ED problem, the effects of multiple fuels resources (coal, nature gas or oil) should also be considered. Each segment of the hybrid cost function implies some information about the fuel being burned or the unit's operation. Since the dispatching units are practically supplied with multi-fuel sources, each unit should be represented with several piecewise quadratic functions reflecting the effects of fuel type changes, and the generator must identify the most economic fuel to burn [2]. The number of non-differentiable points in the objective function increases when multiple fuels are taken into consideration. The incremental cost functions of a generator with multi-fuel options are illustrated in Figure 1. The ED problems with both multiple and fuels valve-point effects can be represented as follows:

$$F_i(P_i) = \begin{cases} a_{i1} + b_{i1}P_i + c_{i1}P_i^2 + \left| e_{i1} \sin(f_{i1})(P_i^{min} - P_{i1}) \right|, & fuel\ 1,\ P_i^{min} < P_i < P_{i1} \\ a_{i2} + b_{i2}P_i + c_{i2}P_i^2 + \left| e_{i2} \sin(f_{i2})(P_{i2}^{min} - P_{i2}) \right|, & fuel\ 2,\ P_{i1} < P_i < P_{i2} \\ a_{ik} + b_{ik}P_i + c_{ik}P_i^2 + \left| e_{i2} \sin(f_{ik})(P_{ik}^{min} - P_{ik}) \right|, & fuel\ k,\ P_{ik-1} < P_i < P_i^{max} \end{cases} \qquad (6)$$

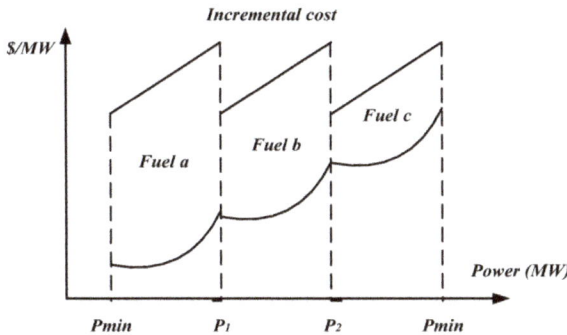

Figure 1. Incremental cost function of a generator with multi-fuel options.

3. The Proposed SQPSO Algorithm

3.1. Conventional Particle Swarm Optimization

PSO is a population-based stochastic optimization algorithm, which is inspired by the social intelligence and movements of fishes or birds in the swarm. In PSO, each potential solution is a point in the search space and is called as 'particle'. Each particle is assumed to have two characteristics: a position and a velocity. The target of the particles is to find the best result of the objective function. Initially, a population of particles is randomly generated within the search space. At each iteration, it stores memory of best position of each individual and best position of the whole population. By taking advantages of the particles' own experience and experience of its neighbours, the particles could fly towards the optimal solution.

For example, in a *n*-dimensional search space, the position and velocity of an individual *i* are represented as the vectors: $X_i = (X_{i1}, X_{i1}, \dots, X_{in})$ and $V_i = (V_{i1}, V_{i2}, \dots, V_{in})$. The best position for each particle is denoted as: $pbest_i = (pbest_{1i}, pbest_{2i}, \dots, pbest_{ni})$ and $gbest_i$ is the best solution found in the whole swarm. In standard PSO, the position and velocity of particles are updated by the following equations:

$$V_i^{(t+1)} = w \times V_i^{(t)} + c_1 \times rand() \times \left(pbest_i - x_i^{(t)}\right) + c_2 \times rand() \times \left(gbest_i - x_i^{(t)}\right) \tag{7}$$

$$x_i^{(t+1)} = x_i^t + v_i^{(t+1)} \tag{8}$$

where:

x_i^t and v_i^t represent the position and velocity of individual *i* at generation *t*;
w is the inertia weight parameter that controls the momentum of particles;
c_1 and c_2 are positive constants, which balance the need for local and global search;
rand() is a random number between 0 and 1.

3.2. Quantum-Behaved Particle Swarm Optimization

In the conventional PSO, a particle moves in the search space by the moments of its position and velocity. In the quantum model of a PSO, the state of a particle is depicted by wave function $\Psi(x,t)$ [8], instead of position and velocity. QPSO introduces the mean best position into the algorithm and uses a strategy based on a quantum delta potential well model to sample around the previous best points Furthermore, QPSO has only one parameter, which is easier to control than PSO algorithm. Employing the Monte Carlo method, particles are updated according to the following equations:

$$\begin{cases} x_{ij}(t+1) = p_{ij}(t) + \beta * \left| Mbest_{ij}(t) - x_{ij}(t) \right| * In(1/u), & \text{if } k \geq 0.5 \\ x_{ij}(t+1) = p_{ij}(t) - \beta * \left| Mbest_{ij}(t) - x_{ij}(t) \right| * In(1/u), & \text{if } k < 0.5 \end{cases} \tag{9}$$

The following gives the explication of the update Equation (9):

(1) $x_{ij}(t+1)$ is denoted as the position of the *j*th dimension of the *i*th particle for the next generation $t+1$.
(2) $P_{ij}(t)$ is the local attractor to make sure SQPSO can converge, which is defined as follows:

$$p_{ij} = \phi * Pbest_{ij} + (1 - \phi) * gbest_j \tag{10}$$

where ϕ is a random number uniformly distributed in (0,1); $Mbest_{ij}$ is a global point, which can be calculated by the mean of the *Pbest* of all particles in the population. The definition is given is as follows:

$$Mbest_{ij}(t) = (\frac{1}{N}\sum_{i=1}^{N} Pbest_{i1}(t-1), \frac{1}{N}\sum_{i=1}^{N} Pbest_{i2}(t-1), ..., \frac{1}{N}\sum_{i=1}^{N} Pbest_{in}(t-1)) \tag{11}$$

where N represents the population size and $Pbest_i$ is the best position of the *i*th particle.

(3)   In this paper, β is called the constriction-expansion coefficient, and it is linearly decreasing when the iteration increases:

$$\beta^t = \beta_{max} - \frac{\beta_{max} - \beta_{min}}{itNum} * t \tag{12}$$

where $itNum$ is the maximum iteration number, t is the current iteration number $\beta_{max} = 1.0$ and $\beta_{min} = 0.5$.

(4) u and k are two random numbers uniformly distributed in (0,1).

3.3. The Proposed Quantum-Behaved Particle Swarm Optimization

In the original QPSO, the local attractor is calculated by Equation (10), which means that the $P_{ij}(t)$ is a random position between the individual best position and the group best position. However, the drawback is the difficulty in maintaining the best information of the swarm, especially when the optimal solution is at the boundary of the problem. In [18], Jong-Bae Park proposed an improved PSO, which introduced a kind of crossover operation. In this operation, particles update the position with the exchange information of previous generation particle position and the individual best position of itself. In this paper, a modified QPSO is proposed, called SQPSO, which introduces a selective probability operator into the update mechanism when calculating the local attractor $P_{ij}(t)$. In SQPSO, the information of global best position and the whole swarm's individual best position are used to update the position for the next generation. The reason behind the inclusion of the selective probability operator is to enable the use of recombination operator into the original QPSO which will help to maintain the best solution and, at the same time, exchange information between individuals in the whole swarm. The pseudo code for the proposed selective probability operator is given in Figure 2.

```
For i=1:PopNum
  For j=1:Dim
    IF rand ≤ SP
      RandPop=floor(rand*popNum)+1;
      P(i,j)=pbest(RandPop,j);
    Else
      P(i,j)=gbest(1,j);
    End
End
```

Figure 2. The pseudo code for the proposed crossover operator of SQPSO.

In Figure 2, *PopNum* is the number of population and *Dim* is the Dimensionality for each individual. *RandPop* is an individual randomly selected from the swarm. *SP* is the selective probability, which can control whether the local attractor $P(i,j)$ is generated from individual best position or global best position. If $rand \leq SP$, then the local attractor $P(i,j)$ will select its value from the *Pbest* of the individual *RandPop* and if $rand > SP$, then the value of $P(i,j)$ will select the point of global best position. Using the *SP*, $P(i,j)$ can not only make use of the previous best swarm information but also increase the population diversity and consequently enhance the global search ability. The principle of the modification is illustrated in Figure 3 and the procedure of the proposed SQPSO is described as follows:

(1) *Initialize the population,* which are generated randomly within the minimum and maximum output of each generator, using the following equations:

$$
\text{population} = \begin{bmatrix} X_1 \\ X_2 \\ ... \\ X_n \end{bmatrix}
\tag{13}
$$

$$
X_i = [x_{i,1}, x_{i,2}, ..., x_{i,n}], x_{ij} = P_{ij}^{\min} + rand * (P_{ij}^{\max} - P_{ij}^{\min})
$$

where X_i is the *ith* individual of the population,(x_{ij} is the *jth* data vector of *ith* individual; $P_{ij}^{\min}$ and $P_{ij}^{\max}$ are the maximum and minimum output limit values of the *jth* control variable.

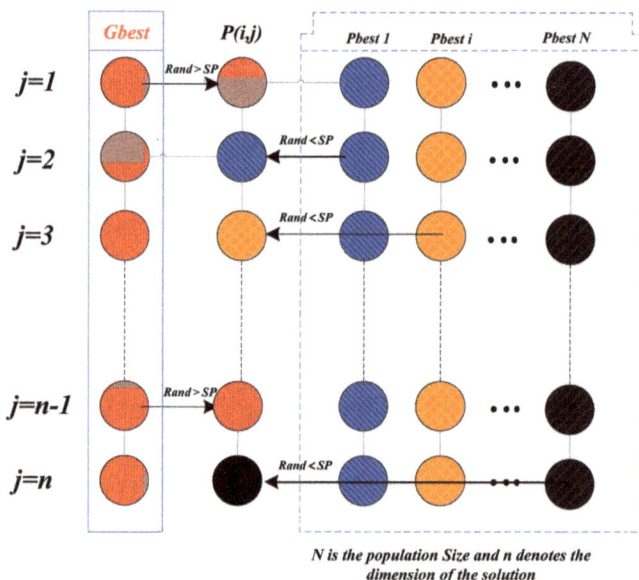

N is the population Size and n denotes the
dimension of the solution

Figure 3. Principle of the modified of SQPSO.

For the multi-fuel ED problem, the relationship between unit output and fuel type is shown in Figure 4, taking a 10-generator problem as an example, each unit has its minimum and maximum output of generation and the sum of the whole power output should satisfy the total output demand, and as shown in Figure 4, different range of unit output corresponds to different type of fuel.

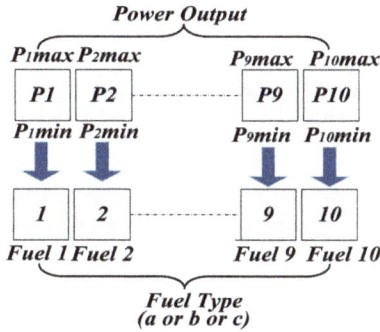

Figure 4. Relationship between unit output and fuel type.

(2) *Constraint handling for real power balance.* Since the individuals of the population are created randomly and with the evolution of particles, newly generated individual may violate the constraints. Therefore, it is important to keep all the individual variables within their feasible ranges. Hence, the following procedure is adopted by the SQPSO to modify the value of new generated variables to satisfy the power balance constraint.

$$x_{ij} = \begin{cases} P_{ij}^{\min} & \text{if } x_{ij} \leq P_{ij}^{\min} \\ P_{ij}^{\max} & \text{if } x_{ij} > P_{ij}^{\min} \\ x_{ij} & \text{otherwise} \end{cases} \tag{14}$$

The amount of power balance violation is calculated by:

$$pd = \sum_{i=1}^{n} P_i - P_D \tag{15}$$

if $pd = 0$, go to step 3; if $pd \neq 0$, the value of pd will be adjusted by allocating it to the output of a unit, which is chosen randomly from the whole set of generating units, so that the generating constraints can be satisfied. If the output of the chosen unit goes outside the feasible boundaries, its value should be modified using Equation (14). The constraints handling procedure is illustrated in Figure 5.

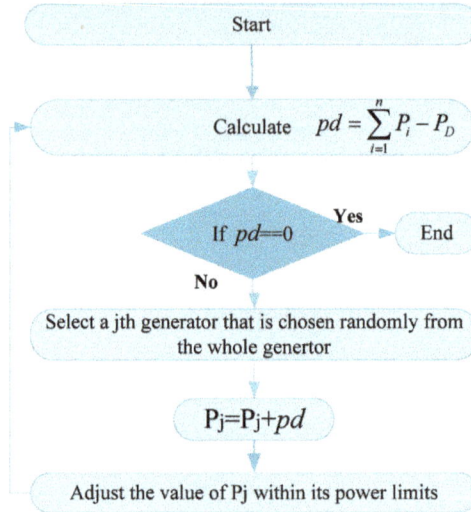

Figure 5. Procedure of constraint handling of the SQPSO algorithm.

(3) *Parameter setting.* There are two parameters in SQPSO, one is the constriction-expansion coefficient which decreases from 1.0 to 0.5 linearly. Another parameter is the introduced selective probability (*SP*). In this paper, the SP for SQPSO increases from 0.5 to 0.8 linearly using the following equation:

$$SP^t = SP_{max} - \frac{SP_{max} - SP_{min}}{itNum} * t \tag{16}$$

where SP^t is the value of SP at iteration t. SP_{max} and SP_{min} are maximum and minimum selective probability. At the early stage, the population will select more vectors from the group best position, which can accelerate the convergence speed. As the iteration number increases, the population will draw more vectors from the individual best positions to enhance the diversity of the whole swarm.

(4) *Evaluate the objective function value of each particle.*

(5) *Update pbest.* Compare each particle's objective function value with its *pbest*. If the current value is better than the *pbest* value, set the *pbest* value to the current value.

(6) *Update gbest.* Determine best *gbest* of the swarm as the minimum *pbest* of all particles.

(7) *Calculate the Mbest, constriction-expansion coefficient β according to Equation (11) and Equation (12), respectively.*

(8) *Calculate the local attractor according to the Selective probability operator proposed in this paper.*

(9) *Update the particle's position using Equation (9)*

(10) *Check if the stop criterion satisfied?*

(11) *If not, then go to step 2.*

(12) *Else, the searching process is stopped.*

4. Experimental Results

4.1. Benchmark Functions

To verify the performance of the proposed SQPSO, five benchmark functions (Sphere, Jason, Griewank, Rosenbrock and Rastrigrin) listed in Table 1 are conducted. These functions are all minimization problems with the minimum value to be zero. The results produced by the proposed SQPSO are compared with that of the EGA, DPSO, HPSO, IPSO and IQPSO in [17]. EGA is a modified genetic algorithm with elitism and adaptive mutation probability control, and DPSO, HPSO, IPSO are three types of revised version of PSO. IQPSO is an improved quantum-inspired particle swarm optimization, which is based on the principle of quantum rotation gates. Additionally, three algorithms are also used in this paper for comparison, which are PSO, DE and QPSO. For PSO, the acceleration coefficients c1 and c2 are set to 2, and the inertia weight decreased from 0.9 to 0.4 linearly [19]. The parameter of DE is set to F = 0.4, CR = 0.8 [20].

Table 1. Benchmark functions.

Name	Function	Dim	Range	Opt
Sphere	$f_1(x) = \sum\limits_{i=1}^{n} x_i^2$	40	$[-100,100]$	0
Jason	$f_2(x) = \sum\limits_{i=1}^{n} (x_i - i)^2$	40	$[-100,100]$	0
Griewank	$f_3(x) = \frac{1}{4000} \sum\limits_{i=1}^{n} x_i^2 - \prod\limits_{i=1}^{n} \cos(\frac{x_i}{\sqrt{i}}) + 1$	40	$[-600,600]$	0
Rosenbrock	$f_4(x) = \sum\limits_{i=1}^{n} [100(x_{i+1} - x_i^2)^2 + (x_i - 1)^2]$	40	$[-2.048,2.048]$	0
Rastrigrin	$f_5(x) = \sum\limits_{i=1}^{n} [x_i^2 - 10\cos(2\pi x_i) + 10]$	40	$[-5.12,5.12]$	0

For QPSO and SQPSO, the coefficient β decreases from 1.0 to 0.5 linearly and the selective probability (SP) for SQPSO increases from 0.5 to 0.8 linearly. To compare the solution quality and convergence characteristics, 50 independent trial runs are performed for each benchmark function and mean function value and best function value are recorded. In order to make a fair comparison, the population size is set to 80 and population dimension is 40 for all the five benchmark functions. The maximum iteration number is set to 5000. All the algorithms are implemented in MATLAB 2008a and executed on an Intel Core2 Duo 1.66 GHz personal computer.

The numerical results in Table 2 show that the proposed SQPSO can achieve satisfactory performance. Specifically, both the sphere and Jason function have only one single optimal solution, so it is usually introduced to test the local search ability of the algorithm. From the results, it can be seen that the SQPSO outperforms all the other algorithms in terms of mean function value and best function value, which indicates SQPSO has strong local search ability. Rosenbrock is a mono-modal function and its optimal solution lies in a narrow area. The experimental results on Rosenbrock show that the mean function value of SQPSO is better than DPSO, HPSO, IPSO, PSO and QPSO. However, the best function value is inferior to other algorithms reported in [21]. Griewank and Rastrigrin are both multi-modal and they are usually used to compare the global search ability of the algorithm. As to Griewank, SQPSO can hit the minimum value zero and the mean function value is superior to other algorithms too. For Rastrigrin, both EGA and IQPSO give a better performance than SQPSO and the results of SQPSO are better than other methods.

Table 2. Mean value and best value for five benchmark functions with different approaches.

Function\\Algorithm	f_1 (Sphere) Mean (Best)	f_2 (Jason) Mean (Best)	f_3 (Griewank) Mean (Best)	f_4 (Rosenbrock) Mean (Best)	f_5 (Rastrigrin) Mean (Best)
EGA [11]	2.743×10^{-10} (0)	8.865×10^{-8} (3.748×10^{-22})	1.042×10^{-4} (7.952×10^{-13})	0.84 (6.537×10^{-4})	2.257 (6.537×10^{-4})
DPSO [11]	5.403×10^{-7} (4.532×10^{-14})	2.595×10^{-6} (1.173×10^{-12})	1.322×10^{-3} (2.167×10^{-10})	28.094 (1.150×10^{-2})	28.826 (19.899)
HPSO [11]	1.319×10^{-6} (2.824×10^{-10})	6.735×10^{-3} (1.503×10^{-10})	2.546×10^{-3} (5.136×10^{-9})	28.995 (2.346×10^{-2})	29.956 (15.393)
IPSO [11]	1.524×10^{-7} (3.406×10^{-11})	1.350×10^{-5} (2.107×10^{-10})	2.224×10^{-3} (1.454×10^{-10})	27.13 (2.339×10^{-2})	31.906 (15.064)
IQPSO [11]	1.085×10^{-23} (0)	2.078×10^{-23} (0)	3.221×10^{-7} (0)	2.19×10^{-2} (2.717×10^{-9})	0.521 (1.075×10^{-4})
PSO	2.885×10^{-21} (1.774×10^{-23})	1.4526×10^{-21} (4.413×10^{-24})	8.0215×10^{-3} (0)	56.1057 (12.4904)	34.6046 (20.8941)
DE	1.3727×10^{-47} (3.9244×10^{-49})	1.0097×10^{-30} (0)	3.4506×10^{-4} (0)	12.4830 (6.6779)	56.7802 (14.9244)
QPSO	5.054×10^{-26} (7.333×10^{-31})	5.3011×10^{-30} (0)	8.1 (0)	48.4957 (25.2717)	25.7895 (13.9294)
SQPSO	$\mathbf{6.5759 \times 10^{-74}}$ ($\mathbf{1.8122 \times 10^{-89}}$)	(0) (0)	2.217×10^{-7} (0)	32.68016 (14.7115)	13.7105 (3.9798)

In addition, compared with original QPSO without selective probability operator, the proposed SQPSO demonstrates good performance for all the five benchmark functions in terms of both the mean function value and best function value, which indicates that the SQPSO is an effective modification of QPSO.

4.2. ED Problem with Valve-Point Effects

A large-scale power system of 40-generating units with quadratic cost function and valve-point effects is being considered here. Transmission losses are ignored and the total load demand of this text system is 10,500 MW. The system data can be found from [1]. One hundred independent runs are made for each method and population size is set to 80. The stopping criterion is set to 500. The result obtained from SQPSO is compared with some methods in the literature including IFEP [1], GA_PS_SQP [22], PC-PSO [23], SOH_PSO [23], NPSO [24] ,NPSO_LRS [24], PSO-GM [25], CBPSO_RVM [25], ICA-PSO [26], ACO [5], APSO(2) [27], HDE [28], ST-HDE [28] and IQPSO [29]. In addition, in order to compare the performance of the crossover operation in [18] with the proposed selective probability operator. The crossover operation [18] is introduced into QPSO, namely CQPSO, and the performance of CQPSO can be seen in the following results. The comparison results of SQPSO with other methods reported in literature are given in Table 3. The best solution of the SQPSO is 121,434.41 $/H, which is comparatively superior to most of the methods and the mean cost is better than other methods as well.

Table 3. Comparison results for ED problem with valve-point effects (40-unit system).

Methods	Generation cost($/H)			Standard Deviation
	Minimum	Mean	Maximum	
IFEP [1]	122,624.35	123,382	125,740.63	NR
GA-PS-SQP [22]	121,458.14	122,039	NR	NR
PC-PSO [23]	121,767.90	122,461.30	122,867.55	NR
SOH-PSO [23]	121,501.14	121,853.57	122,446.3	NR
NPSO [24]	121,704.74	122,221.37	122,995.10	NR
NPSO-LRS [24]	121,664.43	122,209.32	122,981.59	NR
PSO-GM [25]	121,845.98	122,398.38	123,219.22	258.44
CBPSO-RVM [25]	121,555.32	122,281.14	123,094.98	259.99
ICA-PSO [26]	121,422.17	121,428.14	121,453.56	NR
ACO [5]	121,532.41	121,606.45	121,679.64	45.58
APSO(2) [27]	121,663.52	122,153.67	122,912.40	NR
HDE [28]	121,813.26	122,705.66	NR	NR
ST-HDE [28]	121,698.51	122,304.30	NR	NR
IQPSO [21]	121,448.21	122,225.07	NR	NR
FCASO [30]	121,516.47	122,082.59	NR	NR
CASO [30]	121,865.63	122,100.74	NR	NR
CPSO-SQP [31]	121,458.54	122,028.16	NR	NR
CPSO [31]	121,865.23	122,100.87	NR	NR
DE	121,805.56	122,142.97	122,466.75	151.88
PSO	121,956.18	122,459.36	122,785.73	209.12
QPSO	121,487.27	121,750.48	121,991.99	111.68
CQPSO	121,463.39	121,732.98	121,778.74	79.38
SQPSO	**121,434.41**	**121,723.22**	**121,881.51**	**104.29**

The convergence characteristics of the SQPSO in comparison with PSO, DE, QPSO are shown in Figure 6. It is shown that PSO converges fastest among these methods while it suffers the premature convergence. Besides, DE is the slowest among the four methods, as DE involves a series of mutation, crossover and greedy selection operators, which leads to low convergence speed and increases the computational time as well. QPSO and SQPSO converge at nearly the same speed, however the SQPSO can produce a better solution as iteration increases, which indicates stronger searching ability. In addition, compared with CQPSO, SQPSO can outperform it almost in all aspects, which indicates that the proposed elective probability operator is improved compared with the crossover operation in [18].

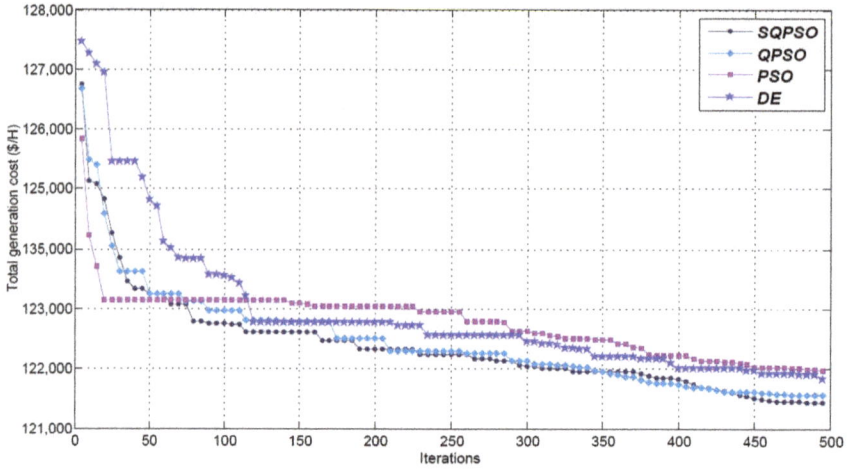

Figure 6. Convergence characteristics for total generation costs (40-uint system).

The distribution of generation costs of the four algorithms for 100 runs is shown in Figure 7 which reflects the robustness of each algorithm. The curve of the SQPSO is at the bottom of the figure and stabilizes at a relatively intensive region, which means the distribution of the solution of SQPSO is much better than other methods. The detailed results of the best solution of DE, PSO, QPSO and SQPSO, for ED problem with valve-point effects are given in Table 4.

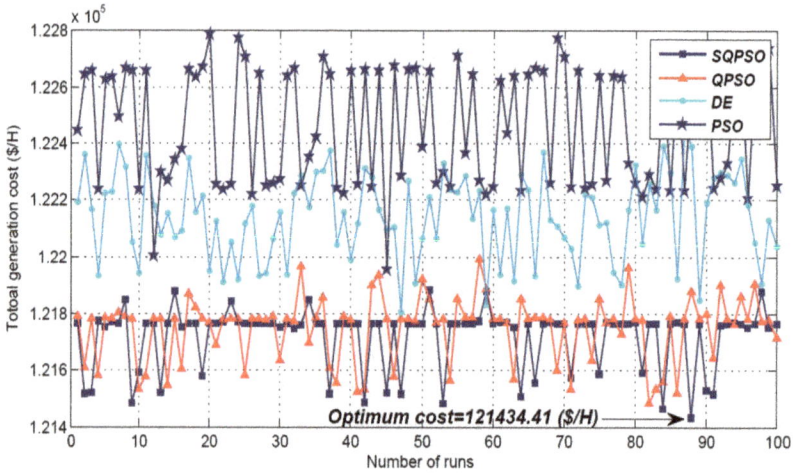

Figure 7. Distribution of generation costs of the four algorithms for 100 runs (40-unit system).

Table 4. Detailed results of the best solution of DE, PSO, QPSO and SQPSO, for ED problem with valve-point effects (40-unit system).

Unit	Methods				
	DE	PSO	QPSO	CQPSO	SQPSO
P_1	111.8012	113.9945	113.6426	113.9999	110.9173
P_2	111.5734	110.9343	111.9581	113.9999	111.7807
P_3	95.79661	100.748	97.56082	120.0000	97.56128
P_4	182.4958	179.1588	179.7457	179.7333	179.7005
P_5	87.27856	97.0000	88.53738	96.9999	93.37496
P_6	140.0000	140.0000	139.9981	140.0000	139.9862
P_7	300.0000	300.0000	299.989	300.0000	259.8548
P_8	285.2077	300.0000	284.9879	299.9999	284.9466
P_9	286.9856	299.9040	284.7968	293.3932	284.5976
P_{10}	130.0000	130.0000	130.0093	130.0000	130.0493
P_{11}	94.25143	94.0000	94.02522	94.0000	168.807
P_{12}	94.61699	94.0000	94.0286	94.0000	94.00315
P_{13}	125.7718	125.0000	125.0323	125.0000	214.7713
P_{14}	393.1819	393.9392	394.2728	394.2794	394.2986
P_{15}	395.1001	394.1116	394.2987	394.2794	304.61
P_{16}	393.7253	304.3765	394.3071	304.5196	394.2632
P_{17}	487.6391	500.0000	489.3179	489.2794	489.363
P_{18}	491.819	490.6004	489.2953	489.2795	489.5688
P_{19}	512.8806	513.8928	511.3082	511.2794	511.2797
P_{20}	511.7995	514.1406	511.3473	511.2794	511.3193
P_{21}	524.2502	524.3505	523.3044	523.2796	523.2616
P_{22}	523.9075	523.4735	523.3182	523.2796	523.3642
P_{23}	519.8336	529.2841	523.3638	523.2796	523.2587
P_{24}	527.6248	547.3133	523.3677	550.0000	523.3996
P_{25}	523.9776	522.9096	523.2928	523.2795	523.2836
P_{26}	523.2693	524.9206	523.3083	523.2798	523.2817
P_{27}	10.3912	10.0000	10.01133	10.0000	10.00975
P_{28}	10.0000	10.0000	10.08587	10.0000	10.0344
P_{29}	10.0335	10.0000	10.00228	10.0000	10.00645
P_{30}	92.73803	91.53567	90.21066	96.9999	88.52085
P_{31}	187.1519	190.0000	189.9984	190.0000	189.9972
P_{32}	189.9415	190.0000	189.9968	190.0000	189.9834
P_{33}	189.4094	190.0000	189.9988	190.0000	189.9822
P_{34}	197.3705	199.9374	199.9794	199.9999	165.321
P_{35}	199.2062	198.4492	199.9942	200.0000	199.9666
P_{36}	198.9157	200.0000	199.9942	200.0000	200.0000
P_{37}	109.5043	110.0000	110.0000	110.0000	110.0000
P_{38}	110.0000	110.0000	109.9926	110.0000	109.9984
P_{39}	108.1849	110.0000	109.9915	110.0000	109.992
P_{40}	512.3655	512.0254	511.3299	511.2794	511.2849
Total Demand	10,500	10,500	10,500	10,500	10,500
Total Cost	121,805.5647	121,956.1827	121,487.2762	121,463.3942	121,434.4071

4.3. The ED Problem with Multi-Fuel Option and Valve-Point Effects

In this section, the proposed SQPSO is applied to multi-fuel economic dispatch problem with valve-point effects. Transmission losses are ignored and system date can be found in [29]. The experimental results are also compared with other algorithms reported in literature, including CGA_MU [2], IGA_MU [2], ACO [5], ED-DE [32], ARCGA [33], PSO-GM [25], NPSO [24], NPSO-LRS [24], PSO-GM [25], CBPSO-RVM [25], APSO [27], GA [34], DSPSO–TSA [34], which are given in Table 5.

Table 5. Comparison of calculation results for multiple fuel ED problems with total demand of 2700 (MW).

Methods	Generation cost ($/H)			Standard Deviation	Average CPU times
	Minimum	Mean	Maximum		
CGA_MU [2]	624.7193	627.6087	633.8652	NR	26.64
IGA_MU [2]	624.5178	625.8692	630.8705	NR	7.32
ACO [5]	623.9000	624.3500	624.7800	NR	8.35
ED-DE [32]	623.8290	623.8807	623.8894	NR	NR
ARCGA [33]	623.8281	623.8495	623.8814	NR	NR
NPSO [24]	624.1624	625.2180	627.4237	NR	NR
NPSO-LRS [24]	624.1273	624.9985	626.9981	NR	NR
PSO-GM [25]	624.3050	624.6749	625.0854	0.1580	NR
CBPSO-RVM [25]	623.9588	624.0816	624.2930	0.0576	NR
APSO [27]	624.0145	624.8185	627.3049	NR	0.52
GA [34]	624.5050	624.7419	624.8169	0.1005	18.3
TSA [34]	624.3078	635.0623	624.8285	1.1593	9.71
DSPSO–TSA [34]	623.8375	623.8625	623.9001	0.0106	3.44
DE	623.9280	624.0068	624.0653	0.0271	0.625
PSO	624.0120	624.2055	624.4376	0.0889	0.308
QPSO	623.8766	623.9639	624.4163	0.0688	0.315
CQPSO	623.8476	623.8652	623.8885	0.0151	0.318
SQPSO	**623.8319**	**623.8440**	**623.8605**	0.0107	0.324

It can be seen that SQPSO can get a minimum generation cost of 623.8319($/H), which is the best solution among all the methods. For the mean cost, SQPSO outperforms most of the methods expect for the ARCGA, which is slightly better than SQPSO, however the CUP times of ARCGA is almost three times that of SQPSO. When considering the average CPU time, the computational time for PSO, QPSO and SQPSO are at the same level, while the results of SQPSO is better than the other two methods. The detailed results of the best solution of DE, PSO, QPSO, CQPSO and SQPSO, for the multiple fuel ED problem with total demand of 2700 MW is given in Table 6.

Table 6. Detailed results of the best solution of DE, PSO, QPSO and SQPSO, for multiple fuel ED problem with total demand of 2700 MW.

Unit	CQPSO		DE		PSO		QPSO		SQPSO	
	Output (MW)	Fuel type	Output (MW)	Fuel type	Output (MW)	Fuel type	Output (MW)	Fuel type	Output (MW)	Fuel type
P_1	217.567	2	220.8058	2	220.8058	2	218.587	2	218.5939	2
P_2	211.7117	1	211.7154	1	211.7154	1	210.4723	1	211.2166	1
P_3	279.6489	1	280.7032	1	280.7032	1	280.7087	1	281.6653	1
P_4	240.5800	3	239.7713	3	239.7713	3	239.3708	3	238.9676	3
P_5	276.3749	1	277.2203	1	277.2203	1	279.6347	1	279.9345	1
P_6	239.6394	3	238.9671	3	238.9671	3	240.7144	3	239.2363	3
P_7	290.0985	1	289.0121	1	289.0121	1	290.1244	1	287.7275	1
P_8	240.8488	3	240.175	3	240.175	3	239.6396	3	239.6394	3
P_9	427.6622	3	425.4145	3	425.4145	3	423.8487	3	427.1502	3
P_{10}	275.8686	1	276.2151	1	276.2151	1	276.8994	1	275.8686	1
Pd	2,700		2,700		2,700		2,700		2,700	
Total Cost	623.8476		623.928		624.012		623.8766		623.8319	

The convergence characteristics and the distribution of generation costs of the SQPSO in comparison with PSO, DE, QPSO are shown in Figures 8 and 9. Clearly, SQPSO converges to the optimal solution faster than other three methods. It can reach the optimal region only in a few iterations, which shows powerful global search ability.

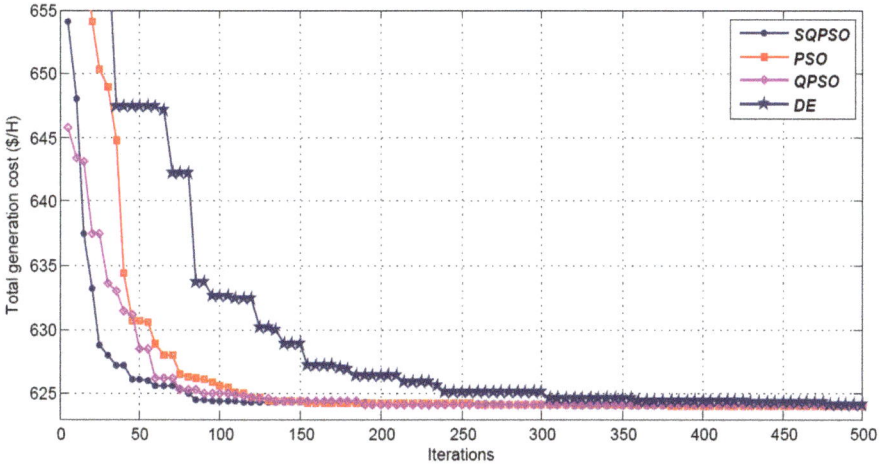

Figure 8. Convergence characteristics for total generation costs (multiple fuel options system).

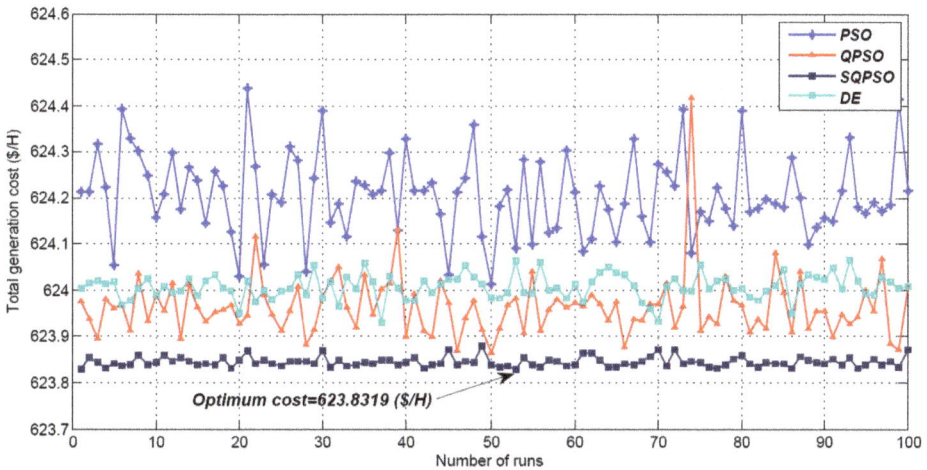

Figure 9. Distribution of generation costs for 100 runs (multiple fuel options system).

The results of different methods for the multiple fuel ED problems with total demand range from 2400 to 2600 MW are summarized in Table 7. It again shows that the SQPSO outperforms all the other methods.

Energies **2012**, *5*, 3655–3673

Table 7. Comparison of calculation results for multiple fuel ED problem with total demand range from 2400–2600 MW.

Demand	Method	Generation cost ($/H)			Standard Deviation	Average CPU times
		Minimum	Mean	Maximum		
2400	DE	481.9030	481.9527	482.0231	0.0285	0.4781
	PSO	482.0807	484.1717	491.5540	2.5598	0.3625
	QPSO	481.9235	483.4540	492.6059	2.2612	0.3562
	CQPSO	481.7469	481.7711	481.7974	0.0180	0.3683
	SQPSO	**481.7320**	**481.7440**	**481.7591**	**0.0068**	**0.3390**
2500	DE	526.4154	526.4771	526.5379	0.0244	0.5156
	PSO	526.4849	527.5594	535.1762	1.3761	0.3578
	QPSO	526.3758	527.5720	534.9611	1.4797	0.3328
	CQPSO	526.2537	526.2839	526.3229	0.0187	0.3453
	SQPSO	**526.2447**	**526.2556**	**526.2897**	**0.0079**	**0.3500**
2600	DE	574.5489	574.6371	574.9653	0.0916	0.5984
	PSO	574.6194	576.0185	589.1900	2.5451	0.3515
	QPSO	574.5857	575.7198	589.1281	2.0467	0.3315
	CQPSO	574.4492	574.6538	574.7928	0.1439	0.3576
	SQPSO	**574.3866**	**574.5076**	**574.7659**	**0.1640**	**0.3484**

5. Conclusions

An improved quantum-behaved particle swarm optimization called SQPSO is proposed in this paper, which introduces selective probability operator into the basic QPSO. The proposed SQPSO has been tested on five classic benchmark functions, as well as two ED problems with valve-point effects and multiple fuel options. It shows superior optimization performance in terms of the convergence rate and the robustness, compared with DE, PSO, CQPSO and QPSO. Additionally, SQPSO also shows competitive ability over other algorithms from the literature.

Acknowledgments: This work is supported by the National Natural Science Foundation of China (61273040), Shanghai Rising-Star Program (12QA1401100), Key Project of Science and Technology Commission of Shanghai Municipality (10JC1405000), the project of Shanghai Municipal Education Commission (12YZ020) and RCUK funded Science Bridge project.

References

1. Sinha, N.; Chakrabarti, R.; Chattopadhyay, P.K. Evolutionary programming techniques for economic load dispatch. *IEEE Trans. Evolut. Comput.* **2003**, *7*, 83–94. [CrossRef]
2. Chiang, C.L. Improved genetic algorithm for power economic dispatch of units with valve-point effects and multiple fuels. *IEEE Trans. Power Syst.* **2005**, *20*, 1690–1699. [CrossRef]
3. Park, J.B.; Lee, K.S.; Shin, J.R. A particle swarm optimization for economic dispatch with non-smooth cost functions. *IEEE Trans. Power Syst.* **2005**, *20*, 34–42. [CrossRef]
4. Balamurugan, R.; Subramanian, S. Hybrid integer coded differential evolution-dynamic programming approach for economic load dispatch with multiple fuel options. *Energy Convers. Manag.* **2008**, *49*, 608–614. [CrossRef]
5. Pothiya, S.; Ngamroo, I.; Kongprawechnon, W. Ant colony optimisation for economic dispatch problem with non-smooth cost functions. *Int. J. Electr. Power Energy Syst.* **2010**, *32*, 478–487. [CrossRef]
6. Basu, M. A simulated annealing-based goal-attainment method for economic emission load dispatch of fixed head hydrothermal power systems. *Electr. Power Energy Syst.* **2005**, *27*, 147–153. [CrossRef]
7. Kennedy, J.; Eberhart, R. Particle Swarm Optimization. In Proceedings of the IEEE International Conference on Neural Networks Proceedings, Perth, WA, Australia, 27 November–1 December 1995; pp. 1942–1948.
8. Sun, J.; Xu, W.B.; Feng, B. A Global Search Strategy of Quantum-Behaved Particle Swarm Optimization. In Proceedings of the IEEE Conference on Cybernetics and Intelligent Systems, Singapore, 1–3 December 2004; pp. 111–116.

9. Fang, W. A review of quantum-behaved particle swarm optimization. *IETE Tech. Rev.* **2010**, *27*, 336–348. [CrossRef]

10. Sun, J.; Wu, X.J.; Palade, V.; Fang, W.; Lai, C.-H.; Xu, W.B. Convergence analysis and improvements of quantum-behaved particle swarm optimization. *Inf. Sci.* **2012**, *193*, 81–103. [CrossRef]

11. Sun, J.; Fang, W.; Palade, V.; Wu, X.J.; Xu, W.B. Quantum-behaved particle swarm optimization with Gaussian distributed local attractor point. *Appl. Math. Comput.* **2011**, *218*, 3763–3775. [CrossRef]

12. Sun, J.; Fang, W.; Wu, X.J.; Palade, V.; Xu, W.B. Quantum-behaved particle swarm optimization: Analysis of the individual particle's behavior and parameter selection. *Evolut. Comput.* **2012**, *20*, 349–393. [CrossRef]

13. Luitel, B.; Venayagamoorthy, G.K. Particle swarm optimization with quantum infusion for system identification. *Eng. Appl. Artif. Intell.* **2010**, *23*, 635–649. [CrossRef]

14. Sun, J.; Liu, J.; Xu, W. Using quantum-behaved particle swarm optimization algorithm to solve non-linear programming problems. *Int. J. Comp. Math.* **2007**, *84*, 261–272. [CrossRef]

15. Sun, C.F.; Lu, S.F. Short-term combined economic emission hydrothermal scheduling using improved quantum-behaved particle swarm optimization. *Expert Syst. Appl.* **2010**, *37*, 4232–4241. [CrossRef]

16. Leandro dos, S.C.; Viviana, C.M. Particle swarm approach based on quantum mechanics and harmonic oscillator potential well for economic load dispatch with valve-point effects. *Energy Convers. Manag.* **2008**, *49*, 3080–3085. [CrossRef]

17. Chakraborty, S.; Senjyu, T.; Yona, A.; Saber, A.Y.; Funabashi, T. Solving economic load dispatch problem with valve-point effects using a hybrid quantum mechanics inspired particle swarm optimization. *IET Gener. Transm. Distrib.* **2011**, *5*, 1042–1052. [CrossRef]

18. Park, J.B.; Jeong, Y.W.; Shin, J.R.; Lee, K.Y. An improved particle swarm optimization for nonconvex economic dispatch problems. *IEEE Trans. Power Syst.* **2010**, *25*, 156–166. [CrossRef]

19. Shi, Y.; Eberhart, R. Empirical Study of Particle Swarm Optimization. In Proceedings of the 1999 Congress on Evolutionary Computation (CEC 99), Washington, DC, USA, 6–9 July 1999; Volume 3, pp. 1945–1950.

20. Coelho, L.S.; Mariani, V.C. Combining of chaotic differential evolution and quadratic programming for economic dispatch optimization with valve-point effect. *IEEE Trans. Power Syst.* **2006**, *21*, 989–996. [CrossRef]

21. Meng, K. Quantum-inspired particle swarm optimization for valve-point economic load dispatch. *IEEE Trans. Power Syst.* **2010**, *25*, 215–222. [CrossRef]

22. Alsumait, J.S.; Sykulski, J.K. A hybrid GA-PS-SQP method to solve power system valve-point economic dispatch problems. *Appl. Energy* **2010**, *87*, 1773–1781. [CrossRef]

23. Chaturvedi, K.T.; Pandit, M.; Srivastava, L. Self-organizing hierarchical particle swarm optimization for nonconvex economic dispatch. *IEEE Trans. Power Syst.* **2008**, *23*, 1079–1087. [CrossRef]

24. Selvakumar, A.I.; Thanushkodi, K. A new particle swarm optimization solution to nonconvex economic dispatch problems. *IEEE Trans. Power Syst.* **2007**, *22*, 42–51. [CrossRef]

25. Lu, H.Y. Experimental study of a new hybrid PSO with mutation for economic dispatch with non-smooth cost function. *Electr. Power Energy Syst.* **2010**, *32*, 921–935. [CrossRef]

26. Vlachogiannis, J.K.; Lee, K.Y. Economic load dispatch—A comparative study on heuristic optimization techniques with an improved coordinated aggregation-based PSO. *IEEE Trans. Power Syst.* **2009**, *24*, 991–1001. [CrossRef]

27. Selvakumar, A.I.; Thanushkodi, K. Anti-predatory particle swarm optimization: Solution to nonconvex economic dispatch problems. *Electr. Power Syst. Res.* **2010**, *78*, 2–10. [CrossRef]

28. Wang, S.K.; Chiou, J.P.; Liu, C.W. Non-smooth/non-convex economic dispatch by a novel hybrid differential evolution algorithm. *IET Gener. Transm. Distrib.* **2007**, *1*, 793–803. [CrossRef]

29. Lee, K.Y.; Sode-Yome, A.; Park, J.H. Adaptive hopfield neural networks for economic load dispatch. *IEEE Trans. Power Syst.* **1998**, *13*, 519–525. [CrossRef]

30. Cai, J.J.; Li, Q.; Li, L.X.; Peng, H.P.; Yang, Y.X. A fuzzy adaptive chaotic ant swarm optimization for economic dispatch. *Electr. Power Energy Syst.* **2012**, *34*, 154–160. [CrossRef]

31. Cai, J.J.; Li, Q.; Li, L.X.; Peng, H.P.; Yang, Y.X. A hybrid CPSO-SQP method for economic dispatch considering the valve-point effects. *Energy Convers. Manag.* **2012**, *53*, 175–181. [CrossRef]

32. Wang, Y.; Li, B.; Weise, T. Estimation of distribution and differential evolution cooperation for large scale economic load dispatch optimization of power systems. *Inf. Sci.* **2010**, *180*, 2405–2420. [CrossRef]

Energies **2012**, *5*, 3655–3673

33. Amjady, N.; Nasiri-Rad, H. Nonconvex economic dispatch with AC constraints by a new real coded genetic algorithm. *IEEE Trans. Power Syst.* **2009**, *24*, 1489–1502. [CrossRef]

34. Khamsawang, S.; Jiriwibhakorn, S. DSPSO-TSA for economic dispatch problem with nonsmooth and noncontinuous cost functions. *Energy Convers. Manag.* **2010**, *51*, 365–375. [CrossRef]

Article

Annual Electric Load Forecasting by a Least Squares Support Vector Machine with a Fruit Fly Optimization Algorithm

Hongze Li [1], Sen Guo [1],*, Huiru Zhao [1], Chenbo Su [2] and Bao Wang [1]

[1]   School of Economics and Management, North China Electric Power University, Beijing 102206, China; lihongze@163.com (H.L.); huiruzhao@163.com (H.Z.); wangbao610@163.com (B.W.)

[2]   School of Electrical and Electronic Engineering, North China Electric Power University, Beijing 102206, China; scb2636@sina.com

*   Author to whom correspondence should be addressed; guosen324@163.com; Tel.: +86-10-5196-3626; Fax: +86-10-8079-6904.

Received: 14 September 2012; in revised form: 18 October 2012; Accepted: 2 November 2012; Published: 8 November 2012

Abstract: The accuracy of annual electric load forecasting plays an important role in the economic and social benefits of electric power systems. The least squares support vector machine (LSSVM) has been proven to offer strong potential in forecasting issues, particularly by employing an appropriate meta-heuristic algorithm to determine the values of its two parameters. However, these meta-heuristic algorithms have the drawbacks of being hard to understand and reaching the global optimal solution slowly. As a novel meta-heuristic and evolutionary algorithm, the fruit fly optimization algorithm (FOA) has the advantages of being easy to understand and fast convergence to the global optimal solution. Therefore, to improve the forecasting performance, this paper proposes a LSSVM-based annual electric load forecasting model that uses FOA to automatically determine the appropriate values of the two parameters for the LSSVM model. By taking the annual electricity consumption of China as an instance, the computational result shows that the LSSVM combined with FOA (LSSVM-FOA) outperforms other alternative methods, namely single LSSVM, LSSVM combined with coupled simulated annealing algorithm (LSSVM-CSA), generalized regression neural network (GRNN) and regression model.

Keywords: annual electric load forecasting; least squares support vector machine (LSSVM); fruit fly optimization algorithm (FOA); optimization problem

---

1. Introduction

With the rapid development of China's electric power industry, electric load forecasting technology has aroused widespread concerns among practitioners and academia. An effective and accurate electric load forecast can provide the basis for the decision-making of electric power system planners. To a certain extent, the annual electric load forecasting can affect the development trends of the electric power industry. With the construction and development of the "Strong Smart Grid" in China, the renewable distributed energy generation capacity is growing rapidly, which may influence the stability of power system operation. In view of this, more accurate annual electric load forecasting is needed for maintaining the secure and stable operation of the electric power grid. However, annual electric loads have complex and non-linear relationships with some factors such as the political environment, human activities, and economic policy [1], making it is quite difficult to accurately forecast annual electric loads.

To improve the accuracy of annual electric load forecasting, many approaches have been proposed by scholars and practitioners in the past decades, such as time series technology and regression models [2–6]. However, it is difficult to achieve significant improvements in terms of forecasting accuracy with these forecasting methods due to their poor non-linear fitting capability. In recent years, many artificial intelligence forecasting techniques have been applied in annual power load forecasting to improve the forecasting accuracy. Niu *et al.* [7] proposed a combined forecasting method based on a particle swarm optimization method, which can improve the forecasting stability and reliability. Wang *et al.* [1] proposed a hybrid model combining support vector regression and a differential evolution algorithm to forecast the annual power load, which was proven to outperform the SVR model with default parameters, regression forecasting model and back propagation artificial neural network (BPNN). Xia *et al.* [8] developed a medium and long term load forecasting model by using a radial basis function neural network (RBFNN), and the computational results indicated that this proposed model has a higher forecasting accuracy and stability. Hsu and Chen [9] formulated an artificial neural network model by collecting empirical data to forecast the regional peak load of Taiwan. Abou El-Ela *et al.* [10] proposed the artificial neural network (ANN) technique for long-term peak load forecasting, which was applied at the Egyptian electrical network based on its historical data. Meng *et al.* [11] applied the partial least squares method which could simulate the relationship between the electricity consumption and its influencing factors to forecast electricity load, and the empirical results revealed that this method is effective. Chen [12] proposed a collaborative fuzzy-neural approach for forecasting Taiwan's annual electricity load, and this approach could improve the forecasting accuracy. Kandil *et al.* [13] implemented a knowledge-based expert system to support the choice of the most suitable load forecasting model, and the usefulness of this method was demonstrated by a practical application. Hong [14] proposed an electric load forecasting model which combined the seasonal recurrent support vector regression model with a chaotic artificial bee colony algorithm, and this method could provide a more accurate forecasting result than the TF-ε-SVR-SA and ARIMA model. Pai *et al.* [15] used support vector machines with a simulated annealing algorithm to forecast Taiwan's electricity load, and the empirical results revealed this model outperforms the general regression neural network model and the autoregressive integrated moving average model. These methods, to a certain extent, all improve the annual electric load forecasting accuracy.

The least squares support vector machine (LSSVM) is a reformulation of the support vector machine (SVM) which leads to solving a linear KKT system [16,17]. The LSSVM can approach the non-linear system with high precision, making it a powerful tool for modeling and forecasting non-linear systems [18]. The LSSVM model has been successfully used to solve forecasting problems in many fields, such as CO concentration [19], gas [20,21], short term electric load [22–24], revenue [25], precipitation [26], wind speed [27], hydropower consumption forecasting [28], and so on. However, it is very regretful to find that the LSSVM model has rarely been applied to annual electric load forecasting. This paper examines the feasibility of using the LSSVM model to forecast annual electric loads. The forecasting performance of the LSSVM model largely depends on the values of its two parameters. Currently, several meta-heuristic algorithms have been used to determine the appropriate values of these two parameters, including particle swarm optimization [20], genetic algorithm [22], chaotic differential evolution approach [29], artificial bee colony algorithm [30], and simulated annealing algorithm [31]. However, these optimization algorithms have the drawbacks of being hard to understand and reaching the global optimal solution slowly. The fruit fly optimization algorithm (FOA) proposed by Pan in 2011 [32], is a novel evolutionary computation and optimization technique. This new optimization algorithm has the advantages of being easy to understand due to the shorter program code compared with other optimization algorithms and of reaching the global optimal solution fast. Therefore, this paper attempts to use the FOA to automatically determine the appropriate values of the two necessary parameters in order to improve the performance of the LSSVM model in annual electric load forecasting.

The rest of this paper is organized as follows: Section 2 introduces the LSSVM model and FOA, then a hybrid annual electric load forecasting model (LSSVM-FOA) that combines LSSVM model and FOA is discussed in detail. Section 3 introduces the sample data processing procedure used in this paper, and the computation, comparison and discussion of a numerical example is presented. Section 4 concludes this paper.

2. Methodology of the LSSVM-FOA Model

2.1. Least Squares Support Vector Machine (LSSVM) Model

The LSSVM is an extension of SVM which applies the linear least squares criteria to the loss function instead of inequality constraints [33]. The basic principle is as follows [34]: given a set of samples $\{x_i, y_i\}_{i=1}^m$, where $x_i \in \mathbf{R}^n$ is the input vector and $y_i \in \mathbf{R}$ is the corresponding output value for sample i. By a nonlinear function φ, the data are mapped from the original feature space to a higher dimensional transformed one, thus, to approximate it in a linear way as follows:

$$f(x) = w^T \varphi(x) + b \tag{1}$$

where w denotes the weight vector; and b denotes the error.

In the primal space, the LSSVM formulation with the equality constraints can be described as:

$$\begin{cases} \min J(w, \xi) = \dfrac{1}{2} w^T w + \dfrac{1}{2} C \sum_{i=1}^m \xi_i^2 \\ s.t. \quad y_i = w^T \varphi(x_i) + b + \xi_i, \quad i = 1, \cdots, m \end{cases} \tag{2}$$

where C is the regularization parameter; and ξ_i is the slack variable.

The Lagrangian function L can be constructed by:

$$L(w, b, \xi, a) = \frac{1}{2} w^T w + \frac{1}{2} C \sum_{i=1}^m \xi_i^2 - \sum_{i=1}^m a_i \{ w^T \varphi(x_i) + b + \xi_i - y_i \} \tag{3}$$

where a_i is the Lagrange multiplier. The Karush–Kuhn–Tucker (KKT) conditions for optimality are given by:

$$\begin{cases} \dfrac{\partial L}{\partial w} = 0 \rightarrow w = \sum_{i=1}^m a_i \varphi(x_i) \\ \dfrac{\partial L}{\partial b} = 0 \rightarrow \sum_{j=1}^m a_i = 0 \\ \dfrac{\partial L}{\partial \xi_i} = 0 \rightarrow a_i = C \xi_i \\ \dfrac{\partial L}{\partial a_i} = 0 \rightarrow w^T \varphi(x_i) + b + \xi_i - y_i = 0 \end{cases} \tag{4}$$

Eliminating the variables w and ξ_i, the optimization problem can be transformed into the following linear solution:

$$\begin{bmatrix} 0 & Q^T \\ Q & K + C^{-1} I \end{bmatrix} \begin{bmatrix} b \\ A \end{bmatrix} = \begin{bmatrix} 0 \\ Y \end{bmatrix} \tag{5}$$

35

where $Q = [1, \dots ,1]^T$, $A = [a_1, a_2, \dots ,a_m]^T$, $Y = [y_1, y_2, \dots ,y_m]^T$. According to the Mercer's condition, the Kernel function can be set as:

$$K(x_i, x_j) = \varphi(x_i)^T \varphi(x_j) \qquad (6)$$

Then, the LSSVM model for regression becomes:

$$f(x) = \sum_{i=1}^{m} a_i K(x, x_i) + b \qquad (7)$$

There are several different types of Mercer kernel function $K(x, x_i)$ such as sigmoid, polynomial and radial basis function (RBF). The RBF is a common option for the kernel function because of fewer parameters that need to be set and an excellent overall performance [35]. Therefore, this paper selected the RBF [as shown in Equation (8)] as the kernel function:

$$K(x, x_i) = \exp\left\{-\|x - x_i\|^2 / 2\sigma^2\right\} \qquad (8)$$

Consequently, there are two parameters that need to be chosen in the LSSVM model, which are the bandwidth of the Gaussian RBF kernel "σ" and the regularization parameter "C". In this paper, the FOA is used to determine the optimal values of these two parameters.

2.2. Fruit Fly Optimization Algorithm (FOA)

The fruit fly optimization algorithm (FOA) is a new swarm intelligence algorithm, which was proposed by Pan [32] in 2011. It is a kind of interactive evolutionary computation method. By imitating the food finding behavior of the fruit fly swarm, the FOA can reach the global optimum.

Fruit flies are a kind of insect, which live in the temperate and tropical climate zones and eat rotten fruit. The fruit fly is superior to other species in vision and osphresis. The food finding process of fruit fly is as follows: it firstly smells the food source with its osphresis organ, and flies towards that location; after it gets close to the food location, its sensitive vision is also used for finding food and other fruit flies' flocking location, and then it flies towards that direction. The FOA has been applied to several fields including traffic incidents [36], export trade forecasting [37], and the design of analog filters [38]. Figure 1 shows the food finding iterative process of a fruit fly swarm.

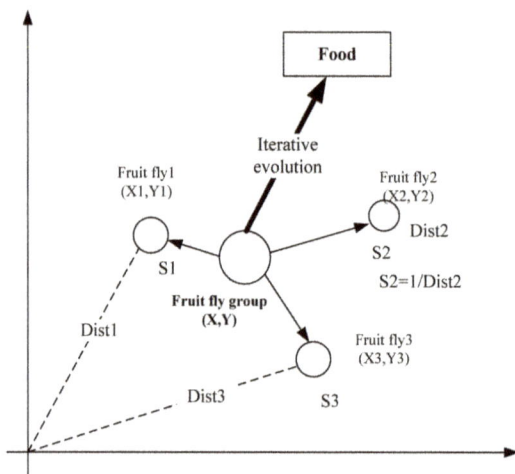

Figure 1. Food finding iterative process of a fruit fly swarm.

According to the food finding characteristics of fruit fly swarm, the FOA can be divided into several steps, as follows:

Step 1: Parameter Initialization

The main parameters of FOA are the maximum iteration number *maxgen*, the population size *sizepop*, the initial fruit fly swarm location (*X_axis,Y_axis*), and the random flight distance range *FR*.

Step 2: Population Initialization

Give the random flight direction and the distance for food finding of an individual fruit fly by using osphresis:

$$X_i = X\_axis + \text{Random Value} \tag{9}$$

$$Y_i = Y\_axis + \text{Random Value} \tag{10}$$

Step 3: Population Evaluation

Firstly, the distance (*Dist*) of the fruit fly to the origin needs to be calculated. Secondly, the smell concentration judgment value (*S*) needs to be calculated. Suppose that *S* is the reciprocal of *Dist*:

$$Dist_i + \left(X_i^2 + Y_i^2 \right)^{1/2} \tag{11}$$

$$S_i = 1/Dist_i \tag{12}$$

Then, we calculate the smell concentration (*Smell_i*) of the individual fruit fly location by substituting the smell concentration judgment value (*S_i*) into the smell concentration judgment function (also called Fitness function). Finally, find out the individual fruit fly with the maximal smell concentration (the maximal value of *Smell_i*) among the fruit fly swarm:

$$Smell_i = \text{Function} (S_i) \tag{13}$$

$$[bestSmell \; bestIndex] = \max (Smell_i) \tag{14}$$

Step 4: Selection Operation

Keep the maximal smell concentration value and *x, y* coordinates. Then, the fruit flies fly towards the location with the maximal smell concentration value by using vision. Enter iterative optimization to repeat the implementation of step 2–3. When the smell concentration is not superior to the previous iterative smell concentration any more, or the iterative number reaches the maximal iterative number, the circulation stops:

$$Smellbest = bestSmell \tag{15}$$

$$X\_axis = X (bestIndex) \tag{16}$$

$$Y\_axis = Y (bestIndex) \tag{17}$$

2.3. LSSVMFOA Forecasting Model

The diagram of procedure structure of the LSSVM-FOA forecasting model is illustrated in Figure 2. The details of FOA for parameter determination of the LSSVM model are as follows:

Step1: Initialization Parameters

The maximum iteration number *maxgen*, the population size *sizepop*, the initial fruit fly swarm location (*X_axis,Y_axis*), and the random flight distance range *FR* should be determined at first. In this study, we suppose *maxgen* = 100, *sizepop* = 20, (*X_axis,Y_axis*) ÿ [−50, 50], *FR* ⊂ [−10,10]. In the LSSVMFOA program, we set *X_axis* = *rands*(1,2), *Y_axis* = *rands*(1,2), where *rands*() denotes the random number generation function.

Step2: Evolution Starts

Set *gen* = 0, and give the random flight direction *rand*() and the flight distance for food finding of an individual fruit fly *i*. In the LSSVMFOA program, we employ two variables [*X(i,:),Y(i,:)*] to represent the flight distance for food finding of an individual fruit fly i, and set $X(i,:) = X\_axis + 20 *$ $rand() - 10$, $Y(i,:) = Y\_axis + 20 * rand() - 10$, respectively.

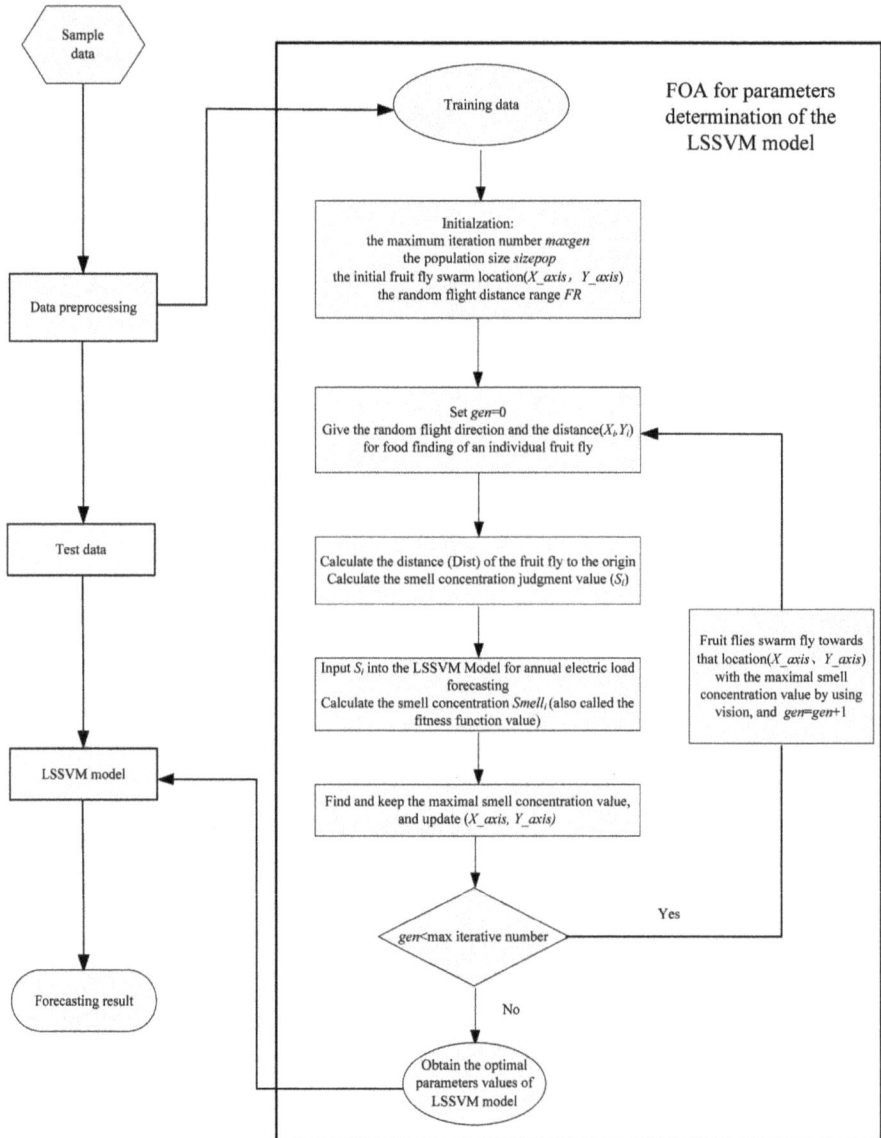

Figure 2. Diagram of the procedure structure of the LSSVM-FOA forecasting model.

Step3: Preliminary Calculations

Calculate the distance $Dist_i$ of the fruit fly i to the origin, and then calculate the smell concentration judgment value S_i. In the LSSVM-FOA program, we employ $(D(i,1),D(i,2))$ to represent $Dist_i$, and set $D(i,1) = (X(i,1)^2 + Y(i,1)^2)^{0.5}$, $D(i,2) = (X(i,2)^2 + Y(i,2)^2)^{0.5}$, respectively. Similarly, we use $(S(i,1), S(i,2))$ to represent S_i in the LSSVM-FOA program, and set $S(i,1) = 1/D(i,1)$, $S(i,2) = 1/D(i,2)$, respectively. Then, input S_i into the LSSVM model for annual electric load forecasting. In the

LSSVM-FOA program, the parameters [*C*,σ] of LSSVM model are represented by [*S*(*i*,1),*S*(*i*,2)], and we set *C* = 20 * *S*(*i*,1) and σ² = *S*(*i*,2), respectively. According to the electric load forecasting result, the smell concentration *Smell*$_i$ (also called the fitness function value) can be calculated. The *Smell*$_i$ is employed by the root-mean-square error (*RMSE*), as shown in Equation (18), which measures the deviations between the forecasting values and actual values:

$$RMSE = \sqrt{\frac{\sum_{i=1}^{n}(f_i - \hat{f_i})^2}{n}}$$

(18)

where *n* is the number of forecasting periods; f_i is the actual value at period *i*; f_i denotes the forecasting value at period *i*.

Step4: Offspring Generation

The offspring generation is generated according to Equations (9–14). Then input the offspring into the LSSVM model and calculate the smell concentration value again. Set *gen* = *gen* + 1.

Step5: Circulation Stops

When *gen* reaches the max iterative number, the stop criterion satisfies, and the optimal parameters of LSSVM model are obtained. Otherwise, go back to Step2.

3. Example Computation and Discussion

3.1. The Preprocessing of Sample Data

The sample data were selected from the annual electricity consumption of China between 1978 and 2011, shown in Table 1. Before the calculation, the sample data were normalized to make them in the range from 0 to 1 using the following formula:

$$Z = \{z_i\} = \frac{x_i - x_{i\min}}{x_{i\max} - x_{i\min}}, \qquad i = 1, 2, 3$$

(19)

where x_{imin} and x_{imax} denote the minimal and maximal value of each input factor, respectively.

The sample data were divided into the training data and testing data. Different from the short term electric load forecasting, the annual electric load forecasting is not suitable for selecting the factors such as temperature, moderate [1]. Therefore, this paper selected the last three load data (L_{n-3}, L_{n-2}, L_{n-1}) as the input variables of the LSSVMFOA model, and the output variable is L_n. Due to using the last three electric load data as the input variables to forecast, the training data started in 1981 and ended in 2005, and the testing data were from 2006 to 2011.

In the training stage, a roll-based data processing procedure was used. Firstly, the top three load data (from 1978 to 1980) of the sample data were substituted into the LSSVM-FOA model, and then the electric load forecasting value of 1981 could be obtained. Secondly, the next roll-top three load data (from 1979 to 1981) were fed into the LSSVM-FOA model, and the forecasting value of 1982 could be produced. In this step, the electric load value of 1981 which was fed into the proposed LSSVM-FOA model should employ the actual electric load value of 1981. Similarly, the forecasting processes were cycling until all the electric load forecasting values (from 1981 to 2005) were obtained. Because of the roll-based data processing procedure, the value of *n* in Equation (18) equals to 25.

Table 1. Annual electricity consumption of China between 1978 and 2011 (unit: 10^9 kWh).

Year	Electricity consumption	Year	Electricity consumption	Year	Electricity consumption
1978	246.53	1990	623.59	2002	1633.15
1979	282.02	1991	680.96	2003	1903.16
1980	300.63	1992	759.27	2004	2197.14
1981	309.65	1993	842.65	2005	2494.03
1982	327.92	1994	926.04	2006	2858.80
1983	351.86	1995	1002.34	2007	3271.18
1984	377.89	1996	1076.43	2008	3454.14
1985	411.90	1997	1128.44	2009	3703.22
1986	451.03	1998	1159.84	2010	4199.90
1987	498.84	1999	1230.52	2011	4690.00
1988	547.23	2000	1347.24	-	-
1989	587.18	2001	1463.35	-	-

Sample data sources: the data of 1978–2010 come from reference [39]; the data of 2011 comes from reference [40].

3.2. The Selection of Comparison Models

To compare the annual electric load forecasting result, several other electric load forecasting models were selected. From Table 1, we can discern that the annual electric load series shows an increasing approximately linear trend. Therefore, the regression forecasting model was employed. In the meantime, the single LSSVM model, LSSVM model combined with coupled simulated annealing algorithm (LSSVM-CSA) [41], and generalized regression neural network (GRNN) model were also employed for comparison. GRNN is a kind of radial basis function (RBF) networks which is based on a standard statistical technique called kernel regression, and it has excellent performances on approximation ability and learning speed [42,43]. In GRNN model, there is only one parameter σ that needs to be determined.

The experimental environment includes Matlab 2010a, LSSVMlabv1.8 toolbox [44,45], GRNN toolbox, self-written MATLAB programs and a computer with an Intel(R) Core(TM)2 T2450 2 GHz CPU, 1.5 GB RAM and the Windows 7 Professional operating system.

3.3. FOA Result for Parameter Determination of the LSSVM Model

In LSSVM-FOA model, the values of the two parameters of LSSVM model were dynamically tuned by the FOA. Figure 3a shows the fruit fly swarm flying route for parameter optimization. It can be seen that the fruit fly swarm flying route is relatively stable, and the fruit fly swarm moves straight to the food location. The fruit fly swarm fixes the food location accurately and fast. The iterative RMSE trend of the LSSVM-FOA model when searching for the optimal parameters is shown in Figure 3b. After 100 evolution iterations, the convergence can be seen in generation 17 with the coordinate of (441,362), and the optimal values of the parameters σ and C are 0.7051, 17.3571, respectively.

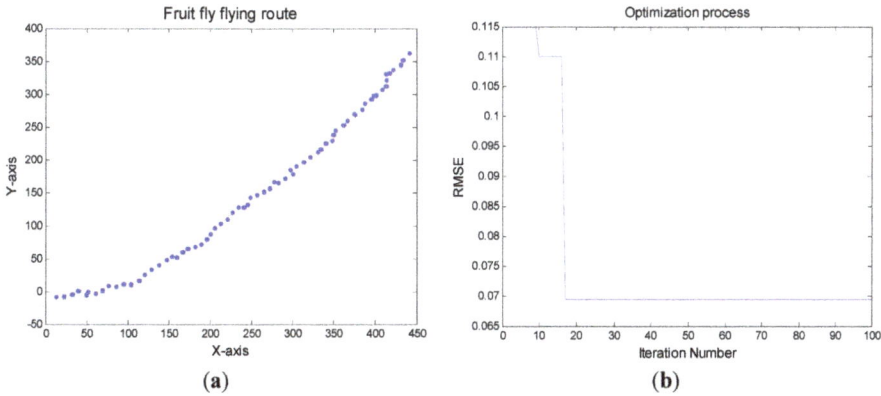

Figure 3. (a) The fruit fly swarm flying route for parameter optimization; (b) The iterative RMSE trend of the LSSVM-FOA model searching for optimal parameters.

3.4. Forecasting Result and Discussion

According to the result of the FOA tuning the parameters of LSSVM model, the values of σ and *C* were chosen as 0.7051 and 17.3571, respectively. In the single LSSVM model, the values of σ and *C* were chosen as 5 and 10, respectively. In the LSSVM-CSA model, radial basis function was chosen as the kernel function. According to the result of CSA optimizing the parameters of LSSVM model, the optimal values of σ and *C* were 10.8494 and 12185.8, respectively. In the GRNN model, the spread parameter value was chosen as 0.2.

With the LSSVM-FOA, single LSSVM, LSSVM-CSA, GRNN and regression model, the training times of the data are 17, 13, 36, 14 and 8 s, respectively. The training time of these five models on disposing of the training data are different. The LSSVM-FOA and LSSVM-CSA use longer times than the single LSSVM, GRNN and regression model because they need to determine the parameters in the each generation. However, the LSSVM-FOA uses 19 s less than the LSSVM-CSA computation.

Table 2 lists the annual electric load forecasting results with the LSSVM-FOA, LSSVM, LSSVM-CSA, GRNN, and regression model. Figure 4 describes the relative errors of the forecasting results of these five models. From Table 2 and Figure 4, the deviations between the forecasting results of these five forecasting models and the actual values can be captured. The relative error ranges [−3%,+3%] and [−1%,+1%] are always considered as a standard to assess the performance of a forecasting model [46]. Firstly, the relative errors of annual electric load forecasting points of LSSVM-FOA model are all in the range [−3%,+3%], and the maximum and minimum relative errors are 2.265% in 2008 and −0.603% in 2009, respectively. In addition, two out of six points means that 33% of the forecasting points are in the scope of [−1%,+1%], which are −0.603% in 2009 and −0.811% in 2011. Secondly, the single LSSVM model has two forecasting points that exceed the relative error range [−3%,+3%], which are 3.139% in 2008 and 4.412% in 2009, respectively. However, all the forecasting points exceed the scope of [−1%,+1%], and the maximum and minimum relative errors are 4.412% in 2009 and −1.863% in 2011, respectively. Thirdly, the LSSVM-CSA model has one forecasting point that exceeds the relative error range [−3%,+3%], which is 3.529% in 2008. For LSSVM-CSA model, there is one forecasting point in the scope of [−1%,+1%], which is −0.632% in 2007, and the maximum and minimum relative errors are 3.529% in 2008 and −0.632% in 2007, respectively. Fourthly, the GRNN model has three forecasting points that exceed the relative error range [−3%,+3%], which are 3.355% in 2006, −3.664% in 2007, and 3.509% in 2009. All the forecasting points of the GRNN model exceed the scope of [−1%,+1%]. Finally, the regression model has four forecasting points that exceed the relative error range [−3%,+3%], which are 7.354% in 2008, 3.017% in 2009, −3.119% in 2010, and 3.477% in

2011, and one forecasting point in the scope of [−1%,+1%], which are −0.410% in 2007. The maximum relative error of regression model is 7.354%, which is the largest among these five forecasting models.

Table 2. Forecasting results of LSSVM-FOA, single LSSVM, LSSVM-CSA, GRNN, and regression model (unit: 10^9 kWh).

Year	Actual value	LSSVM-FOA	LSSVM	LSSVM-CSA	GRNN	Regression
2006	2858.80	2896.83	2914.43	2915.69	2954.72	2794.15
2007	3271.18	3218.18	3180.73	3250.50	3151.32	3257.77
2008	3454.14	3532.38	3562.55	3576.02	3522.79	3708.16
2009	3703.22	3680.91	3866.61	3763.08	3833.16	3591.50
2010	4199.90	4250.26	4282.43	4301.53	4247.11	4068.92
2011	4690.00	4651.95	4602.62	4616.40	4572.24	4853.09

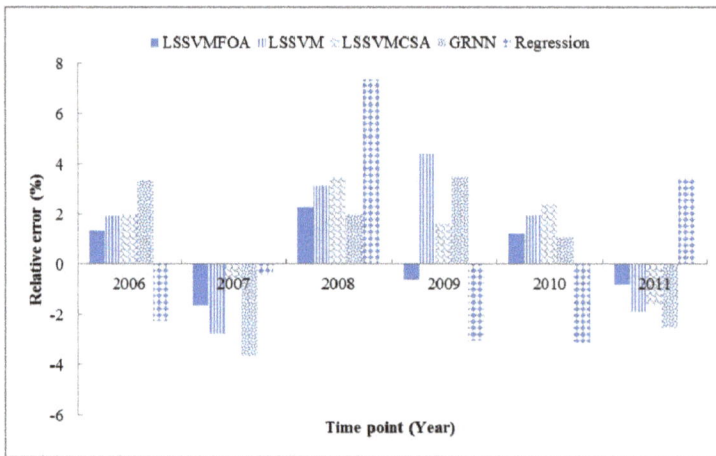

Figure 4. The relative errors of the forecasting results of the different forecasting models.

The mean absolute percentage error (MAPE), mean square error (MSE), and average absolute error (AAE) were also used to assess the performances of different forecasting models in this paper. The values of MAPE, MSE, and AAE can be calculated by:

$$MAPE = \frac{1}{n}\sum_{i=1}^{n}\left|\frac{A(i)-F(i)}{A(i)}\right| \times 100\%$$

(20)

$$MSE = \frac{1}{n}\sum_{i=1}^{n}(A(i)-F(i))^2$$

(21)

$$AAE = \frac{1}{n}\sum_{i=1}^{n}\frac{|A(i)-F(i)|}{\frac{1}{n}\sum_{i=1}^{n}A(i)}$$

(22)

where $A(i)$ is the actual electric load value at time i; and $F(i)$ is the forecasting value at time i.

Comparisons of the values of MAPE, MSE, and AAE for the LSSVM-FOA, LSSVM, LSSVM-CSA, GRNN and regression model are listed in Table 3. It can be seen that the MAPE value of LSSVM-FOA model is 1.305%, which is much smaller than that obtained by single LSSVM, LSSVM-CSA, GRNN and regression model (which are 2.682%, 1.959%, 2.692%, and 3.273%, respectively). The MSE value of LSSVM-FOA model is 2,476, which is dramatically smaller than that obtained by another four models (which are 10,695, 6,308, 10,210, and 20,853, respectively). The AAE value of LSSVM-FOA model is 0.0126, which is much smaller than that obtained by single LSSVM, LSSVM-CSA, GRNN and regression model (which are 0.0265, 0.0196, 0.0261, and 0.0333, respectively). Meanwhile, the values of MAPE, MSE, and AAE of LSSVM-CSA model are much smaller than that of single LSSVM, GRNN and regression models. These indicate that the meta-heuristic algorithms for parameter selection have the potential to be employed for the LSSVM-based annual electric load forecasting model to improve the forecasting accuracy. In this paper, the LSSVM-FOA model has better forecasting performance than the LSSVM-CSA model. Furthermore, because the values of MAPE, MSE, and AAE are the largest, the regression model has the lowest forecasting accuracy, which reveals its poor non-linear fitting capability. The MAPE value of the single LSSVM model is smaller than that of GRNN model, but the MSE and AAE values are much larger. So, it is still unclear when the LSSVM-based annual electric load forecasting model performs better than the GRNN-based annual electric load forecasting model in this paper.

Table 3. The values of MAPE, MSE, and AAE for LSSVM-FOA, single LSSVM, LSSVM-CSA, GRNN and regression model.

Model	LSSVM-FOA	LSSVM	LSSVM-CSA	GRNN	Regression
MAPE (%)	1.305	2.682	1.959	2.692	3.273
MSE	2476	10695	6308	10210	20853
AAE	0.0126	0.0265	0.0196	0.0261	0.0333

In conclusion, the proposed LSSVM-FOA model greatly narrows the deviations between the forecasting values and actual values, and outperforms the single LSSVM, LSSVM-CSA, GRNN, and regression model in the annual electric load forecasting.

4. Conclusions

With the construction of the "Strong Smart Grid" and the increasing generation capacity of renewable distributed energy, accurate electric load forecasting is a guide for effective implementations of energy policies in China of greatly importance. However, the non-linear relationship of annual electric load with its influencing factors makes electric load forecasting very complicated. Thus, how to improve the annual electric load forecasting accuracy is worthy of study. The least squares support vector machine has been widely applied to a variety of fields, but it is regretful to find that the LSSVM have rarely been applied to the problem of annual electric load forecasting. The fruit fly optimization algorithm (FOA) is a new swarm intelligence algorithm which has the advantages of being easy to understand due to its shorter program code compared with other meta-heuristic algorithms, and reaching the global optimal solution fast. In this paper, we hybridized the LSSVM and FOA, in the so-called LSSVM-FOA model, to examine its potential for annual electric load forecasting. To validate the proposed method, four other alternative models (single LSSVM, LSSVM-CSA, GRNN, and regression model) were employed to compare the forecasting performances. Example computation results show that the relative errors of annual electric load forecasting points of LSSVM-FOA model are all in the range [−3%,+3%], and the values of MAPE, MSE and AAE are much smaller than that obtained by single LSSVM, LSSVM-CSA, GRNN, and regression model. These indicate the proposed LSSVM-FOA model has significant superiority over other alternative forecasting models in terms of the annual electric load forecasting accuracy. The hybridization of the least squares support vector machine and fruit fly optimization algorithm is feasible. The LSSVM-FOA model uses 19 s less than

Energies **2012**, *5*, 4430–4445

the LSSVM-CSA computation, which testifies to the FOA's advantage in reaching the global optimal solution fast compared with other meta-heuristic algorithms. Although the LSSVM-FOA model is a little time consuming compared with single LSSVM, some attentions should be paid to this new hybrid forecasting model. The proposed LSSVM-FOA model which uses the FOA to automatically determine the appropriate values of the two parameters for the LSSVM model can effectively improve the annual electric load forecasting accuracy. We also conclude that the artificial intelligence forecasting models have much better performance than the regression models, which reveals that artificial intelligence forecasting models have good non-linear fitting capacity. Meanwhile, the meta-heuristic algorithms for parameter selection have the potential to be employed for the LSSVM-based annual electric load forecasting model to improve the forecasting accuracy.

Acknowledgments: The authors would like to acknowledge the grant from the Beijing Philosophy and Social Science Planning Project (Project number: 11JGB070) and the Humanities and Social Science project of the Ministry of Education of China (Project number: 11YJA790217).

References

1. Wang, J.J.; Li, L.; Niu, D.X.; Tan, Z.F. An annual load forecasting model based on support vector regression with differential evolution algorithm. *Appl. Energy* **2012**, *94*, 65–70. [CrossRef]
2. Pappas, S.S.; Ekonomou, L.; Karamousantas, D.C.; Chatzarakis, G.E.; Katsikas, S.K.; Liatsis, P. Electricity demand loads modeling using auto regressive moving average (ARMA) models. *Energy* **2008**, *33*, 1353–1360. [CrossRef]
3. Dong, R.J.; Pedrycz, W. A granular time series approach to long-term forecasting and trend forecasting. *Phys. A Stat. Mech. Appl.* **2008**, *387*, 3253–3270. [CrossRef]
4. Pappas, S.S.; Ekonomou, L.; Karampelas, P.; Karamousantas, D.C.; Katsikas, S.K.; Chatzarakis, G.E.; Skafidas, P.D. Electricity demand load forecasting of the Hellenic power system using an ARMA model. *Electr. Power Syst. Res.* **2010**, *80*, 256–264. [CrossRef]
5. Ai-Hamadi, H.M.; Soliman, S.A. Long-term/mid-term electric load forecasting based on short-term correlation and annual growth. *Electr. Power Syst. Res.* **2005**, *74*, 353–361.
6. Sorjamaa, A.; Hao, J.; Reyhani, N.; Ji, Y.N.; Lendasse, A. Methodology for long-term prediction of time series. *Neurocomputing* **2007**, *70*, 861–869. [CrossRef]
7. Niu, D.X.; Li, J.C.; Li, J.Y.; Liu, D. Middle-long power load forecasting based on particle swarm optimization. *Comput. Math. Appl.* **2009**, *57*, 1883–1889. [CrossRef]
8. Xia, C.H.; Wang, J.; McMenemy, K. Short, medium and long term load forecasting model and virtual load forecaster based on radial basis function neural networks. *Int. J. Electr. Power* **2010**, *32*, 743–750. [CrossRef]
9. Hsu, C.C.; Chen, C.Y. Regional load forecasting in Taiwan—Applications of artificial neural networks. *Energy Convers. Manag.* **2003**, *44*, 1941–1949. [CrossRef]
10. Abou El-Ela, A.A.; El-Zeftawy, A.A.; Allam, S.M.; Atta, G.M. Long-term load forecasting and economical operation of wind farms for Egyptian electrical network. *Electr. Power Syst. Res.* **2009**, *79*, 1032–1037. [CrossRef]
11. Meng, M.; Niu, D.X. Annual electricity consumption analysis and forecasting of China based on few observations methods. *Energy Convers. Manag.* **2011**, *52*, 953–957. [CrossRef]
12. Chen, T. A collaborative fuzzy-neural approach for long-term load forecasting in Taiwan. *Comput. Ind. Eng.* **2012**, *63*, 66–70.
13. Kandil, M.S.; El-Debeiky, S.M.; Hasanien, N.E. The implementation of long-term forecasting strategies using a knowledge-based expert system: Part-II. *Electr. Power Syst. Res.* **2001**, *58*, 1–5. [CrossRef]
14. Hong, W.C. Electric load forecasting by seasonal recurrent SVR (support vector regression) with chaotic artificial bee colony algorithm. *Energy* **2011**, *36*, 556–578. [CrossRef]
15. Pai, P.F.; Hong, W.C. Support vector machines with simulated annealing algorithms in electricity load forecasting. *Energy Convers. Manag.* **2005**, *46*, 266–688. [CrossRef]
16. Cortes, C.; Vapnik, V. Support-vector networks. *Mach. Learn.* **1995**, *20*, 273–297.
17. Van Gestel, T.; Suykens, J.A.K.; Baesens, B.; Viaene, S.; Vanthienen, J.; Dedene, G.; de Moor, B.; Vandewalle, J. Benchmarking least squares support vector machine classifiers. *Mach. Learn.* **2001**, *54*, 5–32. [CrossRef]

18. Vong, C.M.; Wong, P.K.; Li, Y.P. Prediction of automotive engine power and torque using least squares support vector machines and Bayesian inference. *Eng. Appl. Artif. Intel.* **2006**, *19*, 277–287. [CrossRef]

19. Yeganeha, B.; Motlagh, S.P.; M.; Rashidi, Y.; Kamalan, H. Prediction of CO concentrations based on a hybrid partial least square and support vector machine model. *Atmos. Environ.* **2012**, *55*, 357–365. [CrossRef]

20. Liao, R.J.; Zheng, H.B.; Grzybowski, S.; Yang, L.J. Particle swarm optimization-least squares support vector regression based forecasting model on dissolved gases in oil-filled power transformers. *Electr. Power Syst. Res.* **2011**, *81*, 2074–2080. [CrossRef]

21. Zhao, X.H.; Wang, G.; Zhao, K.K.; Tan, D.J. On-line least squares support vector machine algorithm in gas prediction. *Min. Sci. Technol.* **2009**, *19*, 194–198.

22. Wu, Q. Hybrid model based on wavelet support vector machine and modified genetic algorithm penalizing Gaussian noises for power load forecasts. *Expert Syst. Appl.* **2011**, *38*, 379–385. [CrossRef]

23. Espinoza, M.; Suykens, J.A.K.; de Moor, B. Fixed-size least squares support vector machines: A large Scale application in electrical load forecasting. *Comput. Manag. Sci.* **2006**, *3*, 113–129. [CrossRef]

24. Espinoza, M.; Suykens, J.A.K.; Belmans, R.; de Moor, B. Electric load forecasting—Using kernel based modeling for nonlinear system identification. *IEEE Control Syst.* **2007**, *27*, 43–57. [CrossRef]

25. Lin, K.P.; Pai, P.F.; Lu, Y.M.; Chang, P.T. Revenue forecasting using a least-squares support vector regression model in a fuzzy environment. *Inf. Sci.* **2011**, *14*, 196–209.

26. Kisi, O.; Cimen, M. Precipitation forecasting by using wavelet-support vector machine conjunction model. *Eng. Appl. Artif. Intel.* **2012**, *25*, 783–792. [CrossRef]

27. Zhou, J.Y.; Shi, J.; Li, G. Fine tuning support vector machines for short-term wind speed forecasting. *Energy Convers. Manag.* **2011**, *52*, 1990–1998. [CrossRef]

28. Wang, S.; Yu, L.; Tang, L.; Wang, S.Y. A novel seasonal decomposition based least squares support vector regression ensemble learning approach for hydropower consumption forecasting in China. *Energy* **2011**, *36*, 6542–6554. [CrossRef]

29. Dos Santosa, G.S.; Justi Luvizottob, L.G.; Marianib, V.C.; Dos Santos, L.C. Least squares support vector machines with tuning based on chaotic differential evolution approach applied to the identification of a thermal process. *Expert Syst. Appl.* **2012**, *39*, 4805–4812. [CrossRef]

30. Sulaimana, M.H.; Mustafab, M.W.; Shareefc, H.; Abd-Khalid, S.N. An application of artificial bee colony algorithm with least squares supports vector machine for real and reactive power tracing in deregulated power system. *Int. J. Electr. Power* **2012**, *37*, 67–77. [CrossRef]

31. Li, J.; Liu, J.P.; Wang, J.J. Mid-long term load forecasting based on simulated annealing and SVM algorithm. *Proc. CSEE* **2011**, *31*, 63–66.

32. Pan, W.T. A new fruit fly optimization algorithm: Taking the financial distress model as an example. *Knowl. Based Syst.* **2012**, *26*, 69–74. [CrossRef]

33. Suykens, J.A.K.; Vandewalle, J. Least squares support vector machine classifiers. *Neural Process Lett.* **1999**, *9*, 293–300. [CrossRef]

34. Suykens, J.A.K.; van Gestel, T.; de Brabanter, J.; de Moor, B.; Vandewalle, J. *Least Squares Support Vector Machines*; World Scientific: Singapore, 2002.

35. Keerthi, S.S.; Lin, C.J. Asymptotic behaviors of support vector machines with Gaussian kernel. *Neural Comput.* **2003**, *15*, 1667–1689. [CrossRef] [PubMed]

36. Shi, D.Y.; Lu, J.; Lu, L.J. A judge model of the impact of lane closure incident on individual vehicles on freeways based on RFID technology and FOA-GRNN method. *J. Wuhan Univ. Technol.* **2012**, *34*, 63–68.

37. Xu, Z.H.; Wang, F.L.; Sun, D.D.; Wang, J.Q. A forecast of export trades based on the FOA-RBF neural network [in Chinese]. *Math. Pract. Theor.* **2012**, *42*, 16–21.

38. Xiao, Z.A. Design of analog filter based on fruit fly optimization algorithm. *J Hubei Univ. Educ.* **2012**, *29*, 26–29.

39. China National Bureau of Statistics. *China Energy Statistical Yearbook 2011*; China Statistics Press: Beijing, China, 2011.

40. Electricity Power Supply and Demand of 2011 in China Released by China Electricity Council [in Chinese]. Available online: http://tj.cec.org.cn/fenxiyuce/yunxingfenxi/gongxufenxiyuce/2012-02-03/79721.html (accessed on 1 September 2012).

41. De Souza, X.S.; Suykens, J.A.K.; Vandewalle, J.; Bolle, D. Coupled simulated annealing. *IEEE Trans. Syst. Man Cybern. Part B Cybern.* **2010**, *40*, 320–335. [CrossRef]

42. Specht, D.F. A general regression neural network. *IEEE Trans. Neural Netw.* **1991**, *2*, 568–576. [CrossRef] [PubMed]

43. Amiri, M.; Davande, H.; Sadeghian, A.; Chartier, S. Feedback associative memory based on a new hybrid model of generalized regression and self-feedback neural networks. *Neural Netw.* **2010**, *23*, 69–74. [CrossRef]

44. *LS-SVMlab*, version 1.8; Department of Electrical Engineering (ESAT), Katholieke Universiteit Leuven: Leuven, Belgium, 2011. Available online: http://www.esat.kuleuven.be/sista/lssvmlab/ (accessed on 1 September 2012).

45. De Brabanter, K.; Karsmakers, P.; Ojeda, F.; Alzate, C.; de Brabanter, J.; Pelckmans, K.; de Moor, B.; Vandewalle, J.; Suykens, J.A.K. *LS-SVMlab Toolbox User's Guide Version 1.8*; World Scientific: Singapore, 2002; pp. 10–146.

46. Niu, D.X.; Wang, Y.L.; Wu, D.D. Power load forecasting using support vector machine and ant colony optimization. *Expert Syst. Appl.* **2010**, *37*, 2531–2539. [CrossRef]

energies

MDPI

Article

Day-Ahead Electricity Price Forecasting Using a Hybrid Principal Component Analysis Network

Ying-Yi Hong * and Ching-Ping Wu

Department of Electrical Engineering, Chung Yuan Christian University, 200 Chung Pei Road,
Chung Li 32023, Taiwan; squall59@pchome.com.tw

* Author to whom correspondence should be addressed; yyhong@dec.ee.cycu.edu.tw;
  Tel.: +886-3-265-1200; Fax: +886-3-265-4809.

Received: 5 September 2012; in revised form: 10 November 2012; Accepted: 12 November 2012;
Published: 19 November 2012

Abstract: Bidding competition is one of the main transaction approaches in a deregulated electricity market. Locational marginal prices (LMPs) resulting from bidding competition and system operation conditions indicate electricity values at a node or in an area. The LMP reveals important information for market participants in developing their bidding strategies. Moreover, LMP is also a vital indicator for the Security Coordinator to perform market redispatch for congestion management. This paper presents a method using a principal component analysis (PCA) network cascaded with a multi-layer feedforward (MLF) network for forecasting LMPs in a day-ahead market. The PCA network extracts essential features from periodic information in the market. These features serve as inputs to the MLF network for forecasting LMPs. The historical LMPs in the PJM market are employed to test the proposed method. It is found that the proposed method is capable of forecasting day-ahead LMP values efficiently.

Keywords: locational marginal price; forecasting; principal component analysis

---

1. Introduction

There are two main transaction modes in a deregulated electric power industry, namely, competitive bidding and bilateral contract. Competitive biddings are used in the energy, spot, firm-transmission-right and ancillary service markets while bilateral contract is adopted outside the competitive market for any two individual entities, buyer and seller [1,2]. For either transaction mode, the electricity price information serves as an essential signal for all entities to adjust their offers/bids and/or contract prices. In particular, locational marginal pricing (LMP) is one of the most popular modes for pricing electricity in a deregulated electricity market. LMPs can reflect the electricity value at a node and may be discriminated at different nodes in a power network [3]. LMPs provide information that is helpful to market participants in developing their bidding strategies. It is also a vital indicator for the Security Coordinator to mitigate transmission congestion [4]. LMPs reveal important information for both the spot market and entities with bilateral contracts.

Past studies have investigated short-term System Marginal Price (SMP) forecasting [5,6]. Because the SMP is irrelevant to transmission constraints, forecasting LMPs subject to transmission constraints is more difficult than forecasting Market Clear Prices (MCPs). Current methods for short-term LMP forecasting can be classified at least into three groups: hour-ahead, day-ahead and week-ahead forecastings.

The recurrent neural network integrated with fuzzy-c-means was proposed for hour-ahead LMP forecasting in [7]. Linguistic descriptions in the PJM market were transformed into fuzzy membership functions associated with the recurrent neural network for forecasting volatile hour-ahead LMP variations when contingency occurs [8]. This paper investigates the more difficult problem related to the day-ahead price forecasting, which may be applied to the day-ahead market and will be discussed in the next paragraph.

In recent years, Contreras *et al.* [9] used the ARIMA model and Nogales *et al.* [10] used the dynamic regression approach and transfer function approach to predict the next-day (day-ahead) electricity prices. However, there is no discussion on extracting the market features for usage of these approaches in [9,10]. Li *et al.* [11] integrated the fuzzy inference system with least-squares estimation to conduct the day-ahead electricity price forecasting. The "week day", "yesterday price" and "local demand" were considered in the 18 antecedent (premise or condition) parts of the fuzzy rules in [11]. Giving the membership functions of these three linguistic variables is quite heuristic. Moreover, the "local demand" for the fuzzy rules is not a forecasted but an actual value, which is generally not available in the day-ahead market. Amjady and Keynia [12] combined a mutual information technique (MIT) with the cascaded neuro-evolutionary algorithm (NEA) for the day-ahead electricity price forecasting. In [12], 14 features in the market were selected by MIT for 24 feedforward neural networks trained by the NEA. No reasonable explanation was found for these 14 features. Moreover, many (24) neural networks make the method impractical for industrial application. Garcia *et al.* [13] presented an approach to predicting next-day electricity prices using the Generalized Autoregressive Conditional Heteroskedastic (GARCH) methodology, which is an extended auto-regressive integrated moving average (ARIMA). Amjady [14] presented a fuzzy neural network with an inter-layer and feedforward architecture using a new hypercubic training mechanism. The proposed method predicted hourly market-clearing prices for the day-ahead electricity markets. Again, there is no discussion on extracting the market features for usage of the GARCH in [13,14]. Coelho and Santos [15] proposed a nonlinear forecasting model based on radial basis function neural networks (RBF-NNs) with Gaussian activation functions. Partial autocorrelation functions (PACF), which relies on the mutual linear dependency among studied parameters, was used to identify the market features. However, the relation among power market features is very nonlinear.

The problem of week-ahead price forecasting is generally easier than that of day-ahead price forecasting because the price pattern of a day is similar to that of its corresponding week-ahead day. Catalao *et al.* [16] proposed a wavelet-based Sugeno type fuzzy inference system to predict the electricity price in the electricity market of mainland Spain. However, the selection of numbers of membership functions in [16] is a trade-off between refining and sparseness. Che and Wang [17] presented a method based on support vector regression and ARIMA modeling; however, only the MCPs of California electricity market were used to examine the accuracy of the proposed method. The method has not been applied to forecasting LMPs, whose pattern is more nonlinear than MCPs'.

Because LMPs vary dramatically, it is difficult to analyze the related data with traditional techniques (e.g., regression analysis). Like other forecasting problems [18–20], the LMP forecasting needs feature extraction incorporating a powerful approach. As described above, the neural network is suitable for nonstationary time-series prediction, providing satisfactory results. In this paper, a principal component analysis (PCA) neural network cascaded with the multi-layer-feedforward (MLF) neural network is proposed for day-ahead LMP forecasting. The PCA neural network is used to extract essential features in the electricity market. It also helps reduce high-dimensional data into low-dimensional ones, which serve as inputs for the MLF neural network.

The rest of this paper is organized as follows: the PJM real-time market data will be described in Section 2. The proposed PCA neural network cascaded with the MLF neural network for forecasting day-ahead LMPs will be given in Section 3. Simulation results obtained using the PJM data are presented in Section 4. Concluding remarks are provided in Section 5.

2. Volatile LMPs in a Day-ahead Market

The PJM energy market comprises day-ahead and real-time markets. The day-ahead market is a forward market in which hourly LMPs are calculated for the next operating day using generation offers, demand bids and scheduled bilateral transactions. The real-time market is a spot market in which current LMPs are calculated at five-minute intervals according to actual grid operating conditions. PJM settles transactions hourly and issues invoices to market participants monthly. Figures 1 and 2 illustrate the LMPs in Fisk (4 kV) and Byberry (13 kV), respectively, on 1–7 July 2008. As can be seen, LMPs vary dramatically over a wide range.

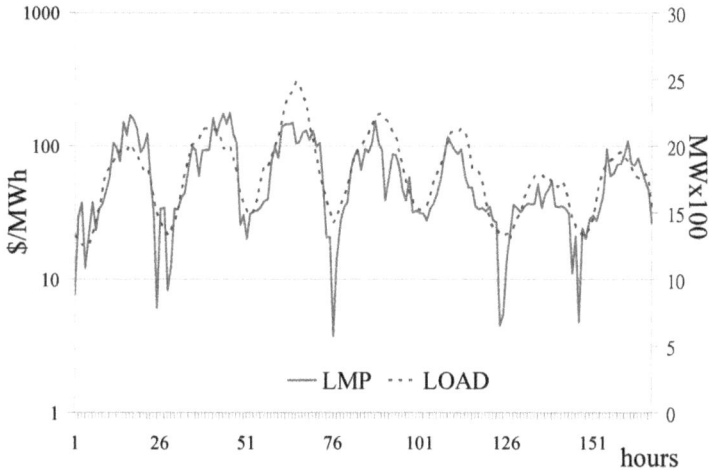

Figure 1. Daily LMPs in Fisk (4 kV) on 1–7 July 2008.

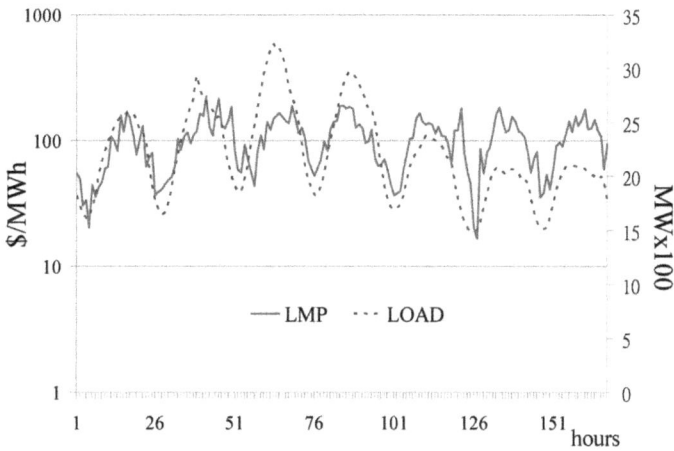

Figure 2. Daily LMPs in Byberry (13 kV) on 1–7 July 2008.

3. The Proposed Method

The hybrid PCA neural network is developed by combining the unsupervised PCA and supervised MLF neural networks to conduct day-ahead LMP forecasting. The PCA neural network is employed to extract essential features in the electricity market. The PCA neural network can also reduce high-dimensional data into low-dimensional ones, which serve as inputs of the MLF neural network to reduce the training CPU time.

3.1. Principal Component Analysis Neural Network

The purpose of the PCA neural network is to find a set of P orthonormal vectors (OVs) in a Q-dimensional space (Q $\geq$ P), such that these OVs will account for as much variance of the input data as possible. OVs are actually P eigenvectors associated with the P largest eigenvalues of the $E(xx^t)$, where x denotes the Q-dimensional input column vector, *i.e.*, $x = (x_1 \, x_2 \, \ldots \, x_Q)^t$. The direction of the q-th principal component will be along the q-th eigenvector, $q = 1, 2, .., Q$.

Let symbol t be the training index. This paper used Sanger's method [21] to update the weightings between the neurons of the PCA network as follows:

$$\Delta w_p(t) = \eta(t) v_p(t) \left(x(t) - \sum_{pj=1}^{p} v_{pj}(t) w_{pj}(t) \right), \; p = 1, 2, \ldots, P \tag{1}$$

where $\eta(t)$ is a parameter of learning rate and $v_{pj}(t) = w_{pj}^t(t) x(t)$ is the output. Equation (1) is employed to train a neural network consisting of P linear neurons so as to find the first P principal components. More specifically, Generalized Hebbian Algorithm was able to make $w_p(t)$, $p = 1, 2, \ldots$, P, converge to the first P principal component directions, in sequential order: $w_p(t) \to \pm v_i$, where v_i is a normalized eigenvector associated with the i-th largest eigenvalue of the correlation matrix. It was shown that if $w_{pj}(t)$, $pj = 1, 2, \ldots, p-1$ have converged to $v_{pj}(t)$, $pj = 1, 2, \ldots, p-1$, respectively, then the maximal eigenvalue λ_p and the corresponding normalized eigenvector v_p of the correlation matrix of x_p, *i.e.*, $C_p \equiv E(x_p x_p^t)$, are exactly the p-th eigenvalue and the p-th normalized eigenvector v_p of the correlation matrix of x, *i.e.*, $C \equiv E(xx^t)$, respectively. Consequently, neuron p can find the p-th normalized eigenvector of C. Detailed explanations can be found in [21].

It was shown that $\eta(t)$ should be smaller than the reciprocal largest eigenvalue of $E(xx^t)$ to ensure the convergence of training a PCA neural network. When the training process is convergent, w_p, $p = 1, 2, \ldots, P$, converges to the p-th eigenvector of $E(xx^t)$.

Figure 3 shows the configuration of the hybrid PCA neural network: The left part of Figure 3 is the unsupervised PCA neural network while the right part is the supervised MLF neural network. Because the training time of unsupervised PCA neural network is trivial while that of supervised MLF is considerable, PCA neural network is adopted to reduce both the dimension of training data and the training time for the cascaded MLF network trained by the back-propagation algorithm, which is well known and ignored here.

In the proposed hybrid PCA, the new hidden layer consists of 20 neurons. The number (p) of orthonormal vectors is 24 or 48, depending on the numbers of inputs. After training the unsupervised PCA, the supervised MLF is trained, using the frozen weights of the unsupervised PCA. The training sets are identical for both unsupervised and supervised NNs.

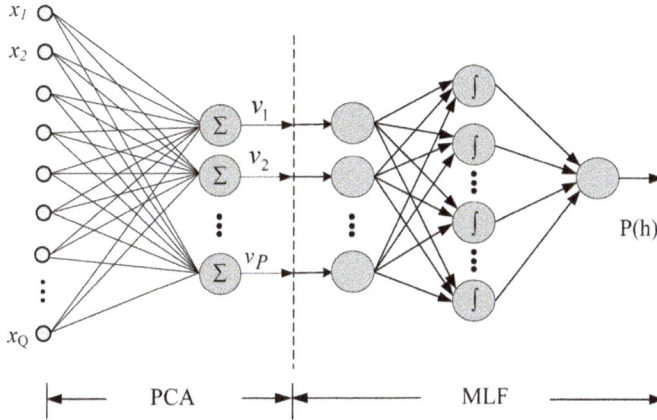

Figure 3. The proposed hybrid PCA neural network.

3.2. Features for Inputs of PCA Neural Network

The performance of a neural network depends strongly on the adopted features at the input layers. As shown in Figures 1 and 2, variations of system load affect LMPs. Assume that the *h*-th LMP is to be forecasted. Let P(h) and L(h) be the LMP and MW demand at hour h, respectively.

Below are 4 alternatives for considering input features $x_1, x_2, \ldots, x_Q$:

(1) The features of the past 2 days: F1(h) ≡ (P(h − 47), P(h − 46), ... , P(h − 25), P(h − 24) and L(h − 47), L(h − 46), ... , L(h − 25), L(h − 24)). That is, Q = 48.

(2) The features of the same day of the last week and those of the past 2 days: F2(h) ≡ (P(h − 168), P(h − 167), ... , P(h − 146), P(h − 145), P(h − 47), P(h − 46), ... , P(h − 25), P(h − 24) and L(h − 168), L(h − 167), ... , L(h − 146), L(h − 145), L(h − 47), L(h − 46), ... , L(h − 25), L(h − 24)). That is, Q = 96.

(3) The features of the past 2 days and the designated day: F3(h) = F1(h) ∪ (D | D is one of the seven days in a week). This implies Q = 49.

(4) The features of the same day of the last week, those of the past 2 days and the designated day: F4(h) = F2(h) ∪ (D | D is one of the seven days in a week). This means Q = 97.

The symbol D for the designated day here means Monday, ... , Saturday or Sunday. Because the neural network cannot deal with symbols, 30, 50, ... , 150 stand for Monday, ... , Saturday and Sunday, respectively, in this paper.

3.3. Moving Data Windows for Forecasting

P(h) at the output layer is paired with F1(h), F2(h), F3(h) or F4(h). More specifically, assume that F1(h) is considered and the 24 LMPs on Wednesday (next day) are forecasted. Figure 4 illustrates the moving data window corresponding to the forecasted LMP. Hence, the paired training data are as follows: (F1(h), P(h)), (F1(h + 1), P(h + 1)), ... , (F1(h + 23), P(h + 23)). In Figure 4, the first data set involves only Monday and Wednesday. The last 23 data on Monday and the first data on Tuesday will be paired with P(h + 1) for the second data set. Restated, forecasting 24 LMPs on Wednesday will be completed at 23:00 p.m. on Tuesday.

When the proposed hybrid PCA neural network is used in the day-ahead market or in the testing stage, the input data for the past day (e.g., Monday in Figure 4) and this day (e.g., Tuesday in Figure 4) are known and output (forecasted) data for the next day (e.g., Wednesday in Figure 4) is unknown.

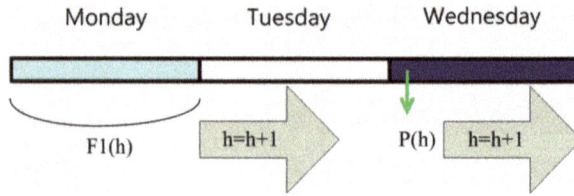

Figure 4. Moving data window (F1(h)) corresponding to forecasted P(h).

Assume that this day is Tuesday and LMPs on Wednesday are to be forecasted. Figure 5 shows the moving data window for F2(h) paired with P(h). The paired training data are as follows: (F2(h), P(h)), (F2(h + 1), P(h + 1)), ... , (F2(h + 23), P(h + 23)). As shown in Figure 5, the first data set involves only the last Wednesday, Monday and Wednesday. The time index h will be increased by one at a time until h + 23. Restated, forecasting 24 LMPs on Wednesday will be completed at 23:00 p.m. on Tuesday.

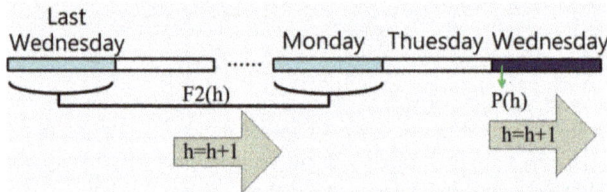

Figure 5. Moving data window (F2(h)) corresponding to forecasted P(h).

3.4. Numbers of Neurons in Different Layers

The numbers of input, output, second, and fourth layers are discussed as follows:

(1) The numbers of input neurons for the hybrid PCA neural networks are 48, 96, 49 and 97 for F1(h), F2(h), F3(h) and F4(h), respectively. That is, subscript Q in Figure 3 can be 48, 96, 49 or 97.
(2) The number of neurons in the MLF output layer is one (*i.e.*, P(h)), regardless of F1(h), F2(h), F3(h) and F4(h) being considered.
(3) Because the purpose of the PCA neural network is to find a set of P orthonormal vectors (OVs) in a Q-dimensional space, P is expected to be smaller than the corresponding number of inputs. It is intuitive to consider P in Figure 3 to be 24 for the studied problem with Q = 48 or 49 because there are 24 hours in a day. Similarly, P = 48 while Q = 96 or 97.
(4) The common number of neurons for the fourth (hidden) layer is (P + number of output neurons)/2 or (P × number of output neurons)$^{0.5}$. The simulation results show no significant difference between these two alternatives.

4. Simulation Results

In order to demonstrate the applicability of the proposed hybrid PCA neural network, the LMPs for the Fisk (4 kV) and Byberry (13 kV) areas in the PJM system were studied. Two sets of 366 × 24 data (1 January–31 December 2008) for Fisk and Byberry from the PJM web site were employed to train/validate and test the proposed hybrid PCA neural network. The entire data set includes four seasons. The data of each season are further divided into three groups: training data, validation data (in total 2/3), and test data (1/3). The training data were used for training the neural network and updating the biases and weights. The validation data were utilized to monitor the training process. The remaining data were employed to test the proposed hybrid PCA neural network after they were well trained. A C++ code was developed using a PC equipped with a Pentium(R) Dual-Core E5200 2.5 GHz CPU and 4-GB RAM for showing the applicability of the proposed method.

4.1. Comparison between Hybrid PCA and Back Propagation-Based Neural Networks

In this subsection, the performance of the proposed hybrid PCA neural network is compared with that of traditional back propagation-based (BP-based) neural network for the Fisk area. The traditional BP-based neural network can be taken as a neural network in which there are no second and third layers as seen in Figure 3. Tables 1–4 display the CPU time (minute: second), correlation coefficient (R2) and mean absolute error (MAE, $/MWh) obtained by these two methods for Fisk. The correlation coefficient represents the resemblance between the actual and the forecasted values. The value of one for the correlation coefficient indicates that the actual values are identical to the forecasted ones. The average value and corresponding standard deviation (sd, $/MWh) of actual LMPs in each season are also shown in the second and third columns of Tables 1–4 . Figure 6 shows the comparisons among actual, BP-based and hybrid PCA-based LMPs for Fisk (1–7 July 2008).

Table 1. Performance comparison between the proposed hybrid PCA and BP-based neural network (Fisk, spring).

Dimension of input vector	LMP ($/MWh)		Hybrid PCA Network			BP-based Network		
Q	average	sd	time	R2	MAE	time	R2	MAE
48			03:50	0.749	14.96	04:21	0.707	15.85
49			03:36	0.814	13.11	04:28	0.777	14.72
96	52.89	25.17	06:20	0.838	14.19	07:05	0.767	15.19
97			06:34	0.838	13.78	08:19	0.769	15.10

Table 2. Performance comparison between the proposed hybrid PCA and BP-based neural network (Fisk, summer).

Dimension of input vector	LMP ($/MWh)		Hybrid PCA Network			BP-based Network		
Q	average	sd	time	R2	MAE	time	R2	MAE
48			04:06	0.793	21.35	04:35	0.759	21.36
49			04:39	0.823	20.68	04:50	0.800	21.42
96	55.61	40.52	08:48	0.826	21.15	12:00	0.824	21.16
97			08:53	0.843	20.51	12:59	0.829	20.74

Table 3. Performance comparison between the proposed hybrid PCA and BP-based neural network (Fisk, fall).

Dimension of input vector	LMP ($/MWh)		Hybrid PCA Network			BP-based Network		
Q	average	sd	time	R2	MAE	time	R2	MAE
48			03:45	0.840	15.57	04:50	0.840	16.42
49			03:55	0.858	15.34	04:55	0.847	15.81
96	53.23	32.43	13:41	0.901	13.06	14:34	0.875	16.22
97			13:55	0.904	13.66	14:44	0.876	15.42

Table 4. Performance comparison between the proposed hybrid PCA and BP-based neural network (Fisk, winter).

Dimension of input vector	LMP ($/MWh)		Hybrid PCA Network			BP-based Network		
Q	average	sd	time	R2	MAE	time	R2	MAE
48			03:53	0.706	11.02	04:16	0.683	12.19
49			03:57	0.765	10.69	04:24	0.725	12.58
96	42.36	19.15	08:03	0.818	10.13	12:26	0.786	10.71
97			08:21	0.822	10.03	12:40	0.780	11.28

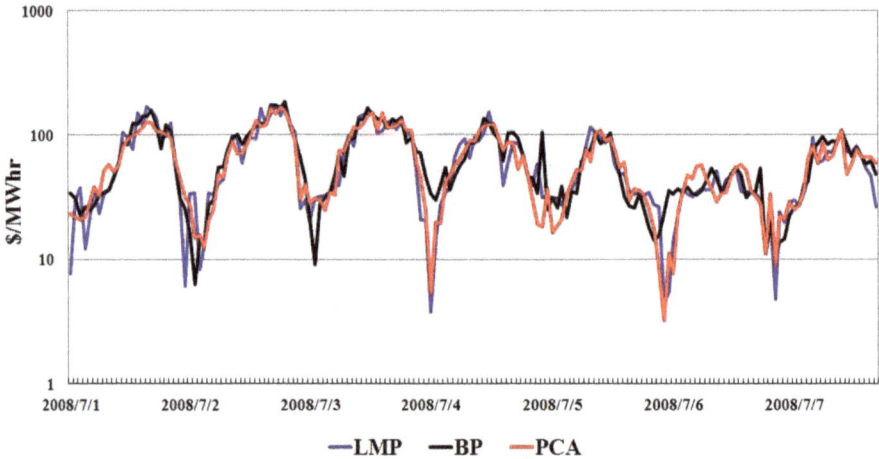

Figure 6. Comparisons among actual, BP-based and hybrid PCA-based LMPs for Fisk.

According to Tables 1–4 , the following comments can be made:

(1) The most volatile LMPs with an average of 55.61 $/MWh and a standard deviation of 40.52 $/MWh occurred in summer. The most steady LMPs with an average of 42.36 $/MWh and a standard deviation of 19.15 $/MWh occurred in winter.

(2) For the same neural network, the R2 obtained by 49 inputs is better (larger) than that by 48 inputs; for the same reason, the neural network with 97 inputs has better performance than that with 96 inputs in terms of R2.

(3) For the same neural network, the R2 obtained by 96 (97) inputs is much better (larger) than that by 48 (49) inputs; however, the CPU times required by 96 (97) inputs are longer.

(3) For the same number of inputs, the R2 and MAE $/MWh obtained by the hybrid PCA neural network are better than those obtained by the BP-based neural network.

(4) For the same number of inputs, the CPU time required by the hybrid PCA neural is shorter than that required by the BP-based neural network.

Tables 5–8 show the comparison of performance between the proposed hybrid PCA neural network and the traditional BP-based neural network for the Byberry area. The same conclusions can be made for both Byberry and Fisk areas. Figure 7 shows the comparisons among actual, BP-based and hybrid PCA-based LMPs for Byberry (1–7 July 2008).

Table 5. Performance comparison between the proposed hybrid PCA and BP-based neural network (Byberry, spring).

Dimension of input vector	LMP ($/MWh)		Hybrid PCA Network			BP-based Network		
Q	average	sd	time	R2	MAE	time	R2	MAE
48			03:28	0.681	25.89	04:01	0.634	28.15
49			03:39	0.734	25.41	04:05	0.669	28.56
96	73.95	42.05	06:00	0.784	24.60	07:05	0.673	28.40
97			06:07	0.793	24.22	07:35	0.681	29.22

Table 6. Performance comparison between the proposed hybrid PCA and BP-based neural network (Byberry, summer).

Dimension of input vector	LMP ($/MWh)		Hybrid PCA Network			BP-based Network		
Q	average	sd	time	R2	MAE	time	R2	MAE
48			03:41	0.797	28.06	04:32	0.789	28.15
49			03:52	0.842	26.36	04:58	0.827	26.96
96	87.22	52.73	08:25	0.859	25.10	14:47	0.852	25.73
97			08:30	0.868	24.37	14:51	0.848	26.50

Table 7. Performance comparison between the proposed hybrid PCA and BP-based neural network (Byberry, fall).

Dimension of input vector	LMP ($/MWh)		Hybrid PCA Network			BP-based Network		
Q	average	sd	time	R2	MAE	time	R2	MAE
48			03:08	0.765	25.01	04:15	0.725	25.66
49			03:15	0.769	24.73	04:36	0.761	25.24
96	88.41	50.30	08:06	0.848	23.48	11:41	0.823	23.59
97			08:11	0.874	23.17	11:54	0.864	23.62

Table 8. Performance comparison between the proposed hybrid PCA and BP-based neural network (Byberry, winter).

Dimension of input vector	LMP ($/MWh)		Hybrid PCA Network			BP-based Network		
Q	average	sd	time	R2	MAE	time	R2	MAE
48			03:28	0.725	14.72	04:29	0.688	14.98
49			03:58	0.779	13.94	04:53	0.709	15.24
96	56.73	25.26	09:53	0.830	12.96	11:22	0.754	13.62
97			09:58	0.837	12.99	11:54	0.757	16.03

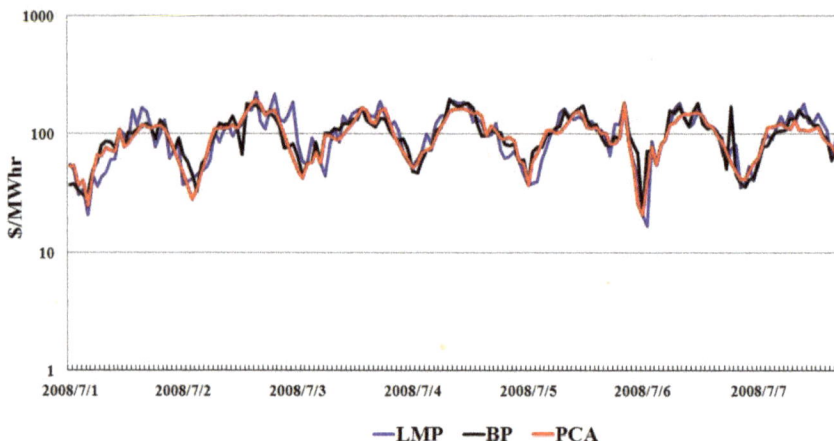

Figure 7. Comparisons among actual, BP-based and hybrid PCA-based LMPs for Byberry.

4.2. Investigation of Number of Output Neurons for PCA Network

The second layer of the proposed hybrid PCA neural network shown in Figure 3 denotes the features of the electricity market. The number (*i.e.*, P) of neurons at this layer hence plays a crucial role in developing the proposed method. Tables 9 and 10 show the impact of different P's at the second layer on R2 and MAE for Fisk and Byberry areas, respectively, in fall. In order to show the effectiveness of the proposed method, only 97 inputs (*i.e.*, Q) in Figure 3 were studied. The following remarks can be made according to Tables 9 and 10:

(1)   The larger the P, the longer the CPU time is required due to the supervised MLF neural network at the third, fourth and fifth layers in Figure 3.
(2)   A larger P will result in a better performance in terms of R2 and MAE obtained. Hence, there is a trade-off between performance and CPU time. In general, performance is more important.

Table 9. Comparison between different P's when Q = 97 (Fisk, fall).

P	CPU	R2	MAE
48	13:55	0.904	13.66
36	04:28	0.896	13.86
24	03:37	0.890	13.91
12	02:48	0.863	15.01

Table 10. Comparison between different P's when Q = 97 (Byberry, fall).

P	CPU	R2	MAE
48	08:11	0.874	23.17
36	07:47	0.840	23.94
24	06:25	0.826	23.95
12	05:52	0.811	23.99

4.3. Comparison between Hybrid PCA Network and ARIMA

Traditional nonstationary time-series prediction method using ARIMA [9] is employed to study the same PJM day-ahead market. Because the hybrid PCA neural network with 97 inputs gained the best performance as described in Section 4.1, it was compared with the ARIMA. The general ARIMA formulation was given as follows [9]: φ(B)P(h) = θ(B)ε(h) where P(h) is the LMP at time h, φ(B) and θ(*B*) are functions of the backshift operator B: B_kP(h) ≡ P(h − k), and ε(h) () is the error term. This paper adopted the functions φ(B) and θ(B) given in [9] for comparisons. In [9], the load factor was not considered as a regressor in the ARIMA. Twenty four LMPs were used as lagged regressors in the ARIMA. Tables 11 and 12 show the correlation coefficient (R2) and mean absolute error (MAE, $/MWh) obtained for all four seasons by the two methods.

The following comments could be made according to the results shown in Tables 11 and 12:

(1) For either Fisk or Byberry, the performances of the proposed hybrid PCA neural network are always better than those of the ARIMA in terms of both R2 and MAE obtained.
(2) The R2's obtained by the ARIMA for both Fisk and Byberry in winter are very low (0.488 and 0.419) while those obtained by the proposed method are much higher (0.822 and 0.837).
(3) The LMPs in the Byberry area are more volatile than those in the Fisk area in terms of average R2 (0.566 with respect to 0.725). However, the proposed method is more reliable regardless of the studied areas; that is, 0.843 for Byberry is close to 0.852 for Fisk.

Table 11. Comparison between the proposed hybrid PCA and ARIMA [9] (Fisk).

Seasons	Hybrid PCA Network		ARIMA [9]	
	R2	MAE	R2	MAE
Spring	0.838	13.78	0.808	15.89
Summer	0.843	20.51	0.757	21.33
Autumn	0.904	13.66	0.846	16.31
Winter	0.822	10.03	0.488	11.68
Average	0.852	14.50	0.725	16.30

Table 12. Comparison between the proposed hybrid PCA and ARIMA [9] (Byberry).

Seasons	Hybrid PCA Network		ARIMA [9]	
	R2	MAE	R2	MAE
Spring	0.793	24.22	0.578	28.10
Summer	0.868	24.37	0.743	27.81
Autumn	0.874	23.17	0.522	26.86
Winter	0.837	12.99	0.419	15.04
Average	0.843	21.19	0.566	24.45

4.4. Diebold and Mariano Test

Diebold and Mariano proposed and evaluated explicit tests of the null hypothesis of no difference in the accuracy of two competing forecasts [22]. The loss function does not need to be quadratic, and even to be symmetric, and forecast errors can be non-Gaussian, nonzero mean, serially correlated, and contemporaneously correlated in this method. This subsection utilizes Diebold and Mariano test to evaluate the performance of the proposed hybrid PCA network, the BP-based network, and ARIMA. The loss function used in this paper is based on "mean squared error" (MSR) [23].

Let H_0 be the null hypothesis of no difference in the accuracy of the proposed hybrid PCA network and the BP-based network. The alternative hypothesis is the union of H_1 and H_2, which mean that the proposed hybrid PCA network is significantly better than the BP-based network and that the BP-based network is significantly better than the proposed hybrid PCA network, respectively. Under the null hypothesis, the test statistic S_1 defined in [22] and used to test H_0, H_1 and H_2 has an asymptotic standard normal distribution. Let the confidence level be 95%. If S_1 is greater than 1.96, than H_1 is accepted and H_0 is declined. If S_1 is smaller than −1.96, than H_2 is accepted and H_0 is declined. When

S_1 is within $[-1.96, 1.96]$, H_0 is accepted and there is no significant difference in forecasting accuracy between the two models. According to Tables 13 and 14, the proposed hybrid PCA network has better performance in 9 out of 16 tests while 7 tests accept H_0. H_2 has never been accepted.

Similarly, Diebold and Mariano test is conducted to compare the performance between the proposed hybrid PCA network and ARIMA. Based on the same comparisons given in Tables 11 and 12, Table 15 shows that the proposed hybrid PCA network is significantly better than ARIMA.

Table 13. Diebold and Mariano test between the proposed hybrid PCA and BP-based neural network (Fisk).

Seasons	Number of Input Q	S_1	Results
spring	48	2.8701	accept H_1, decline H_0
	96	3.2559	accept H_1, decline H_0
summer	48	1.5524	accept H_0
	96	0.2173	accept H_0
fall	48	0.7640	accept H_0
	96	3.4976	accept H_1, decline H_0
winter	48	3.1032	accept H_1, decline H_0
	96	3.1033	accept H_1, decline H_0

Table 14. Diebold and Mariano Test between the proposed hybrid PCA and BP-based neural network (Byberry).

Seasons	Number of Input Q	S_1	Results
spring	48	2.7228	accept H_1, decline H_0
	96	3.2615	accept H_1, decline H_0
summer	48	0.7169	accept H_0
	96	1.5594	accept H_0
fall	48	0.5016	accept H_0
	96	0.8247	accept H_0
winter	48	2.7702	accept H_1, decline H_0
	96	4.1360	accept H_1, decline H_0

Table 15. Diebold and Mariano Test between the proposed hybrid PCA and ARIMA [9].

Seasons	FISK		Byberry	
	S_1	Results	S_1	Results
Spring	3.2055	accept H_1, decline H_0	6.5186	accept H_1, decline H_0
Summer	3.5084	accept H_1, decline H_0	4.2310	accept H_1, decline H_0
Autumn	3.1485	accept H_1, decline H_0	10.000	accept H_1, decline H_0
Winter	7.3961	accept H_1, decline H_0	11.678	accept H_1, decline H_0

5. Conclusions

In this paper, a new method using the hybrid principal component analysis (PCA) neural network for the day-ahead LMP forecasting in a deregulated market is proposed. The purpose of the PCA neural network is to find a set of 24 or 48 orthonormal vectors in a Q-dimensional space (24 for Q = 48, 49, and 48 for Q = 96 and 97 in this paper). The PCA can extract more essential features of the power market and hence reduce the training time required for the cascaded multi-layer feedforward neural network.

Simulation results show that the features of the same day of the last week and of the designated day provide crucial information serving as inputs of the PCA neural network. Simulation results also show that the performance of the proposed method is always better than that of the back-propagation-based neural network and ARIMA by evaluating R2 and MAE. The results of the Diebold and Mariano test

Energies **2012**, *5*, 4711–4725

show that the proposed method is better than the back-propagation-based neural network for most of the studied cases. The proposed hybrid PCA network is significantly better than the ARIMA according to the Diebold and Mariano test.

Acknowledgments: The authors would like thank the financial support from Department of Industrial Technology, Ministry of Economic Affairs under the Grant 98-EC-17-A-07-S2-0029. This study was also conducted under the "III Innovative and Prospective Technologies Project" of the Institute for Information Industry which is subsidized by the Ministry of Economy Affairs of the Republic of China. Comments from Chi-Cheng Chuang are appreciated.

References

1. Cristian, C.; Andrea, L. A dynamic approach for the optimal electricity dispatch in the deregulated market. *Energy* **2004**, *29*, 2273–2287. [CrossRef]
2. Salem, Y.A.A. Supply curve bidding of electricity in constrained power networks. *Energy* **2010**, *35*, 2886–2892. [CrossRef]
3. Ott, A. PJM: A full service ISO market evolution. In Proceedings of IEEE Power Engineering Society 1999 Summer Meeting, Albert, Canada, 18–22 July 1999; pp. 746–748.
4. *Comparison of System Redispatch Methods for Congestion Management*; Congestion Management Working Group of NERC Market Interface Committee: Atlanta, GA, USA, 1999.
5. Wang, B.; Zhang, S.; Xue, Q.L.; Shen, P. Prediction of power system marginal price based on chaos characteristics. In Proceedings of IEEE International Conference on Industrial Technology, Chengdu, China, 21–24 April 2008.
6. Chen, Y.Q.; Xiao, X.Q.; Song, Y.H. The short-term forecast of system marginal price based on artificial neural network. In Proceedings of International Conference on Industrial and Information Systems, Haikou, China, 24–25 April 2009.
7. Hong, Y.Y.; Hsiao, C.Y. Locational marginal price forecasting in deregulated electricity markets using artificial intelligence. *IEE Proc. Gen. Trans. Dist.* **2002**, *149*, 621–626. [CrossRef]
8. Hong, Y.Y.; Lee, C.F. A neuro-fuzzy LMP forecasting approach in deregulated electric markets. *Electr. Power Syst. Res.* **2005**, *73*, 151–157. [CrossRef]
9. Contreras, J.; Espinola, R.; Nogales, F.J.; Conejo, A.J. ARIMA models to predict next-day electricity prices. *IEEE Trans. Power Syst.* **2003**, *18*, 1014–1020. [CrossRef]
10. Nogales, F.J.; Contreras, J.; Conejo, A.J.; Espinola, R. Forecasting next-day electricity prices by time-series models. *IEEE Trans. Power Syst.* **2002**, *17*, 342–348. [CrossRef]
11. Li, G.; Liu, C.C.; Mattson, C.; Lawarrée, J. Day-ahead electricity price forecasting in a grid environment. *IEEE Trans. Power Syst.* **2007**, *22*, 266–274. [CrossRef]
12. Amjady, N.; Keynia, F. Day-ahead price forecasting of electricity markets by mutual information technique and cascaded neuro-evolutionary algorithm. *IEEE Trans. Power Syst.* **2009**, *24*, 306–318. [CrossRef]
13. Garcia, R.C.; Contreras, J.; Akkeren, M.V.; Garcia, J.B.C. A GARCH forecasting model to predict day-ahead electricity prices. *IEEE Trans. Power Syst.* **2005**, *20*, 867–874. [CrossRef]
14. Amjady, N. Day-ahead price forecasting of electricity markets by a new fuzzy neural network. *IEEE Trans. Power Syst.* **2006**, *21*, 887–896. [CrossRef]
15. Coelho, L.D.S.; Santos, A.A.P. A RBF neural network model with GARCH errors: Application to electricity price forecasting. *Electr. Power Syst. Res.* **2012**, *81*, 74–83. [CrossRef]
16. Catalao, J.P.S.; Pousinhoa, H.M.I.; Mendes, V.M.F. Short-term electricity prices forecasting in a competitive market by a hybrid intelligent approach. *Energy Convers. Manag.* **2011**, *52*, 1061–1065. [CrossRef]
17. Che, J.X.; Wang, J.Z. Short-term electricity prices forecasting based on support vector regression and auto-regressive integrated moving average model. *Energy Convers. Manag.* **2010**, *51*, 1911–1917. [CrossRef]
18. Iranmanesh, H.; Abdollahzade, M.; Miranian, A. Mid-term energy demand forecasting by hybrid neuro-fuzzy models. *Energies* **2012**, *5*, 1–21. [CrossRef]
19. Wang, F.; Mi, Z.Q.; Su, S.; Zhao, H.S. Short-term solar irradiance forecasting model based on artificial neural network using statistical feature parameters. *Energies* **2012**, *5*, 1355–1370. [CrossRef]
20. Meng, M.; Niu, D.X.; Sun, W. Forecasting monthly electric energy consumption using feature extraction. *Energies* **2011**, *4*, 1495–1507. [CrossRef]

21. Sanger, T.D. Optimal unsupervised learning in a single-layer linear feedforward neural network. *Neural Netw.* **1989**, *2*, 459–473. [CrossRef]

22. Diebold, F.X.; Mariano, R.S. Comparing predictive accuracy. *J. Bus. Econ. Stat.* **1995**, *13*, 253–263.

23. Ibisevic, S. *Diebold-Mariano Test Statistic*; MathWorks Inc.: Natick, MA, USA, 1995; Available online: http://www.mathworks.com/matlabcentral/fileexchange/33979-diebold-mariano-test-statistic/content/dmtest.m (accessed on 15 October 2012).

energies

MDPI

Article

A Bayesian Method for Short-Term Probabilistic Forecasting of Photovoltaic Generation in Smart Grid Operation and Control

Antonio Bracale [1], **Pierluigi Caramia** [1], **Guido Carpinelli** [2], **Anna Rita Di Fazio** [3,]* and **Gabriella Ferruzzi** [4]

[1]   Department for Technologies, University Parthenope of Napoli, Centro Direzionale di Napoli, Is. C4,
      80143 Napoli, Italy; antonio.bracale@uniparthenope.it (A.B.); pierluigi.caramia@uniparthenope.it (P.C.)
[2]   Department of Electrical Engineering and of Information Technologies, University Federico II of Napoli,
      via Claudio 21, 80125 Napoli, Italy; guido.carpinelli@unina.it
[3]   Department of Electrical and Information Engineering, University of Cassino and Southern Lazio,
      via Di Biasio 43, 03042 Cassino, Italy
[4]   Department of Economical-Management Engineering, University Federico II of Napoli,
      Piazzale V. Tecchio 80, 80125 Napoli, Italy; gabriella.ferruzzi@unina.it
*   Author to whom correspondence should be addressed; a.difazio@unicas.it; Tel.: +39-0776-299-4366.

Received: 28 December 2012; in revised form: 24 January 2013; Accepted: 25 January 2013;
Published: 6 February 2013

Abstract: A new short-term probabilistic forecasting method is proposed to predict the probability density function of the hourly active power generated by a photovoltaic system. Firstly, the probability density function of the hourly clearness index is forecasted making use of a Bayesian auto regressive time series model; the model takes into account the dependence of the solar radiation on some meteorological variables, such as the cloud cover and humidity. Then, a Monte Carlo simulation procedure is used to evaluate the predictive probability density function of the hourly active power by applying the photovoltaic system model to the random sampling of the clearness index distribution. A numerical application demonstrates the effectiveness and advantages of the proposed forecasting method.

Keywords: smart grid; photovoltaic generation; clearness index; forecasting; probability density functions; autoregressive models; Bayesian inference

1. Introduction

In recent years, power systems have been undergoing radical changes and in the near future their planning and operation will be undertaken according to the Smart Grid (SG) vision. The SG initiatives aim at introducing new technologies and services in power systems to make the electrical networks more reliable, efficient, secure and environmentally-friendly [1].

Increasing the exploitation of renewable energy sources (such as wind and solar energy) is certainly one of the most important goals of SGs. Indeed, the random behavior of such energy sources introduce challenging issues in the design of advanced tools and techniques for the optimal SG operation and control. In tackling these issues, forecasting is a fundamental task for an efficient utilization of the available distributed energy resources and for a secure and economic behaviour of the power system [2].

In general, the power system operator can use accurate forecast information about renewable power generation and load consumption to guarantee a balance between generation and demand at all the time with reduced capacity and costs of the operating reserves [3,4]. From the perspective of the producers, forecasting the renewable power output can be very useful for decision making on the energy market. In this way, not only the deviation between scheduled and actual generation can be minimized, but also the revenues are increased, thus reducing the penalties related to regulation costs and enhancing the competitiveness of renewable energies in comparison with dispatchable energy sources [5]. Finally, prosumers can use prediction models to plan their consumption patterns so as to match the power they generate on-site thus maximizing their benefits [6].

In the relevant literature, various forecasting methods have been proposed to estimate the expected power generated from a renewable energy source, which essentially differ in the type of the information characterizing the predicted output and in the time horizon of their application.

Concerning the type of information, two main forecasting methods can be adopted, referred to as deterministic and probabilistic forecasting. In the former one, a single value of the renewable power generation is provided and no uncertainty of the prediction is considered. In the latter one, the output value is accompanied with information on its intrinsic unpredictability and, then, it is more appropriate to solve problems of management and control in future SGs [3,5,7]. Probabilistic forecasting methods can be distinguished in two further categories according to the adopted approach: the prediction error or the direct approach. While the first one provides the uncertainty of the error deriving from the application of a deterministic forecasting method, the second one directly yields the statistic representation of the predicted output.

Concerning the time horizon, renewable generation forecasting can basically be divided into different time intervals, depending on the time frames corresponding to the tasks of grid operation and control and to the sessions of electricity markets. Short-term forecasting covers time intervals ranging from less than 1 hour to few hours ahead and is very useful for frequency regulation and load balancing. Medium term forecasting, up from several hours to few days ahead, is needed for unit commitment and energy trading. Finally, long-term forecasting can be required to support system planning and economic analyses in seasonal and annual horizons. However, recent renewable integration studies have shown that it is the short term forecasting that gains the most in a SG [3].

One of the most promising renewable energy conversion system to be integrated in SGs is the PhotoVoltaic (PV) power generation, due to the expected cost reduction and the increased efficiency of both PV panels and converters [8]. The power generated by a PV power system varies according to the solar radiation on the earth's surface, which mainly depends on the installation site and the weather conditions. While the dependence on the specific location can be essentially predicted on a deterministic way, the atmospheric conditions (such as cloud cover, ambient temperature, relative humidity) are the main causes of the randomness of the solar radiation and it is very important to consider them when short-term forecasting is concerned [9,10].

Several methods have been proposed in the relevant literature for forecasting the PV power generated in a short time horizon. In [11] a recurrent neural network has been proposed to perform a short term forecasting of the PV power production using meteorological data of the last 16 days and has been compared with a feed-forward neural network. A method to predict PV power output has been presented in [12] by deriving hourly site-specific irradiance forecasts from data provided by a weather forecasts center. In [13] an advanced Grey-Markov chain model has been applied to predict the daily power production of grid-connected PV systems using operating data collected at 15 minute intervals. A two-stage method to predict hourly value of the PV power for time horizon up to 36 hours has been proposed in [14]. In [15] Kalman filters are applied to predict sub-hourly and hourly PV power production using solar irradiance as input. First studies on the application of Bayesian theory for PV power production forecasting are shown in [16,17].

In this paper, extending and improving the approach based on the Bayesian theory outlined in [16–18], a new method for short-term probabilistic forecasting is proposed, that directly yields to the statistic representation of the predicted PV power output. The proposed method forecasts at the generic hour h the probability density function (pdf) of the active power produced by a PV power system at the hour $h + k$ with $k = 1, ..., K$, starting from the evaluation of samples of the pdf of the hourly clearness index at hour $h + k$. The forecast of the pdf of the hourly clearness index is obtained firstly selecting for the pdf an analytical expression and, then, evaluating the pdf parameters by applying the Bayesian Inference (BI). To this aim, an Auto Regressive (AR) time series model, representing the relationship between the pdf parameters, the clearness index and some explanatory meteorological variables, is used together with appropriate sets of historical measurements of the random variables involved in the AR model. Finally, a Monte Carlo (MC) simulation procedure is applied to generate the predicted pdf of the PV active power: a random sampling of the pdf of the hourly clearness index is performed and, using the PV system model, the PV power samples are obtained.

The key steps of the proposed method are: (i) the choice of the analytical expression of the pdf modeling the hourly clearness index; (ii) the definition of an adequate AR time series model so as to consider only the meteorological variables that most affect the hourly clearness index behaviour; and (iii) the selection of appropriate data vectors from historical measurements collected before the time of the forecast.

The peculiarity of the method is that it takes into account the dependence of the terrestrial solar radiation on some explanatory atmospheric variables and combines probabilistic techniques, such as BI and MC simulation, to provide a probabilistic forecasting of the PV power generation useful for optimal SG operation and control.

This paper is organized as follows: Section 2 briefly recalls the probabilistic forecasting method based on the Bayesian approach. In Section 3 the probabilistic method is applied to forecast the power production of a PV system. Finally, numerical simulations are reported in Section 4 to give evidence of the effectiveness of the proposed approach.

2. Probabilistic Forecasting method based on the Bayesian approach

The probabilistic forecasting method based on the Bayesian approach predicts at the generic hour $t = h$ the pdf of a random variable X_t at the hour $t = h + k$, with $k = 1, \ldots,$ K. For the sake of simplicity, in the following the analysis is referred to the case of $k = 1$.

In applying this method, the starting point is the knowledge of the analytical expression of the pdf $f_{X_t}(X_t)$ of the random variable to be forecasted. Usually, the analytical expression of the pdf is characterized by some distribution parameters and it is modeled as a conditional pdf. For the sake of conciseness, reference is made to only one distribution parameter (*i.e.*, the mean value), generically referred to as ϑ_t, and the conditional pdf is indicated as $f_{X_t}(X_t | \vartheta_t)$.

Forecasting at $t = h$ the pdf $f_{X_{hh+1}}(X_{h+1} | \vartheta_{h+1})$ requires an estimation of ϑ_{h+1}. To this aim, a first order AR time series model can be used, representing the relationship between ϑ_{h+1} and both the measurements x_h and $(v_{1,h}, \ldots, v_{M,h})$ collected at the hour h of, respectively, the random variable X_t to be forecasted and the M explanatory random variables $V_{1,t}, \ldots, V_{M,t}$ influencing X_t:

$$\vartheta_{h+1} = \alpha_1 x_h + \beta_1 v_{1,h} + \ldots + \beta_M v_{M,h} + \alpha_0 \tag{1}$$

where $\alpha_0, \alpha_1, \beta_1, \ldots, \beta_M$ are the coefficients of the AR model. Explanatory variables are variables such that changes in their value are thought to cause changes in another variable.

In the classical statistics, $\alpha_0, \alpha_1, \beta_1, \ldots, \beta_M$ are assumed to be constant. Indeed, when Bayesian approach are adopted, such coefficients are modeled as random variables, known as *prior random parameters,* and the BI [19] allows to estimate the conditional pdf $p(\alpha_0, \alpha_1, \beta_1, \ldots, \beta_M | S_{X_h})$ of the parameters $\alpha_0, \alpha_1, \beta_1, \ldots, \beta_M$ given the set $S_{X_h} = (x_{s_1}, \ldots, x_{s_{N_h}})$ composed of N_h measurements of X_t observed before the hour h. The pdf $p(\alpha_0, \alpha_1, \beta_1, \ldots, \beta_M | S_{X_h})$ is known as *a posteriori distribution* of the prior random parameters and it is very difficult to obtain its expression in closed form. Actually, only a simplified expression, known as *unnormalized a posteriori distribution* of the prior random parameters, and indicated as $q(\alpha_0, \alpha_1, \beta_1, \ldots, \beta_M | S_{X_h})$, can be provided. Fortunately, the knowledge of the unnormalized *a posteriori* distribution of the prior random parameters is sufficient for developing algorithms that provide information about the normalised *a posteriori* distributions.

The unnormalized a posteriori distribution of the prior random parameters is derived from the application of the Bayes' rule assuming the independency of the prior random parameters so that:

$$q(\alpha_0, \alpha_1, \beta_1, \ldots, \beta_M \mid S_{X_h}) = p(S_{X_h} \mid \alpha_0, \alpha_1, \beta_1, \ldots, \beta_M) \prod_{i=0}^{1} p(\alpha_i) \prod_{j=1}^{M} p(\beta_j) \qquad (2)$$

where $p(S_{X_h} \mid \alpha_0, \alpha_1, \beta_1, \ldots, \beta_M)$ is the likelihood function; and $p(\alpha_i)$ and $p(\beta_j)$ are the *a priori* distributions of the prior random parameters.

The *a priori* distributions are the initial pdfs of the prior random parameters which are not conditional on observed data. Their expressions can be vague or informative and reflects the knowledge that we have in advance about the pdfs that we are interested in.

The likelihood function is the conditional data distribution, that is the pdf modeling X_t, whose realizations are contained in S_{X_h}, given the prior random parameters. Its expression can be derived making use of the $f_{X_t}(X_t \mid \vartheta_t)$ for the set S_{X_h} and assuming that $x_{s_1}, \ldots, x_{s_{N_h}}$ are independent realizations of X_t. Substituting for ϑ_t the AR time series model and using the vectors $S_{V_{1,h}} = \left(v_{1,s_1}, \ldots, v_{1,s_{N_h}}\right), \ldots, S_{V_{M,h}} = \left(v_{M,s_1}, \ldots, v_{M,s_{N_h}}\right)$ of the N_h measurements of the explanatory variables $V_{1,t}, \ldots, V_{M,t}$ corresponding to S_{X_h}, it obtains:

$$p(S_{X_h} \mid \alpha_0, \alpha_1, \beta_1, \ldots, \beta_M) = \prod_{i=2}^{N_h} f_{X_t}(x_{s_i} \mid \vartheta_{s_i} = \alpha_1 x_{s_i-1} + \beta_1 v_{1,s_i-1} + \ldots + \beta_M v_{M,s_i-1} + \alpha_0) \qquad (3)$$

Once the unnormalized *a posteriori* distribution of the prior random parameters is known, it is trivial to evaluate the normalised *a posteriori* distributions of each parameter by applying the theory of the joint pdfs [20]. Then, the Monte Carlo Markov Chain (MCMC) simulation method based on the Metropolitan-Hasting algorithm [21] can be directly applied to the unnormalized distributions of every parameter to obtain samples of their *a posteriori* distributions. In the MCMC approach, a Markov chain is constructed, characterized by a transition probability matrix reflecting the *a posteriori* distributions of the prior random parameters. Then, the Markov chain is simulated until the samples are representative of the *a posteriori* distributions of every parameter.

Eventually, incorporating the AR time series model in this procedure, the samples derived from the *a posteriori* pdfs of $\alpha_0, \alpha_1, \beta_1, \ldots, \beta_M$ can be used together with the measurements x_h and $v_{1,h}, \ldots, v_{M,h}$ collected at the hour h to obtain samples of ϑ_{h+1}. Finally, for each simulated sample of ϑ_{h+1}, the samples of the random variable X_t are drawn from the analytical expression of the pdf so as to provide the full predictive distribution $f_{X_{hh+1}}(X_{h+1} \mid \vartheta_{h+1})$.

3. Probabilistic Forecasting of the Photovoltaic Generation

In the following, the probabilistic forecasting method described in Section 2 is applied to predict at hour h the pdf of the PV power production at hour $h + 1$, starting from an estimation of the terrestrial hourly solar radiation, expressed in terms of the pdf of the hourly clearness index at hour $h + 1$. The next four subsections dealt with:

- The description of the adopted model for the PV system;
- The description of the pdf modeling the hourly clearness index;
- The definition of the AR time-series model including meteorological variables; and
- The probabilistic characterization of the prior random parameters.

3.1. PV System Model

The hourly active power produced by a PV system depends on the availability of the solar radiation at the installation site. The solar radiation in a given locality cannot be exactly predicted owing mainly to the irregular presence of clouds. The sky conditions are often taken into account by representing the terrestrial solar radiation in terms of clearness index that is defined as the ratio of the surface radiation to the extraterrestrial radiation for a given period [22].

When the PV system is equipped with a maximum power point tracker, an analytical relationship exists between the PV active power P_{PV_t} at the hour t and the corresponding hourly clearness index K_t, [22–24] that is defined as the ratio of the hourly total solar radiation on an horizontal plane I_t to the extra-terrestrial hourly total solar radiation I_0; it results:

$$P_{PV_t} = S_C \eta \left(T K_t - T' K_t^2 \right)$$ (4)

where S_C is the array surface area; η is the efficiency of the PV system; and T and T' are defined as:

$$T = \left[\left(R_b + \rho \frac{1 - \cos \gamma}{2} \right) + \left(\frac{1 + \cos \gamma}{2} - R_b \right) p \right] r_d \frac{H_0}{3600}$$ (5)

$$T' = \left(\frac{1 + \cos \gamma}{2} - R_b \right) q r_d \frac{H_0}{3600}$$ (6)

where R_b is the ratio of beam radiation on a tilted surface to that on a horizontal surface at any time; ρ is the reflectance of the ground; γ is the inclination of the array surface to the horizontal plane; r_d is the ratio between diffuse radiation in hours and diffuse radiation in a day; H_0 is the extra-terrestrial total solar radiation; and p, q are coefficients reported in [23], which link the diffuse fraction of the hourly total solar radiation on horizontal plane with the hourly clearness index.

The analysis of the relationship (4) clearly reveals that K_t is the only variable affecting P_{PV_t}, once the hour of the day, the installation site and the technical characteristics of the PV system are assigned. The hourly clearness index K_t is a random variable modelling the uncertain behaviour of the terrestrial solar radiation. The hourly PV active power P_{PV_t}, as function of K_t, is itself a random variable and its pdf can be determined by the pdf of the hourly clearness index.

In this paper, a MC simulation procedure is used to generate at the hour $t = h$ samples of the predictive pdf of the hourly PV active power $f_{P_{PV_{h+1}}}\left(P_{PV_{h+1}}\right)$ by performing a random sampling of the pdf of the hourly clearness index $f_{K_{h+1}}(K_{h+1})$ estimated for the hour $t = h + 1$ and applying Equation (4).

3.2. Probability Density Function of the Hourly Clearness Index

The analytical expression of the pdf of the hourly clearness index can be experimentally obtained by a statistical analysis of historical solar measurements collected in the site in which the PV system is installed. In the literature, starting from the fitting of the hourly clearness index data collected in a specific location, the investigation has often resulted in the individuation of pdfs characterized by standard distributions. In [25] the hourly clearness index measurements recorded at various locations in Algeria are conveniently described by Beta distributions. In [26] bimodal distributions are considered more adequate to model clear and cloudy sky conditions of hourly clearness index measurements collected in different cities in the U.S.A. On the other hand, several attempts have been made to find universal standard pdfs that are independent of the location and the time period used to define the clearness index [27–29]. Following the latter approach, in this paper the model proposed in [29] has been adopted and the following modified Gamma distribution $f_{K_t}(K_t|C_t, \lambda_t)$ is used to model the pdf of the hourly clearness index K_t:

$$f_{K_t}(K_t|C_t, \lambda_t) = C_t \frac{\overline{k}_u - K_t}{\overline{k}_u} e^{\lambda_t K_t}$$ (7)

where $\bar{k}_u$ is the upper bound of the observed values of K_t; and C_t and λ_t are the distribution parameters, defined as:

$$C_t = \frac{\lambda_t^2 \bar{k}_u}{\left(e^{\lambda_t \bar{k}_u} - 1 - \lambda_t \bar{k}_u\right)} \tag{8}$$

$$\lambda_t = \frac{(2F_t - 17.519e^{-1.3118F_t} - 1062e^{-5.0426F_t})}{\bar{k}_u} \tag{9}$$

with:

$$F_t = \frac{\bar{k}_u}{\bar{k}_u - \mu_{K_t}} \tag{10}$$

where μ_{K_t} is the mean value of the hourly clearness index K_t at hour t. Assuming the knowledge of $\bar{k}_u$, the distribution parameters C_t and λ_t only depend on the mean value μ_{K_t} and the pdf in Equation (7) can be rewritten as:

$$f_{K_t}(K_t \mid \mu_{K_t}) = C_t(\mu_{K_t}) \frac{\bar{k}_u - K_t}{\bar{k}_u} e^{\lambda_t(\mu_{K_t})K_t} \tag{11}$$

3.3. AR Time-Series Model

To predict at the hour $t = h$ the pdf $f_{K_{hh+1}}(K_{h+1} \mid \mu_{K_{h+1}})$ at the hour $t = h+1$ an estimation of the mean value $\mu_{K_{h+1}}$ is required, as shown in the Subsection 3.2. To this aim, an AR time series model can be used to define the relationship among such mean value and the measurements of the clearness index and of some meteorological variables influencing the solar radiation, such as the ambient temperature AT, the relative humidity RH, the wind speed WS and the cloud cover CC , where the cloud cover is defined as the ratio in % of the sky hidden by all visible cloud. In this paper, the following first order AR time series model is adopted:

$$\mu_{K_{h+1}} = \alpha_1 k_h + \beta_1 at_h + \beta_2 rh_h + \beta_3 ws_h + \beta_4 cc_h + \alpha_0 \tag{12}$$

where $k_h, at_h, rh_h, ws_h, cc_h$ are the measurements of, respectively, the clearness index, the ambient temperature, the relative humidity, the wind speed and the cloud cover, which are collected at the hour $t = h$.

The inclusion of meteorological variables in the AR model significantly increases the computational efforts in the application of the proposed forecasting method. Despite of the highest complexity of the procedure, taking into account the dependence of the clearness index on the meteorological variables allows to perform a more accurate forecasting [10,30]. To reduce computational efforts, a correlation analysis can help to individuate the meteorological variables presenting the highest influence on the solar radiation. Such analysis is performed "off-line" and correlates historical measurements collected in the specific site in which the PV system is installed. In this way, only the meteorological variables presenting the highest correlation value with the clearness index are selected so as to found a good compromise between results' accuracy and computational efforts.

3.4. Probabilistic Characterization of the Prior Random Parameters

In this paper, the coefficients α_0, α_1, $\beta_1, \beta_2, \beta_3$, β_4 of the AR time series model in Equation (12) are assumed to be the prior random parameters of the BI. The *a priori* pdfs of the prior random parameters $p(\alpha_0)$, $p(\alpha_1), p(\beta_1), p(\beta_2), p(\beta_3)$, $p(\beta_4)$ are usually chosen with a large variance so that the data, rather than the *a priori* distributions, determine the relevant parameters values in the *a posteriori* distributions [19–21]. In this paper the *a priori* pdfs are assumed to be Gaussian with a mean value $\mu = 0.5$ and a standard deviation $\sigma = 0.5$ so as:

$$p(\alpha_i) = \frac{1}{\sigma\sqrt{2\pi}}e^{\frac{(\mu-\alpha_i)^2}{2\sigma^2}} = \frac{1}{0.5\sqrt{2\pi}}e^{\frac{-\alpha_i^2}{2(0.5)^2}} \quad i = 0,1 \tag{13}$$

$$p(\beta_j) = \frac{1}{\sigma\sqrt{2\pi}}e^{\frac{(\mu-\beta_j)^2}{2\sigma^2}} = \frac{1}{0.5\sqrt{2\pi}}e^{\frac{-\beta_j^2}{2(0.5)^2}} \quad j = 1, \dots, 4 \tag{14}$$

The likelihood function $p(S_{K_h} \mid \alpha_0, \alpha_1, \beta_1, \beta_2, \beta_3, \beta_4)$ is the pdf in Equation (11) specified for the set $S_{K_h} = \left(k_{s_1}, \dots, k_{s_{ih}}, \dots, k_{s_{N_h}}\right)$ of N_h measurements of k_t observed before the hour h:

$$p(S_{K_h} \mid \alpha_0, \alpha_1, \beta_1, \beta_2, \beta_3, \beta_4) = \prod_{i=2}^{N_h} f_{K_t}(k_{s_i} \mid \mu_{K_{si}}) = \prod_{i=2}^{N_h} C\left(\mu_{K_{s_i}}\right)\frac{\overline{k}_u - k_{s_i}}{\overline{k}_u}e^{\lambda(\mu_{K_{s_i}})k_{s_i}} \tag{15}$$

where:

$$\mu_{K_{s_i}} = \alpha_1 k_{s_i-1} + \beta_1 at_{s_i-1} + \beta_2 rh_{s_i-1} + \beta_3 ws_{s_i-1} + \beta_4 cc_{s_i-1} + \alpha_0 \tag{16}$$

Relationship (15) is obtained by substituting for $\mu_{K_{s_i}}$ the AR time series model in Equation (16). The generic measurements k_{s_i-1}, at_{s_i-1}, rh_{s_i-1}, ws_{s_i-1}, cc_{s_i-1} are contained in the sets $S_{K_h} = \left(k_{s_1}, \dots, k_{s_i}, \dots, k_{s_{N_h}}\right)$, $S_{AT_h} = \left(at_{s_1}, \dots, at_{s_i}, \dots, at_{s_{N_h}}\right)$, $S_{RH_h} = \left(rh_{s_1}, \dots, rh_{s_i}, \dots, rh_{s_{N_h}}\right)$, $S_{WS_h} = \left(ws_{s_1}, \dots, ws_{s_i}, \dots, ws_{s_{N_h}}\right)$, $S_{CC_h} = \left(cc_{s_1}, \dots, cc_{s_i}, \dots, cc_{s_{N_h}}\right)$, including N_h measurements collected before the hour h of, respectively, the clearness index and the meteorological variables.

According to Equation (2), the *a posteriori* unnormalized distribution $q(\alpha_0, \alpha_1, \beta_1, \beta_2, \beta_3, \beta_4, S_{K_h})$ of the prior random parameters is equal to

$$q(\alpha_0, \alpha_1, \beta_1, \beta_2, \beta_3, \beta_4, S_{K_h}) = \left[\prod_{i=2}^{N_h} C(\mu_{K_{s_i}})\frac{\overline{k}_u - k_{s_i}}{\overline{k}_u}e^{\lambda(\mu_{K_{s_i}})k_{s_i}}\right]\prod_{i=0}^{1}\frac{1}{0.5\sqrt{2\pi}}e^{\frac{-\alpha_i^2}{2(0.5)^2}}\prod_{j=0}^{4}\frac{1}{0.5\sqrt{2\pi}}e^{\frac{-\beta_j^2}{2(0.5)^2}} \tag{17}$$

and the samples of the individual a posteriori distributions are evaluated by applying the MCMC simulation method based on the Metropolitan-Hasting algorithm. The samples of the a posteriori pdfs of $\alpha_0, \alpha_1, \beta_1, \beta_2, \beta_3, \beta_4$ are used in Equation (12) together with the measurements k_h, t_h, r_h, w_h, cc_h collected at the hour h to provide samples of the mean value $\mu_{K_{h+1}}$. Finally, for each samples of $\mu_{K_{h+1}}$, the samples of the hourly clearness index X_{h+1} are drawn from the analytical expression of the pdf in Equation (11). The simulated samples of X_{h+1} describes the predictive distribution $f_{K_{h+1}}(K_{h+1} \mid \mu_{K_{h+1}})$.

It should be noted that the choice of the measurements contained in the sets $S_{K_h}, S_{AT_h}, S_{RH_h}, S_{WS_h}, S_{CC_h}$ represent a key issue in the BI, since they are used to make inference about the prior random parameters $\alpha_0, \alpha_1, \beta_1, \beta_2, \beta_3, \beta_4$. In general, these sets contain N_h measurements recorded before the hour h of the forecast. Actually, these data are not necessarily the ones collected from the hour $h - 1$ to $h - N_h - 1$, but can be selected with adequate criteria thus improving the accuracy of the proposed forecasting method. In [16] the *homologue* and the *coded group* criteria have been proposed. In the first one, the sets contain measurements at the hour h which are collected N_h days before the forecast (e.g., if the forecast has to be performed at h = 10:00, the sets include measurements recorded N_h days before at 10:00). In the second one, the sets contain a coded group of measurements around the hour h collected some days before the forecast (e.g. if the forecast has to be performed at h = 10:00, the sets include measurements from the 8:00 to 10:00 recorded N_h/3 days before). In addition, the measurements contained in $S_{K_h}, S_{AT_h}, S_{RH_h}, S_{WS_h}, S_{CC_h}$ can be collected at time intervals different from an hour. In [17] the data of the clearness index and of the meteorological variables contained in the vectors are extracted from measurements registered at time intervals of 15 minutes. If this is the case, the application of the proposed forecasting method will provide at the hour h the predictive distribution of the clearness index at the first 15 minutes of the hour $h + 1$. To estimate the pdf at the hour $h + 1$ the following approach is adopted in this paper:

$$f_{K_{h+1}}(K_{h+1} \mid \mu_{K_{h+1}}) = f_{K_{h+15'}}(K_{h+15'} \mid \mu_{K_{h+15'}}) \tag{18}$$

that is the pdf forecasted at $h + 1$ is assumed to be equal to the pdf forecasted at the first 15 minutes of the hour $h + 1$. Eventually, Figure 1 shows a block diagram describing the main steps applied in the proposed Bayesian approach.

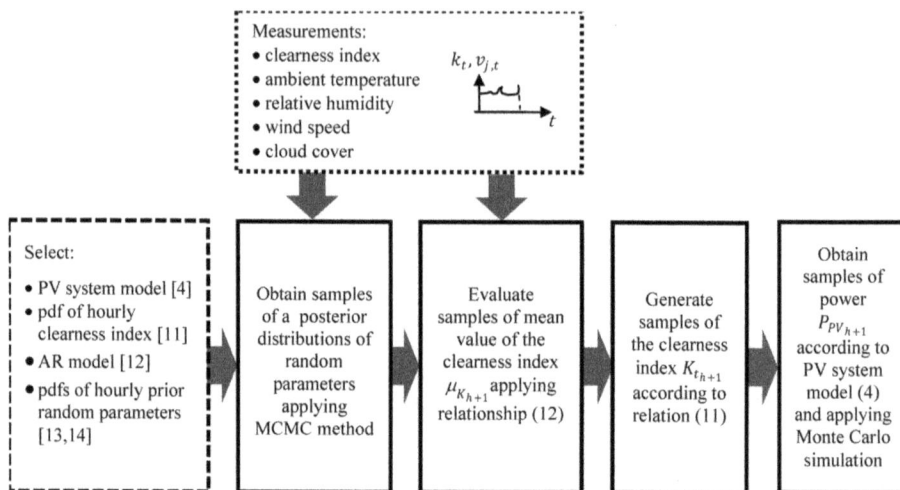

Figure 1. Block diagram describing the main steps applied in the proposed Bayesian approach.

4. Experimental Section

In this section, the proposed Bayesian forecasting method is applied to a 75-kWp PV system, presenting an array surface $S_C = 600$ m^2 and an efficiency $\eta = 0.09$. Measurements of the clearness indexand of the meteorological variables cited in the Section 3.4 (air temperature, relative humidity, wind speed and total cloud cover) are available at the website of the National Renewable Energy Laboratory. In particular, a meteorological station in Colorado (39.742° N, 105.18° W) has been selected and measurements referred to the time interval [8 a.m., 8 p.m.] and collected every 15 minutes are chosen.

In the following, at first a correlation analysis is performed to individuate the meteorological variables to be included in the time series AR model; then, the proposed method is used to forecast the pdf of the hourly active power produced by the PV system.

To individuate the most suitable AR time series model, an "off-line" correlation analysis between the clearness index and the meteorological variables is carried out, on the basis of measurements recorded from January to December 2010. Figure 2 reports the time evolution of the correlation coefficient between the clearness index and the meteorological variables. To avoid excessive computational efforts, only the meteorological variables furnishing the highest values of the correlation coefficient are taken into account. As such, the analysis of the Figure 2 clearly reveals that the total cloud cover and the relative humidity are the meteorological variables presenting the highest influence on the clearness index; consequently, the AR time-series model in Equation (12) reduces to:

$$\mu_{K_{h+1}} = \alpha_1 k_h + \beta_1 rh_h + \beta_2 cc_h + \alpha_0 \tag{19}$$

To make inference about the prior random parameters α_0, α_1, β_1 and β_2, the sets S_{K_h}, S_{CC_h}, S_{RH_h} (see Section 3.4) contain $N_h = 144$ measurements collected at time intervals of 15 minutes recorded before the hour h of the forecast.

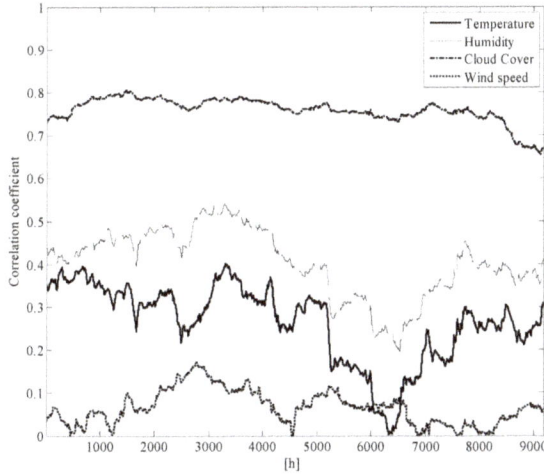

Figure 2. Time evolution of the correlation coefficient between the clearness index and the selected meteorological variables.

The application of the proposed approach to forecast the hourly PV active power is performed referring to the four seasons of the 2011. Figures 3–6 show, respectively, the actual measured values of the PV active powers (red lines), the mean value (blue line) and the range between the 5th and 95th percentile values of the forecasted pdfs of hourly PV active power. In particular, the results refer to the Recommended Average days of winter (Figure 3), spring (Figure 4), summer (Figure 5) and autumn (Figure 6). In [22] Recommended Average Days are days which have the extraterrestrial radiation closest to the average value in the month. A similar behavior characterizes the vast majority of the considered days (in almost all considered days).

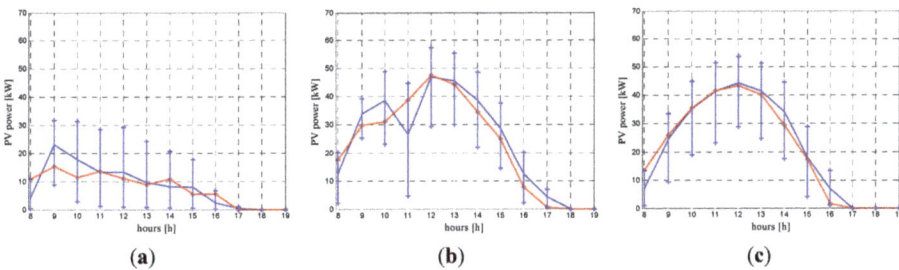

Figure 3. Actual measures of the hourly PV active power (red line); mean values (blue line) and range between 5th and 95th percentile values of the forecasted pdfs of the hourly PV active power. (a) 17 January; (b) 16 February; and (c) 10 December.

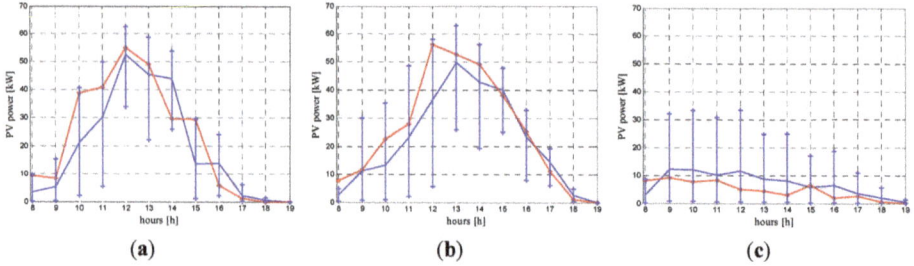

Figure 4. Actual measures of the hourly PV active power (red line); mean values (blue line) and range between 5th and 95th percentile values of the forecasted pdfs of the hourly PV active power. (a) 16 March; (b) 15 April; and (c) 15 May.

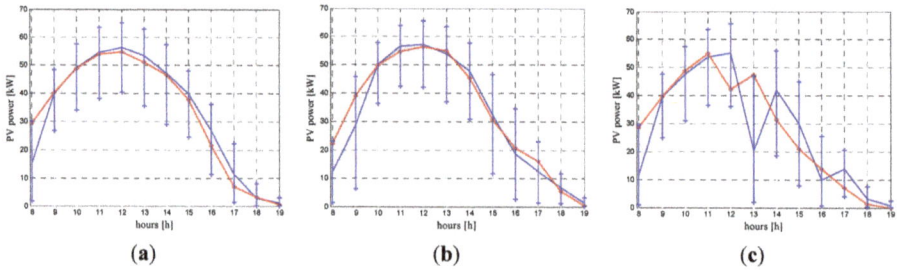

Figure 5. Actual measures of the hourly PV active power (red line); mean values (blue line) and range between 5th and 95th percentile values of the forecasted pdfs of the hourly PV active power. (a) 11 June; (b) 17 July and; (c) 16 August.

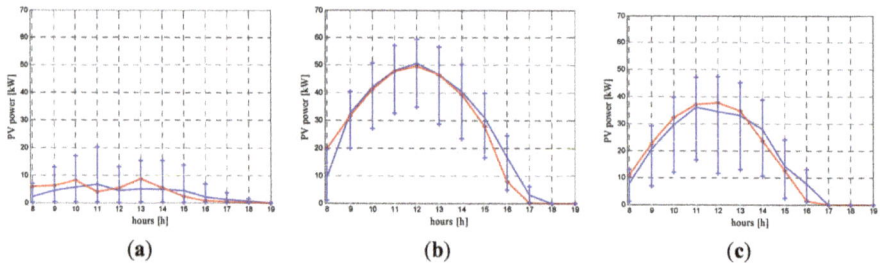

Figure 6. Actual measures of the hourly PV active power (red line); mean values (blue line) and range between 5th and 95th percentile values of the forecasted pdfs of the hourly PV active power. (a) 15 September; (b) 15 October and; (c) 14 November.

From the analysis of the figures, it appears that the actual values of the hourly PV active power are always comprised between the 5th and 95th percentile values. In addition, it should be noted that the mean value appears in most cases a good estimator for the forecasted pdfs, particularly in the range of hours between 11 a.m. and 3 p.m. At the beginning (end) of the period characterized by the presence of solar radiation, usually higher (lower) percentiles appear the most adequate estimators.

Anyway, if the mean value of forecasted pdf would be used as the only estimator of the forecasted PV power, the Mean Absolute Relative Error (MARE), defined as:

$$MARE = \frac{1}{N} \sum_{h=1}^{N} \frac{\left| P_{PV_h} - P_{PV_h}^* \right|}{P_{PV_h}} \tag{20}$$

is estimated between 14.5% (winter season) and 18.0% (autumn season).

Finally, Figures 7 and 8 show the forecasted (represented by a blue histogram) and the analytical (represented by a continuous red line) pdfs of the hourly PV active power in March (Figure 7a), April (Figure 7b), August (Figure 8a) and September (Figure 8b). The analytical pdf is obtained applying the fundamental theorem for the function of a random variable to Equation (9) proposed in [31]. From the analysis of Figures 7 and 8 it is evident that in different conditions of solar radiation the forecasted pdfs are close to the analytical distributions.

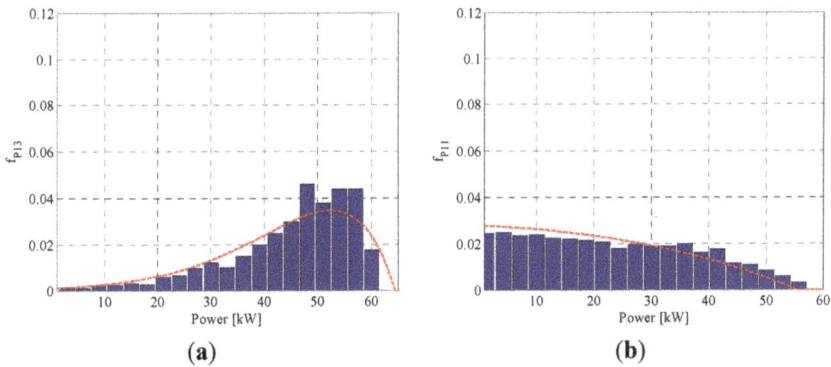

(a) (b)

Figure 7. Forecasted (histogram) and analytical (red line) pdfs of PV power. (**a**) h = 1 p.m. of 16 March; and (**b**) h = 11 a.m. of 15 April.

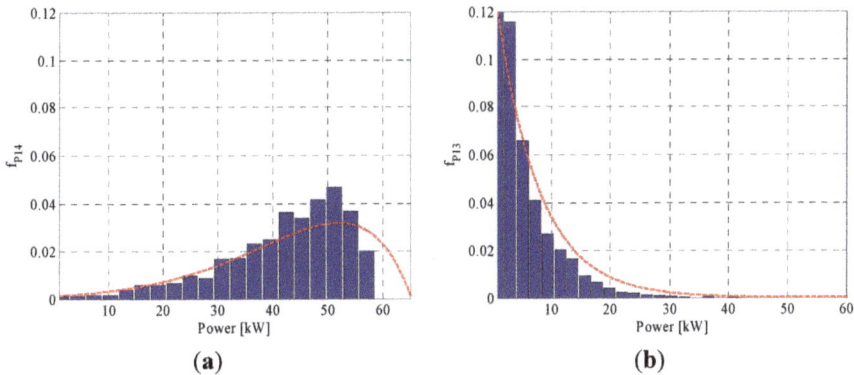

(a) (b)

Figure 8. Forecasted (histogram) and analytical (red line) pdfs of PV power. (**a**) h = 2 p.m. of 16 August and; (**b**) h = 1 p.m. of 15 September.

5. Conclusions

A new method based on the Bayesian inference has been proposed to perform a short-term forecasting of the active power produced by a photovoltaic system starting from an estimation of

Energies **2013**, *6*, 733–747

the hourly clearness index. It takes into account the dependence of the terrestrial solar radiation on some explanatory atmospheric variables, including the cloud cover and humidity. The combination of probabilistic techniques, such as Bayesian inference and Monte Carlo simulation, allows to provide the predictive probability density function of the photovoltaic generated power, which is very useful for the optimal operation and control of the smart grids of the future.

However, if only a value is requested as estimator of the forecasted photovoltaic power, arises the problem of individuate which pdf parameter (mean value, percentiles ...) is the most representative for the distribution. Moreover, the non-linear relationship between the clearness index and the photovoltaic power output can reflect in not negligible errors in the forecasted distributions of the photovoltaic power. Then, future works will investigate the direct application of the proposed probabilistic forecasting method to the active power produced by the PV system, even if the application of the Bayesian inference in this case seems to be arduous. The research will also focus on the choice of the best parameter to be extracted from the predicted probability distribution so as to test the performance of the proposed method in terms of traditional measures of the forecasting accuracy; in this case comparison with ARIMA and neural networks methods will be affected.

Acknowledgments: This paper has been developed in the context of the Italian project PON Research and Competitiveness 2007–2013, Action II-PON 01_02864 "FC SMART GEN"—Fuel cell and smart hybrid generation from fossil and renewable sources.

References

1. Bouhafs, F.; Mackay, M.; Merabti, M. Links to the future: Communication requirements and challenges in the smart grid. *IEEE Power Energy Mag.* **2012**, *10*, 24–28. [CrossRef]
2. Pilo, F.; Pisano, G.; Soma, G.G. Optimal coordination of Energy resources with a two-stage online active management. *IEEE Trans. Ind. Electron.* **2011**, *58*, 4526–4537. [CrossRef]
3. Potter, C.M.; Archambault, A.; Kenneth, W. Building a smarter smart grid to better renewable energy information. In Proceedings of Power Systems Conference and Exposition (PSCE 09), Seattle, WA, USA, 15–18 March 2009.
4. Smith, J.C.; Milligan, M.R.; De Meo, E.A.; Parson, B. Utility wind integration and operating impact state of art. *IEEE Trans. Power Syst.* **2007**, *22*, 900–907. [CrossRef]
5. Pinson, P.; Chevallier, C.; Kariniotakis, G.N. Trading wind generation from short-term probabilistic forecasts of wind power. *IEEE Trans. Power Syst.* **2007**, *22*, 1148–1156. [CrossRef]
6. Sharma, N.; Sharma, P.; Irwin, D.; Shenoy, P. Predicting solar generation from weather forecast using machine learning. In Proceedings of IEEE International Conference on Smart Grid Communications (SmartGridComm), Bruxelles, Belgium, 17–20 October 2011; pp. 528–533.
7. Watson, R. *IEA Expert Meeting on Wind Forecasting Techniques*; NREL Report; NREL (National Renewable Energy Laboratory): Golden, CO, USA, 2000.
8. Liserre, M.; Sauter, T.; Hung, J.Y. Future energy systems: Integrating renewable energy sources into the smart power grid through industrial electronics. *IEEE Ind. Electron. Mag.* **2010**, *4*, 18–37. [CrossRef]
9. Mellit, A.; Massi Pavan, A. A 24-h forecast of solar irradiance using artificial neural network: Application for performance prediction of a grid-connected PV plant at Trieste, Italy. *Sol. Energy* **2010**, *84*, 807–821.
10. Al Riza, D.F; Gilani, S.I.; Aris, M.S. Hourly solar radiation estimation using ambient temperature and relative humidity data. *Int. J. Environ. Sci. Dev.* **2011**, *2*, 188–193. [CrossRef]
11. Yona, A.; Senjyu, T.; Funabashi, T. Application of recurrent neural network to short-term-ahead generating power forecasting for photovoltaic system. In Proceedings of IEEE Power Engineering Society General Meeting, Tampa, FL, USA, 24–28 June 2007.
12. Lorenz, E.; Hurka, J.; Heinemann, D.; Beyer, H.G. Irradiance forecasting for the power prediction of grid-connected photovoltaic systems. *IEEE J. Sel. Top. Appl. Earth Obs. Remote Sens.* **2009**, *2*, 2–10. [CrossRef]
13. Li, Y.; He, L.; Nie, R. Short-term forecast of power generation for grid-connected photovoltaic system based on advanced Grey-Markov chain. In Proceedings of IEEE International Conference on Energy and Environment Technology (ICEET '09), Guilin, China, 16–18 October 2009; Volume 2, pp. 275–278.

14. Bacher, P.; Madsen, H.; Nielsen, P. Online short-term solar power forecasting. *Sol. Energy* **2009**, *83*, 1772–1783. [CrossRef]

15. Hassanzadeh, M.; Etezadi-Amoli, M.; Fadali, M.S. Practical approach for sub-hourly and hourly prediction of PV power output. In Proceedings of IEEE Conferences North American Power Symposium, Arlington, TX, USA, 1–5 September 2010.

16. Bracale, A.; Caramia, P.; Fantauzzi, M.; Di Fazio, A.R. A Bayesian-based approach for photovoltaic power forecast. In Proceedings of Cigrè International Symposium on Smart Grid, Bologna, Italy, 13–15 September 2011.

17. Bracale, A.; Caramia, P.; De Martinis, U.; Di Fazio, A.R. An improved Bayesian-based approach for short term photovoltaic power forecasting in smart grids. In Proceedings of International Conference on Renewable Energies and Power Quality (ICREPQ 2012), Santiago De Compostela, Spain, 28–30 March 2012.

18. Bracale, A.; Caramia, P.; Carpinelli, G.; Di Fazio, A.R.; Varilone, P. A Bayesian-based approach for very short-term steady-state analysis of a smart grid. *IEEE Trans. Smart Grids* **2013**, in press.

19. Gelman, A.; Carlin, J.B.; Stern, H.S.; Rubin, D.B. *Bayesian Data Analysis*; Chaoman & Hall: London, UK, 1995.

20. Papoulis, A. *Probability, Random Variables and Stochastic Processes*; McGraw-Hill: New York, NY, USA, 1991.

21. Zhang, J.; Pu, J.; McCalley, J.D.; Stern, H.; Gallus, W.A., Jr. A Bayesian approach to short term transmission line thermal overload risk assessment. *IEEE Trans. Power Deliv.* **2002**, *17*, 770–778. [CrossRef]

22. Duffie, J.A.; Beckman, W.A. *Solar Engineering of Thermal Processes*, 2nd ed.; Wiley Interscience: New York, NY, USA, 1991.

23. Orgill, J.F; Hollands, K.T.G. Correlation equation for hourly diffuse radiation on a horizontal surface. *Sol. Energy* **1977**, *19*, 357–359. [CrossRef]

24. Kroposki, B.; Emery, K.; Myers, D.; Mrig, L. A comparison of photovoltaic module performance evaluation methodologies for energy ratings. In Proceedings of First World Conference on Photovoltaic Energy Conversion (WPEC 1994), Hawaii, HI, USA, 5–9 December1994; pp. 858–862.

25. Ettoumi, F.Y.; Mefti, A.; Adane, A.; Bouroubi, M.Y. Statistical analysis of solar measurements in Algeria using beta distributions. *Renew. Energy* **2002**, *124*, 28–33.

26. Ibanez, M.; Beckman, W.A.; Sanford, A.K. Frequency distributions for hourly and daily clearness indices. *J. Sol. Energy Eng.* **2002**, *26*, 47–67.

27. Liu, B.Y.H.; Jordan, B.C. The interrelationship and characteristic distribution of direct, diffuse and total solar radiation. *Sol. Energy* **1960**, *4*, 1–19. [CrossRef]

28. Bendt, P.; Collares-Pereira, M.; Rabl, A. The frequency distribution of daily insolation values. *Sol. Energy* **1981**, *27*, 1–19. [CrossRef]

29. Hollands, K.T.G.; Huget, R.G. A probability density function for clearness index, with applications. *Sol. Energy* **1983**, *30*, 195–209. [CrossRef]

30. Huang, Y.; Lu, J.; Liu, C.; Xu, X.; Wang, W.; Zhou, X. Comparative study of power forecasting methods for PV stations. In Proceedings of International Conference on Power System Technology (POWERCON 2010), Hangzhou, China, 24–28 October 2010; pp. 1–6.

31. Conti, S.; Raiti, S. Probabilistic load flow using Monte Carlo techniques for distribution networks with photovoltaic generators. *Sol. Energy* **2007**, *81*, 1473–1481. [CrossRef]

energies

MDPI

Article

A Fuzzy Group Forecasting Model Based on Least Squares Support Vector Machine (LS-SVM) for Short-Term Wind Power

Qian Zhang [1,*], Kin Keung Lai [2,3], Dongxiao Niu [3], Qiang Wang [3] and Xuebin Zhang [1]

[1]   School of Economics and Management, North China Electric Power University, Baoding 071003, China
[2]   Department of Management Science, City University of Hong Kong, Kowloon, Hong Kong;
      msKklai@cityu.edu.hk
[3]   School of Economics and Management, North China Electric Power University, Beijing 102206, China;
      niudx@126.com (D.N.); jshdw@126.com (Q.W.)
*   Author to whom correspondence should be addressed; hdzhq@yeah.net;
     Tel.: +86-137-80226160; Fax: +86-0312-7525111.

Received: 20 April 2012; in revised form: 15 August 2012; Accepted: 21 August 2012;
Published: 5 September 2012

Abstract: Many models have been developed to forecast wind farm power output. It is generally difficult to determine whether the performance of one model is consistently better than that of another model under all circumstances. Motivated by this finding, we aimed to integrate groups of models into an aggregated model using fuzzy theory to obtain further performance improvements. First, three groups of least squares support vector machine (LS-SVM) forecasting models were developed: univariate LS-SVM models, hybrid models using auto-regressive moving average (ARIMA) and LS-SVM and multivariate LS-SVM models. Each group of models is selected by a decorrelation maximisation method, and the remaining models can be regarded as experts in forecasting. Next, fuzzy aggregation and a defuzzification procedure are used to combine all of these forecasting results into the final forecast. For sample randomization, we statistically compare models. Results show that this group-forecasting model performs well in terms of accuracy and consistency.

Keywords: wind power forecasting; LS-SVM; ARIMA; fuzzy group

---

1. Introduction

Along with science and technology in general, wind power technology has also developed rapidly. Because wind power technology is mature, many medium- and large-sized wind farms have been built and put into operation. Wind power has become an important source of the entire power system; worldwide, the installed wind power capacity was 157.9 GW in 2009, representing an annual growth of 20% over the preceding 10 years. Wind energy resources available in China are estimated at 1000 GW, ranking the country third after Russia and the U.S. In recent years, wind power has experienced rapid development in China, as the capacity increased from 0.34 to 25.8 GW between 2000 and 2009. In 2020, the total installed capacity of wind power is expected to reach 150 GW [1].

Wind power is always fluctuating because wind is volatile and intermittent. When the power output exceeds a certain value, it significantly affects power quality, power system security and the stability of operations. If an accurate short-term wind power output forecast is available, the power dispatching department can adjust scheduling in accordance with changes in wind power output to ensure power quality and reduce the system's excess capacity and power system cost. Therefore, short-term wind power forecasts are of key importance [2–4].

Modern wind farms usually incorporate remote monitoring systems in wind turbines so that all turbines can capture and record all signals. The real-time output data from wind generators can be used directly for wind power forecasts without any additional cost, which reduces the cost and improves the quality of data collection, as well as increases forecast accuracy. The existing forecasting methods can be classified into two groups. The first group consists of univariate forecasting models based on historical and real-time power data, in which changes in wind speed are not considered. The second group consists of multivariate models, in which forecasts are based on the relationship between weather data and output power [5]. The numerical weather prediction (NWP) model is popular for short-term wind power prediction with advantages in accuracy, but, it needs more weather information [6]. Detailed algorithms include time series methods, such as the auto-regressive moving average (ARMA) and the auto-regressive conditional heteroskedasticity (ARCH) models [7,8], the linear regression model [9], the grey theory model [10,11], the support vector machine (SVM) [12,13], adaptive fuzzy logic algorithms [14,15] and artificial neural networks (ANNs) [16,17], among others [18].

In the above-mentioned individual models, it is difficult to determine whether the performance of one model is consistently better than that of another model under all circumstances. Typically, a number of different models are utilised, and the model with the most accurate results is selected. However, the selected model may not necessarily be the best for future use because of potentially influential factors, such as sampling variation, model uncertainty and structure change. It is almost universally agreed upon in the forecasting literature that no single method is best in every situation, primarily because a real-world problem is often complex in nature and because any single model may not be able to capture different patterns equally well. Therefore, there is a certain optimal combination of forecasts to be studied, such as an adaptive combination of forecasts [19] and an optimal combination of wind power forecasts [20]. Motivated by this finding, we aimed to integrate multiple models into an aggregated model to obtain further performance improvement. Therefore, certain intelligent SVM forecasting models were developed. The models are selected by a decorrelation maximisation method, and the remaining models can be regarded as experts in forecasting. Then, the fuzzy theory is used to combine all of these forecasting results into the final forecast.

The remainder of this paper is organised as follows: Section 2 describes three group models. In Section 3, real datasets are statistically used for the testing of these models. Finally, conclusions are presented in Section 4.

2. The Forecasting Model

2.1. Principle of Least Squares SVM (LS-SVM)

In this study, SVM was selected as the basic algorithm with which to construct forecasting models because this algorithm is often viewed as a "universal approximator". It has been proven to provide a good arbitrary approximation of any continuous function. Therefore, the model is used here to simulate mutual relationships between historical data and the forecast power output. The models have the ability to provide flexible mapping between inputs and outputs. The SVM model of a data set is given by the formula described below.

Consider an n set of data$\{(x_1, y_1), \dots, (x_N, y_N)\}$, where x_i is the i_{th} input vector and y_i is the corresponding desired output. Because $i = 1, 2, \dots, N$, where N is the size of the sample, the estimating function assumes the following form:

$$f(x) = w \cdot \phi(x) + b \tag{1}$$

where w is the weight vector, b is the bias and $\phi(x)$ is the high-dimensional feature space nonlinearly mapped from the input space, and $(\cdot)$ represents the inner product.

This leads to the optimisation problem associated with standard SVM:

$$minR_{str} = \frac{1}{2}\|w\|^2 + \gamma R_{emp} \tag{2}$$

where γ is a positive real constant that determines the penalty for estimation errors and $R_{emp}(w, b) = \frac{1}{N} \sum_{i=1}^{N} |y_i - f(x_i)|_\delta$ is the estimation error measured by the experimental risk and loss function. Usually, the ε- insensitive loss function is adopted because of its excellent sparsity:

$$|y - f(x)|_\varepsilon = \begin{cases} 0, & |y - f(x)| \leq \varepsilon \\ |y - f(x)| - \varepsilon, & \text{elsewhere} \end{cases}$$

(3)

For least-squares SVM (LS-SVM), the two norms of the estimation error are adopted as the loss function in the objective function and equality constraints instead of inequality constraints. Therefore, the optimisation problem is described as:

$$\begin{cases} \min\limits_{w,b,\xi_i} \dfrac{1}{2} w^T w + \dfrac{1}{2} \gamma \sum_{i=1}^{N} \xi_i^2 \\ s.t. \quad y_i = w^T \phi(x_i) + b + \xi_i \quad i = 1, 2, \cdots, N \end{cases}$$

(4)

where ξ_i is a slack variable, $\xi_i \geq 0$. It is a variable added to an inequality constraint to transform it to equality. It is non-negative number in this paper.

After the introduction of Lagrange multipliers α_i, the Lagrange function is constructed as:

$$L = \frac{1}{2} w^T w + \frac{1}{2} \gamma \sum_{i=1}^{N} \xi_i^2$$
$$- \sum_{i=1}^{N} \alpha_i \left\{ w^T \phi(x_i) + b + \xi_i - y_i \right\}$$

(5)

According to *KKT conditions* which can transform inequality constraints into equality constraints, defined as:

$$\alpha_i \left[y_i \left(w^T \phi(x_i) + b \right) - 1 + \xi_i \right] = 0, i = 1, 2 \ldots, N$$
$$(\gamma - \alpha_i) \xi_i = 0, i = 1, 2 \ldots, N$$

(6)

The following equation can then be obtained:

$$\begin{cases} \dfrac{\partial L}{\partial w} = 0 \\ \dfrac{\partial L}{\partial b} = 0 \\ \dfrac{\partial L}{\partial \xi_i} = 0 \\ \dfrac{\partial L}{\partial \alpha_i} = 0 \end{cases} \Rightarrow \begin{cases} w = \sum_{i=1}^{N} \alpha_i \phi(x_i) \\ \sum_{i=1}^{N} \alpha_i = 0 \\ \alpha_i = \gamma \xi_i \\ w^T \phi(x_i) + b + \xi_i - y_i = 0 \end{cases}$$

(7)

After eliminating w and γ, we obtain:

$$
\begin{bmatrix} 0 & \Theta^T \\ \Theta & \Omega + \dfrac{1}{\gamma} I \end{bmatrix} \begin{bmatrix} b \\ \alpha \end{bmatrix} = \begin{bmatrix} 0 \\ y \end{bmatrix}
$$

(8)

where $\Theta = [1, \dots, 1]_{1 \times N}$, I is a unit matrix, Ω is a square matrix and the element of Ω is expressed as: $\Omega_{ij} = \phi(x_i)^T \phi(x_j)$. In the equation (8), $\alpha = [\alpha_1, \dots, \alpha_N]$, $y = [y_1, \dots, y_N]$.

By solving Equation (7), values of α and b are obtained. According to Mercer's condition, there exists a kernel function with a value that is equal to the inner product of the two vectors x_i and x_j in the feature spaces $\phi(x_i)$ and $\phi(x_j)$; that is, $K(x_i, x_j) = \phi(x_i)^T \phi(x_j)$. Then, the LS-SVM model for regression is expressed as:

$$
y(x) = \sum_{i=1}^{N} \alpha_i K(x, x_i) + b
$$

(9)

2.2. Group Model Based on LS-SVM

2.2.1. Group 1: Diversified Univariate LS-SVM Model

The first group is the univariate forecasting model. It is based on historical and real-time power data; other weather data, such as wind speed, are not considered. Many experimental results have shown that the generalisation of individual networks is not unique. Even for some simple problems, different SVMs with different settings (e.g., different network architectures and different initial conditions) may result in different generalisation results. Diverse models are generated by selecting different core learning algorithms, such as the steep-descent algorithm, the Levenberg-Marquardt algorithm and other training algorithms [21]. Finally, 10 different univariate least squares support vector machine (LS-SVM) models are formulated [22,23]. All of these models use the Gaussian function as the kernel function, and the output is the one-hour-ahead forecasted wind power output. Other parameters are shown in Table 1.

Table 1. Ten univariate LS-SVM models.

Models	Inputs	γ	σ^2
LS-SVM-1	3 previous observations	10	5
LS-SVM-2	4 previous observations	20	5
LS-SVM-3	5 previous observations	30	5
LS-SVM-4	6 previous observations	40	5
LS-SVM-5	7 previous observations	50	5
LS-SVM-6	3 previous observations	50	2
LS-SVM-7	4 previous observations	50	4
LS-SVM-8	5 previous observations	50	6
LS-SVM-9	6 previous observations	50	8
LS-SVM-10	7 previous observations	50	10

2.2.2. Group 2: Diversified Univariate Hybrid Model of ARIMA and the SVM Model

2.2.2.1. Brief Introduction of the Hybrid Model

Because real-world time series are rarely purely linear or nonlinear, researchers have revealed that hybrid models that hybridise two or more different algorithms can produce forecasts of higher accuracy than those produced by individual models. ARIMA and LS-SVM models have different capabilities of capturing data characteristics in linear and nonlinear domains; therefore, the hybrid model proposed in this study is composed of an ARIMA component and an LS-SVM component. Thus, the hybrid model is expected to capture linear and nonlinear patterns with improved overall forecasting performance. Experimental results with real data sets indicate that the hybrid model can be an effective means by which to improve forecasting accuracy over that achieved by either of the models separately. In this section, a type of hybrid approach using both ARIMA and LS-SVM models is proposed. Because ARIMA is a linear model [24] and LS-SVM [22,25] is a nonlinear model, the hybrid approach is expected to capture both linear and nonlinear patterns in wind park power time series.

Based on the structure proposed by [26], the hybrid model (y_t) can be represented as:

$$y_t = L_t + N_t \tag{10}$$

where L_t denotes the linear component and N_t denotes the nonlinear component.

These two components must be estimated from the data. First, ARIMA is used to model the linear component, resulting in the residuals from the linear model containing only the nonlinear relationship. The residual at time t (from the linear model) is denoted as e_t, and then:

$$e_t = y_t - \hat{L}_t \tag{11}$$

where $\hat{L}_t$ is the forecast value at time t from the ARIMA models. Specifications of the $(1, 0, 0) \times (0, 1, 1)$ model are as described in Equation (11):

$$Y_t = \delta + Y_{T-4} + \phi_1(Y_{t-1} - Y_{t-5}) \tag{12}$$

Residuals are also important. By modelling residuals using LS-SVM, nonlinear relationships can be discovered. With n input nodes, the LS-SVM model for residuals will be:

$$e_t = f(e_{t-1}, e_{t-2}, \ldots e_{t-n}) + \Delta_t \tag{13}$$

where f is a nonlinear function determined by the LS-SVM model and Δ_t is its corresponding random error. Therefore, the forecast of the hybrid model is:

$$\hat{y}_t = \hat{L}_t + \hat{e}_t \tag{14}$$

2.2.2.2. Generating the Diversified Hybrid Model from the ARIMA and LS-SVM Models

The proposed hybrid method is applied to forecast wind power output, *i.e.*, the LS-SVM model is used to model the nonlinearity of residuals obtained from the ARIMA models. As mentioned in Section 2.1, to generate the diverse models, the structure of the above LS-SVM can be varied by changing the number of nodes in the input layer and the second layer. Because the number of input layers is changed, there should be different training data. These data can be acquired by re-sampling and pre-processing the data. There are many techniques that can be used to obtain diverse training data sets, such as bagging noise injection, cross-validation and stacking. With these different training datasets and structures, 10 diverse hybrid models are generated using ARIMA and LS-SVM models as described in Table 2. For all of these models, the linear parts use ARIMA ($Y_t = \delta + Y_{T-4} + \phi_1(Y_{t-1} - Y_{t-5})$) and the nonlinear parts use different LS-SVMs. All of these LS-SVM models use the Gaussian function as the kernel function, and the output is the forecasted error. Other parameters are shown in Table 2.

Table 2. Ten diverse hybrid models using ARIMA and LS-SVM.

Models	Inputs	γ	σ^2
H-AR-LS-1	3 previous observations	10	5
H-AR-LS-2	4 previous observations	20	5
H-AR-LS-3	5 previous observations	30	5
H-AR-LS-4	6 previous observations	40	5
H-AR-LS-5	7 previous observations	50	5
H-AR-LS-6	3 previous observations	50	2
H-AR-LS-7	4 previous observations	50	4
H-AR-LS-8	5 previous observations	50	6
H-AR-LS-9	6 previous observations	50	8
H-AR-LS10	7 previous observations	50	10

2.2.2.3. Group 3: Diversified Multivariate LS-SVM model

In this group of multivariate methods, the relationship between weather data and power output is considered. There are five fundamental variables that impact wind power output. The first, w_1, is the wind speed, measured in metres/second (m/s); the second, w_2, is the wind direction, measured as the angle between the incoming wind and the north; the third, w_3, is the air temperature, measured in °C; the fourth, w_4, is the atmospheric pressure in Pa; and the fifth, w_3a, is the relative humidity. These five fundamental variables are used as input data, and the wind power output is the output of the LS-SVM model.

To generate the diverse models, the structure of the above LS-SVM model is varied by changing the number of nodes in the second layer. Different initial conditions can also create diversity in models; these initial conditions include random weights, learning rates and momentum rates from which each network is trained. With these different initial conditions and structures, 10 diverse LS-SVMs are generated. All of these models use the Gaussian function as the kernel function, and the output is the one-hour-ahead forecasted wind power output. Other parameters are shown in Table 3.

Table 3. Ten diverse multivariate LS-SVMs.

Models	Inputs	γ	σ^2
DLS-SVM-1	$w_1; w_2; w_3; w_3; w_4; w_5;$ 2 previous observations	10	5
DLS-SVM-2	$w_1; w_2; w_3; w_3; w_4; w_5;$ 2 previous observations	20	5
DLS-SVM-3	$w_1; w_2; w_3; w_3; w_4; w_5;$ 2 previous observations	30	5
DLS-SVM-4	$w_1; w_2; w_3; w_3; w_4; w_5;$ 2 previous observations	40	5
DLS-SVM-5	$w_1; w_2; w_3; w_3; w_4; w_5;$ 2 previous observations	50	5
DLS-SVM-6	$w_1; w_2; w_3; w_3; w_4; w_5;$ 3 previous observations	50	2
DLS-SVM-7	$w_1; w_2; w_3; w_3; w_4; w_5;$ 3 previous observations	50	4
DLS-SVM-8	$w_1; w_2; w_3; w_3; w_4; w_5;$ 3 previous observations	50	6
DLS-SVM-9	$w_1; w_2; w_3; w_3; w_4; w_5;$ 3 previous observations	50	8
DLS-SVM-10	$w_1; w_2; w_3; w_3; w_4; w_5;$ 3 previous observations	50	10

2.3. Group Model Based on LS-SVM

As mentioned above, each group consists of 10 forecasting models. We need to select a subset of representatives to improve ensemble efficiency. It is clear that it is a necessary requirement of diverse models for making fuzzy group decisions. In this study, a decorrelation maximisation method was used to select the appropriate number of ensemble members. As noted previously, the basic starting point of the decorrelation maximisation algorithm is the principle of ensemble model diversity; that is, the correlations between the selected models should be as small as possible. If there are p models $(f_1, f_2, \ldots, f_p)$ with n forecast values, an error matrix $(e_1, e_2, \ldots, e_p)$ of p predictors can be represented by:

$$E = \begin{bmatrix} e_{11} & e_{12} & \cdots & e_{1p} \\ e_{21} & e_{22} & \cdots & e_{2p} \\ \vdots & \vdots & & \vdots \\ e_{n1} & e_{n2} & \cdots & e_{np} \end{bmatrix}_{n \times p}$$

(15)

From the matrix, the mean, variance and covariance of E can be calculated as:

$$\text{Mean: } \bar{e} = \frac{1}{n}\sum_{k=1}^{n} e_{ki} \ \ (i = 1, 2, \ldots, p)$$

(16)

$$\text{Variance: } V_{ii} = \frac{1}{n}\sum_{k=1}^{n}(e_{ki} - \bar{e}_i)^2 \ \ (i = 1, 2, \ldots, p)$$

(17)

$$\text{Covariance: } V_{ij} = \frac{1}{n}\sum_{k=1}^{n}(e_{ki} - \bar{e}_i)(e_{ki} - \bar{e}_j) \ \ (i,j = 1, 2, \ldots, p)$$

(18)

Considering Equations (17) and (18), we can obtain a variance covariance matrix:

$$V_{p \times p} = (V_{ij})$$

(19)

Based on the variance-covariance matrix, correlation matrix R can be calculated using the following equations:

$$R = (r_{ij})$$

(20)

$$r_{ij} = \frac{V_{ij}}{\sqrt{V_{ii}V_{jj}}}$$

(21)

where r_{ij} is the correlation coefficient, representing the degrees of correlation classifiers f_i and f_j.

Subsequently, the plural-correlation coefficient $\rho f_i \mid (f_1, f_2, \ldots, f_{i-1}, f_{i+1}, \ldots, f_p)$ between classifier f_i and other $p - 1$ classifiers can be computed based on the results of Equations (20) and (21). For convenience, $\rho f_i \mid (f_1, f_2, \ldots, f_{i-1}, f_{i+1}, \ldots, f_p)$ is abbreviated as ρ_i, representing the degree of correlation between f_i and $(f_1, f_2, \ldots, f_{i-1}, f_{i+1}, \ldots, f_p)$. To calculate the plural-correlation coefficient, the correlation matrix R can be represented by a block matrix; that is:

$$R \xrightarrow{\text{after transformation}} \begin{bmatrix} R_{-i} & r_i \\ r_i^T & 1 \end{bmatrix}$$

(22)

where $R - i$ denotes the deleted correlation matrix. It should be noted that $r_{ii} = 1 (i = 1, 2, \ldots, p)$. Next, the plural-correlation coefficient can be calculated by:

$$\rho_i^2 = r_i^T R_{-i}^T r_i (i = 1, 2, \ldots p) \tag{23}$$

For a pre-specified threshold θ, if $\rho_i{}^2 > \theta$, then model f_i should be removed from p models. Otherwise, model f_i should be retained. Generally, the decorrelation maximisation algorithm can be summarised in the following steps:

Computing the variance-covariance matrix V_{ij} and the correlation matrix R with Equations (19) and (20). For the i_{th} classifier ($i = 1, 2, \ldots, p$), the plural-correlation coefficient ρ_i can be calculated using Equation (23).

For a pre-specified threshold θ, if $\rho_i < \theta$, then the i_{th} classifier should be deleted from the ρ classifiers. Conversely, if $\rho_i > \theta$, then the i_{th} classifier should be retained. For each group of models, we select eight as the representative for the subsequent step.

2.4. Fuzzy Group Prediction

For a specified forecasting problem, different experts usually give different estimations based on a set of criteria $X = (c_1, c_2, \ldots, c_m)$. Some experts give optimistic estimates, some prefer pessimistic estimates, and others present the most likely estimates. To incorporate these different judgements into the final forecasting result and to make full use of the different estimates, a process of fuzzification is used. In this paper, a typical triangular fuzzy number can be used to describe the forecasting results provided by the experts; that is:

$\tilde{Z}_i = (z_{i1}, z_{i2}, z_{i3}) = $ (the lowest forecast value; the most likely forecast value; the highest forecast value), where i represents the numerical index of experts.

Like human experts, individual LS-SVM forecasting groups can also generate different forecasting results by using different parameter settings and training sets. For example, the first forecasting group (univariate LS-SVM model group) generates eight different forecasting results from the eight models (selected from the first 10 models; Section 2.3) of different hidden neurons or different initial weights. The entire first group can be considered an expert in forecasting. Assume that this expert produces k different results, $f_1^i(X_A), f_2^i(X_A), \ldots f_k^i(X_A)$, for a specified applicant "A" over a set of models of different hidden neurons or different initial weights in this group. To make full use of all of the information provided by these results, without loss of generalisation, we use the triangular fuzzy number to construct the fuzzy opinion for consistency; that is the smallest, average and largest of the k forecasting results are used as the left-, medium- and right-membership degrees, respectively. In other words, the smallest and largest scores are seen as optimistic and pessimistic evaluations, respectively, and the average forecasting result is considered to be the most likely score. Of course, the median can also be used as the most likely score to construct the triangular fuzzy number. However, that approach can cause the loss of certain useful information because some other scores are ignored. Therefore, the average is selected as the most likely power output to incorporate the full information from all of the models into the fuzzy judgement. Using this fuzzification method, the expert can make a fuzzy forecast for each point. More precisely, the triangular fuzzy number used for forecasting can be represented as:

$$\tilde{Z}_i = (z_{i1}, z_{i2}, z_{i3}) = \left\{ \left[\min(f_1^i(X_A), f_2^i(X_A), \ldots, f_K^i(X_A)) \right], \right.$$
$$\left. \left[\sum_{j=1}^{k} f_j^i(X_A)/k \right], \left[\max(f_1^i(X_A), f_2^i(X_A), \ldots, f_K^i(X_A)) \right] \right\} \tag{24}$$

Suppose there are p experts, and let $\tilde{Z}_i = \psi(\tilde{Z}_1, \tilde{Z}_2, \ldots \tilde{Z}_p)$ be the aggregation of p fuzzy judgements, where $\psi()$ is an aggregation function. Many methods have been developed to determine the aggregation function. Usually, fuzzy judgements of the p group members are aggregated by using a common linear additive procedure; that is:

$$\tilde{Z} = \sum_{i=1}^{p} w_i \tilde{Z}_i = \left(\sum_{i=1}^{p} w_i z_{i1}, \sum_{i=1}^{p} w_i z_{i2}, \sum_{i=1}^{p} w_i z_{i3} \right)$$

(25)

where w_i is the weight of the i_{th} fuzzy judgement, $i = 1, 2, \ldots, p$. The weights usually satisfy the following normalisation condition:

$$\sum_{i=1}^{p} w_i = 1$$

(26)

At this point, the goal is to determine the optimal weight w_i of the i_{th} fuzzy expert. In this study, three groups of models are used as experts, and we give them the same weight of 1/3 each. After completing aggregation, a fuzzy group consensus can be obtained using Equation (25). To obtain a crisp value of the credit score, we use a defuzzification procedure to obtain the crisp value for decision-making purposes. According to Bortolan and Degani, the defuzzified value of a triangular fuzzy number $\tilde{Z}_i = (z_1, z_2, z_{i3})$ can be determined by its centroid, which is computed by:

$$z = \frac{\int_{z_1}^{z_3} x \mu_{\tilde{z}}(x) dx}{\int_{z_1}^{z_3} \mu_{\tilde{z}}(x) dx} = \frac{z_1 + z_2 + z_3}{3}$$

(27)

At this point, a final group consensus has been computed using the above process. To summarise, the proposed intelligent-agent-based fuzzy group forecasting model is comprised of five steps:

(1) Three forecasting groups are presented, and each group has eight models with varied structures and initial data, for example.
(2) Based on the datasets, each forecasting group can produce eight different forecasting results from the different models.
(3) For the different forecasting results, Equation (25) is used to fuzzify the judgements of intelligent agents into fuzzy opinions.
(4) The fuzzy opinions are aggregated into a group consensus, using the optimisation method proposed above, in terms of the maximum agreement principle.
(5) The aggregated fuzzy group consensus is defuzzified into a crisp value. This defuzzified value can be used as the final forecasting result.

To illustrate and verify the proposed intelligent-agent-based fuzzy group forecasting model, the following section presents an illustrative numerical example of real-world data. The flow chart of the entire procedure is shown in Figure 1.

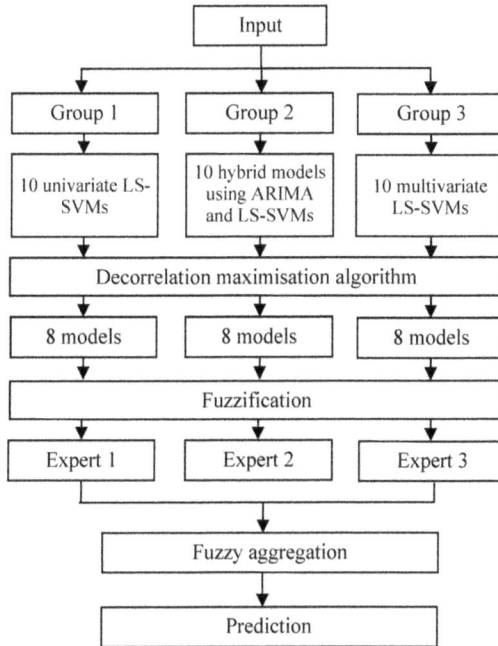

Figure 1. Procedure flow chart.

3. Empirical Analyses

3.1. Forecasting Results

In this study, we collected wind power output data from the Changshun wind park in Huade County, Inner Mongolia Autonomous Region, China. This wind park is located on the slopes of hills and mountains within an area of 260 km^2. Details of the park's geographical information are provided in Table 4. This wind park was completed in May 2010 and has a capacity of 49.5 MW. Its wind power-out data from 1 January 2011, to 31 December 2011, were collected as shown in Figure 2. The short-term forecasting model for predicting hourly power output over a 24-hour horizon was tested. Other input data, such as the actual climate information, were collected from local environmental stations.

Table 4. Wind park geographical information.

Latitude (North)	Longitude (East)	Elevation (m)	Wind speed (m/s)		Temperature (°C)		
			Mean	Max	Mean	Min	Max
41°10'–41°45'	113°49'–114°03'	1500	4.8	29	2.2	−35.9	35.5

Note: The very low minimum temperature is the extremely low temperature in this area, the lowest temperatures in this wind park is −27 °C in January. There is no stop in 2011 due to low temperature.

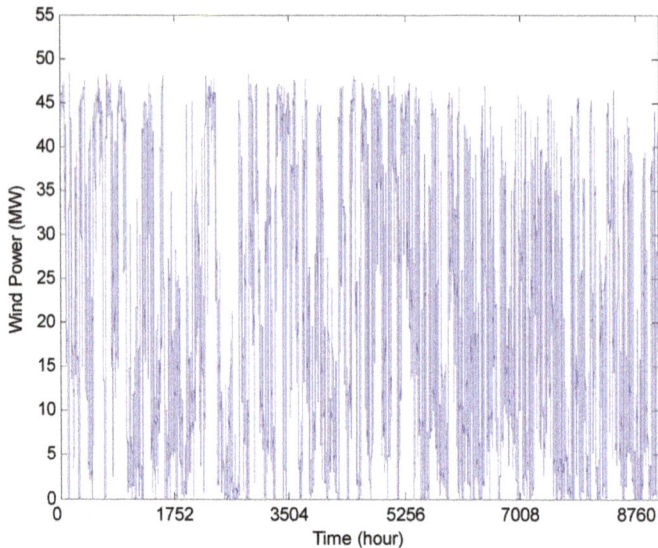

Figure 2. Time series plots of hourly wind power output.

The data from 1 January 2011, to 31 October 2011, are used for constructing and training the models. The data from November 2011 are used to test the models and select the group modes according to Section 2.3. The results are presented in Table 5.

The data from December 2011 are used in the testing of the models and in the model analysis. There are 24 points for each day. To judge the accuracy of the model, individual models and the combined fuzzy forecasting model are compared using the following MAPE:

$$MAPE = \frac{1}{N}\sum_{i=1}^{N}\left|\frac{p_i - \hat{p}_i}{p_i}\right| \times 100\%$$

(28)

where $\hat{p}_i$ is the forecast data, p_i is the real-time data, and N is the number of time points used in determining the forecast.

Also the relative error is adopted to evaluate the models performance. The error is calculated as the follows:

$$RE = \frac{p_i - \hat{p}_i}{p_i} \times 100\%$$

(29)

The MAPEs of the individual models and the combined fuzzy forecasting model are calculated. The results are shown in Table 5.

Table 5. The MAPEs of individual models and the combined fuzzy forecasting model.

Group 1		Group 2		Group 3	
model	MAPE	model	MAPE	model	MAPE
LS-SVM-1	19.71%	H-AR-LS-1	17.26%	DLS-SVM-1	18.06%
LS-SVM-2	24.03%	H-AR-LS-2	21.22%	DLS-SVM-2	21.91%
LS-SVM-3	24.75%	H-AR-LS-3	21.85%	DLS-SVM-3	20.65%
LS-SVM-4	19.52%	H-AR-LS-4	18.05%	DLS-SVM-4	17.62%
LS-SVM-6	18.94%	H-AR-LS-5	17.46%	DLS-SVM-5	20.85%
LS-SVM-7	25.36%	H-AR-LS-6	18.50%	DLS-SVM-6	16.71%
LS-SVM-9	22.45%	H-AR-LS-7	16.99%	DLS-SVM-9	18.31%
LS-SVM-10	18.08%	H-AR-LS-10	19.01%	DLS-SVM-10	22.35%
Average	21.61%	Average	18.79%	Average	19.56%
GFSVM	15.27%				

3.2. Statistical Test

The best individual model is DLS-SVM-6, and the second best is H-AR-LS-7 in terms of MAPE, Statistical test is carried out among the GFSVM model and those two models. According to the methods mentioned in reference [27], comparison in made between the GFSVM model and the best individual model DLS-SVM-6.

$\{y_{it}\}_{t=1}^{T}$ is the history data series, $\{\hat{y}_{it}\}_{t=1}^{T}$ is the results from the GFSVM model, $\{\hat{y}_{jt}\}_{t=1}^{T}$ is the result from the DLS-SVM-6 model. $\{e_{it}\}_{t=1}^{T}$ is the error of GFSVM model and $\{e_{jt}\}_{t=1}^{T}$ is the error of DLS-SVM-6 model. The loss function will be a direct function of the forecast error, that is $g(y_t, \hat{y}_{it}) = g(e_{it})$. The loss differential is $d_t = [g(e_{it}) - g(e_{jt})]$. Empirically, the forecast error has many features: 1. zero mean 2, Gaussian 3. Serially correlated 4 contemporaneously correlated. The null hypothesis is a positive median loss differential: $\text{med}(g(e_{it}) - g(e_{jt})) < 0$. So, we introduce two test statistics in reference [27], S_1 and S_{2a} as the follows:

$$\sqrt{T}\left(\bar{d} - \mu\right) \xrightarrow{d} N\left(0, 2\pi f_d(0)\right) \tag{30}$$

$$\bar{d} = \frac{1}{T}\sum_{t=1}^{T}\left[g\left(e_{it}\right) - g\left(e_{jt}\right)\right] \tag{31}$$

$$f_d(0) = \frac{1}{2\pi}\sum_{\tau=-\infty}^{\infty}\gamma(\tau) \tag{32}$$

$$\gamma(\tau) = E\left[\left(d_t - \mu\right)\left(d_{t-\tau} - \mu\right)\right] \tag{33}$$

$$S_1 = \frac{\bar{d}}{\sqrt{\dfrac{2\pi \hat{f}_d(0)}{T}}} \tag{34}$$

where $\hat{f}_d(0)$ is a consistent estimate of $f_d(0)$:

$$S_2 = \sum_{t=1}^{T} I_+\left(d_t\right)$$

$$\tag{35}$$

$$S_2 = \sum_{t=1}^{T} I_+\left(d_t\right)$$

$$\tag{36}$$

where $I_+(d_t) = 1$ if $d_t > 0$; $I_+(d_t) = 0$ otherwise:

$$S_{2a} = \frac{S_2 - 0.5T}{\sqrt{0.25T}} \xrightarrow{a} N(0,1)$$

$$\tag{37}$$

The comparison result between the GFSVM model and DLS-SVM-6 the model is shown as Figure 3 and Table 6.

The same comparison is made between the GFSVM model and the H-AR-LS-7 model, and the result is shown as Figure 4 and Table 7.

In Tables 6 and 7, T is sample size, ρ is the contemporaneous correlation, and θ is the serial correlation. All tests are at the 10% level. We perform 260 replications.

For comparison between the GFSVM model and the DLS-SVM-6 model, we obtain $S_1 = 11.74$, $S_{2a} = 10.67$ which implying a p-value= 0.089, 0.076. Thus, for sample at hand we do not reject at conventional level the hypothesis of the accuracy of the GFSVM model is better than the DLS-SVM-6 model. In the similar way, we can also statistically conclude that the GFSVM model is better than the H-AR-LS-7 model.

From above, we can draw a statistical conclusion that the GFSVM model is better than the DLS-SVM-6 model and the H-AR-LS-7 model.

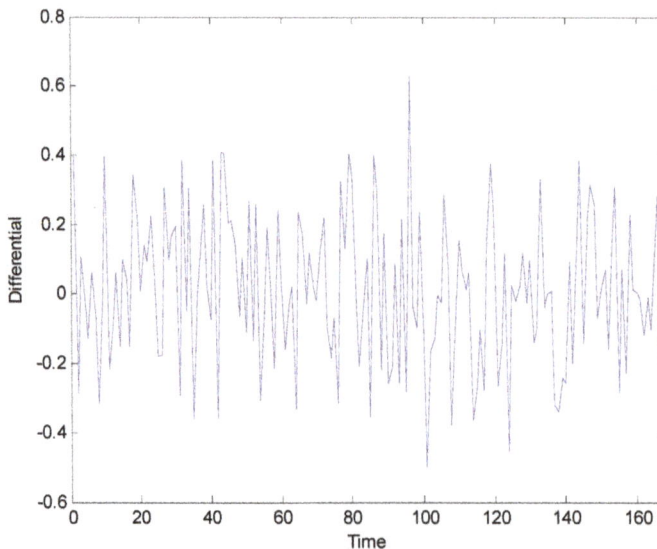

Figure 3. Loss Differential (GFSVM - DLS-SVM-6).

Table 6. Empirical Size under Quadratic Loss, Test Statistic S_1, S_{2a} (GFSVM—DLS-SVM-6).

T	ρ	S_1			S_{2a}		
		$\theta = 0$	$\theta = 0.5$	$\theta = 0.9$	$\theta = 0$	$\theta = 0.5$	$\theta = 0.9$
168	0	11.47	11.72	11.89	10.93	10.96	11.06
168	0.5	11.26	11.62	11.41	10.84	10.94	11.11
168	0.9	11.53	11.08	11.17	10.41	11.03	10.92

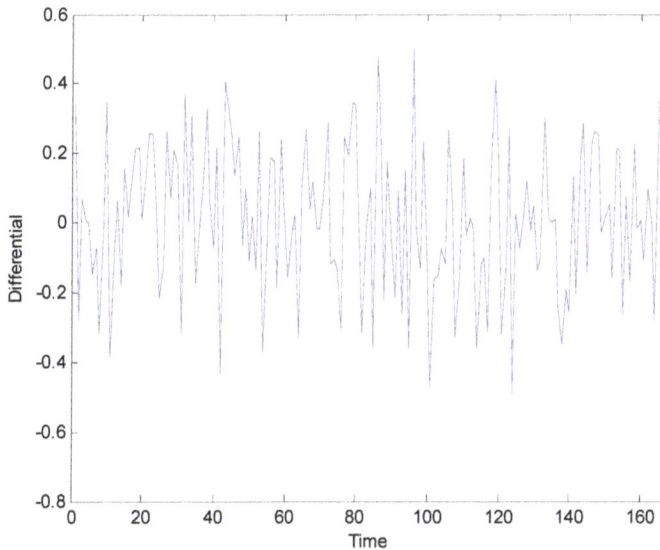

Figure 4. Loss Differential (GFSVM - H-AR-LS-7).

Table 7. Empirical Size under Quadratic Loss, Test Statistic S_1, S_{2a} (GFSVM—H-AR-LS-7).

T	ρ	S_1			S_{2a}		
		$\theta = 0$	$\theta = 0.5$	$\theta = 0.9$	$\theta = 0$	$\theta = 0.5$	$\theta = 0.9$
168	0	11.45	11.69	11.78	10.87	10.91	11.13
168	0.5	11.23	11.61	11.37	10.81	10.97	11.12
168	0.9	11.54	11.11	11.15	10.38	10.92.	10.97

3.3. Result Discussions

From Table 5, it can be observed that the fuzzy group forecasting model (GFSVM) performs best in terms of MAPE, with a MAPE of only 15.27%. The average MAPEs of these 8 models for groups 1, 2 and 3 are 21.61, 18.79 and 19.6%, respectively; all of these MAPEs are higher than those of the GFSVM. The best and second best individual models are DLS-SVM-6 and H-AR-LS-7, and their relative errors for total testing points are shown in Figures 5 and 6 respectively. From these two figures, it can be observed that the range of the relative errors from the fuzzy group forecasting model GFSVM is smaller than that for DLS-SVM-6 and H-AR-LS-7. This means that the GFSVM is much more reliable than the other models. Table 8 represents the number of predictions between ±10%, ±20%, ±30% and ±40% for DLS-SVM-6, H-AR-LS-7 and GFSVM. For example, for the GFSVM model, 47.3% of the predictions have errors between ±10%, whereas for the DLS-SVM-6 model, 34.1% of the errors are in the same error margin, and for H-AR-LS-7 model, only 30.5% of the errors are in the same error margin. Obviously, the accuracy of GFSVM model is the best among these three models. From Figure 7, we know that the GFSVM can imitate the actual wind power output with high accuracy.

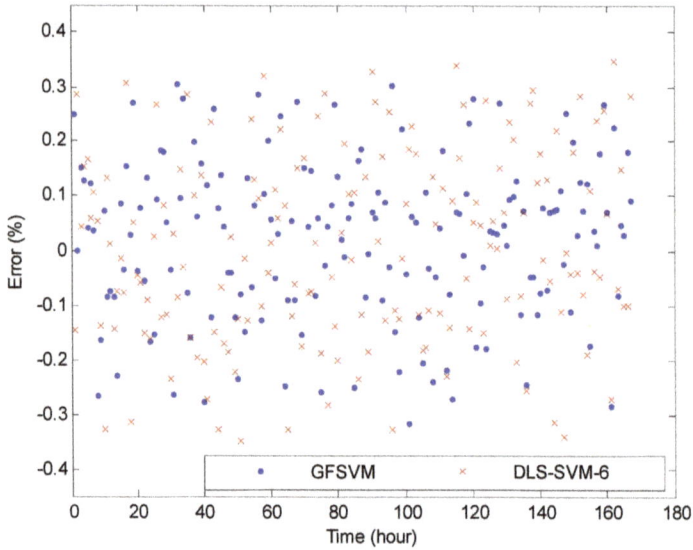

Figure 5. Wind power forecast relative errors of GFSVM model and DLS-SVM-6 model.

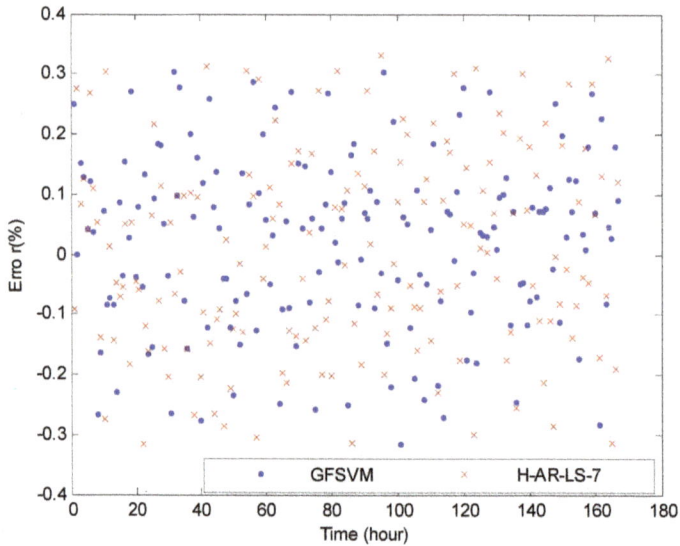

Figure 6. Wind power forecast relative errors of GFSVM model and H-AR-LS-7model.

Table 8. Wind power forecast errors distribution for three models (% of errors in each margin).

	GFSVM	DLS-SVM-6	H-AR-LS-7
±10%	47.3%	34.1%	30.5%
±20%	81.4%	76.6%	74.3%
±30%	98.2%	92.2%	91.6%
±40%	100.0%	100.0%	100.0%

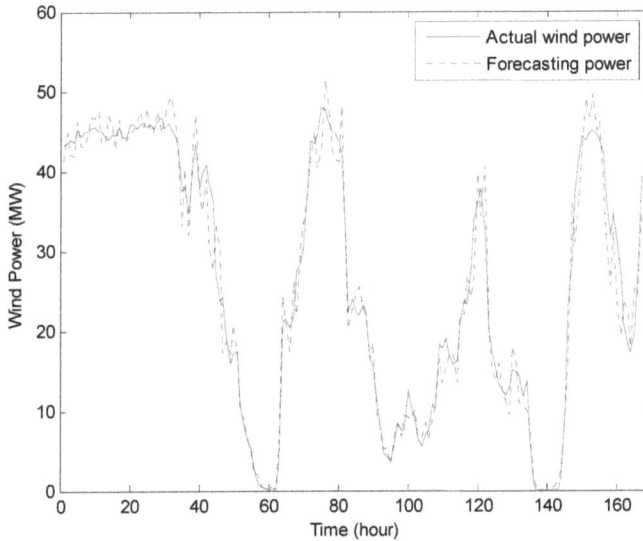

Figure 7. Forecasts derived from the fuzzy model (2011, 12, 01-2011, 12, 07).

It is found that there is correlations among the current wind power output and those 1 h before and later. It is feasible to use them for predicting. From the Statistical test, it can be proved that the performance of the GFSVM model is better than that of DLS-SVM-6 model and H-AR-LS-7 model. It is the best in terms of accuracy and reliability among the models of these three groups. Also its Robustness is higher than those of the LS-SVM, ARIMA LS-SVM, DLS-SVM models. The overall prediction of the proposed method is better, but there is still individual prediction with large error, which needs further research.

4. Conclusions

In this study, we integrated groups of models into an aggregated model by using fuzzy theory to improve forecasting performance. The fuzzy group model overcame the intrinsic defects of single models, obtained information from various single models, and then created the optimum combination. Therefore, in most cases, we can achieve the purpose of improving forecasting results by combination forecasting, which obviously improves accuracy. Combination forecasting can be used to forecast wind power output over short time horizons. Through imitation computation and comparison, we proved that the forecasting accuracy is improved. Our approach thus offers a new and effective method for wind power forecasting.

Acknowledgments: This study was supported by the Fundamental Research Funds for the Central Universities (12MS135), the Hebei Social Science Research Project and the Research on the Special Rules in Project Network, Postdoctoral Science Foundation, China.

References

1. Liu, Y.; Kokko, A. Wind power in China: Policy and development challenges. *Energy Policy* **2010**, *38*, 5520–5529. [CrossRef]
2. Ramirez-Rosado, I.J.; Fernandez-Jimenez, L.A.; Monteiro, C.; Sousa, J.; Bessa, R. Comparison of two new short-term wind-power forecasting systems. *Renew. Energy* **2009**, *34*, 1848–1854. [CrossRef]
3. Kavasseri, R.G.; Seetharaman, K. Day-ahead wind speed forecasting using f-ARIMA models. *Renew. Energy* **2009**, *34*, 1388–1393. [CrossRef]
4. Costa, A.; Crespo, A.; Navarro, J.; Lizcano, G.; Madsen, H.; Feitosa, E. A review on the young history of the wind power short-term prediction. *Renew. Sustain. Energy Rev.* **2008**, *12*, 1725–1744. [CrossRef]
5. De Giorgi, M.G.; Ficarella, A.; Tarantino, M. Assessment of the benefits of numerical weather predictions in wind power forecasting based on statistical methods. *Energy* **2011**, *36*, 3968–3978. [CrossRef]
6. Ernst, B. Wind Power Prediction. In *Wind Power in Power Systems*, 2nd ed.; John Wiley & Sons, Ltd.: Chichester, UK, 2012; pp. 753–766.
7. Tol, R. Autoregressive conditional heteroscedasticity in daily wind speed measurements. *Theor. Appl. Climatol.* **1997**, *56*, 113–122. [CrossRef]
8. Torres, J.; Garcia, A.; de Blas, M.; de Francisco, A. Forecast of hourly average wind speed with ARMA models in Navarre (Spain). *Sol. Energy* **2005**, *79*, 65–77. [CrossRef]
9. Riahy, G.H.; Abedi, M. Short term wind speed forecasting for wind turbine applications using linear prediction method. *Renew. Energy* **2008**, *33*, 35–41. [CrossRef]
10. Atwa, Y.M.; El-Saadany, E.F. Annual wind speed estimation utilizing constrained grey predictor. *IEEE Trans. Energy Conver.* **2009**, *24*, 548–550. [CrossRef]
11. El-Fouly, T.H.M.; El-Saadany, E.F.; Salama, M.M.A. Grey predictor for wind energy conversion systems output power prediction. *IEEE Trans. Power Syst.* **2006**, *21*, 1450–1452. [CrossRef]
12. Lei, M.; Shiyan, L.; Chuanwen, J.; Hongling, L.; Yan, Z. A review on the forecasting of wind speed and generated power. *Renew. Sustain. Energy Rev.* **2009**, *13*, 915–920. [CrossRef]
13. Mohandes, M.A.; Halawani, T.O.; Rehman, S.; Hussain, A.A. Support vector machines for wind speed prediction. *Renew. Energy* **2004**, *29*, 939–947. [CrossRef]
14. Ul Haque, A.; Meng, J.L. Short-term wind speed forecasting based on fuzzy artmap. *Int. J. Green Energy* **2011**, *8*, 65–80. [CrossRef]
15. Hong, Y.Y.; Chang, H.L.; Chiu, C.S. Hour-ahead wind power and speed forecasting using simultaneous perturbation stochastic approximation (SPSA) algorithm and neural network with fuzzy inputs. *Energy* **2010**, *35*, 3870–3876. [CrossRef]
16. Fan, G.F.; Wang, W.S.; Liu, C.; Dai, H.Z. *Wind Power Prediction Based on Artificial Neural Network*; Electric Power Research Institute: Beijing, China, 2008; pp. 118–123.
17. Öztopal, A. Artificial neural network approach to spatial estimation of wind velocity data. *Energy Convers. Manag.* **2006**, *47*, 395–406. [CrossRef]
18. Ackermann, T. Wind power in power systems. *Wind Eng.* **2006**, *30*, 447–449.
19. Sánchez, I. Adaptive combination of forecasts with application to wind energy. *Int. J. Forecast.* **2008**, *24*, 679–693.
20. Nielsen, H.A.; Nielsen, T.S.; Madsen, H.; Pindado, M.J.; Marti, I. Optimal combination of wind power forecasts. *Wind Energy* **2007**, *10*, 471–482. [CrossRef]
21. Saini, L.; Soni, M. Artificial Neural Network based Peak Load Forecasting Using Levenberg-Marquardt and Quasi-Newton Methods. *IEE Proc. Gener. Transm. Distrib.* **2002**, *149*, 578–584. [CrossRef]
22. Du, Y.; Lu, J.; Li, Q.; Deng, Y. Short-term wind speed forecasting of wind farm based on least square-support vector machine. *Power Syst. Technol.* **2008**, *32*, 62–66.
23. Zhao, D.; Pang, W.; Zhang, J.S.; Wang, X. Based on Bayesian theory and online learning SVM for short term load forecasting. *Proc. Chin. Soc. Electr. Eng.* **2005**, *25*, 8–13.
24. Chen, P.Y.; Pedersen, T.; Bak-Jensen, B.; Chen, Z. ARIMA-based time series model of stochastic wind power generation. *IEEE Trans. Power Syst.* **2010**, *25*, 667–676. [CrossRef]
25. Eristi, H.; Demir, Y. A new algorithm for automatic classification of power quality events based on wavelet transform and SVM. *Expert Syst. Appl.* **2010**, *37*, 4094–4102. [CrossRef]

Energies **2012**, *5*, 3329–3346

26. Zhang, G.P. Time series forecasting using a hybrid ARIMA and neural network model. *Neurocomputing* **2003**, *50*, 159–175. [CrossRef]
27. Diebold, F.X.; Mariano, R.S. Comparing predictive accuracy. *J. Bus. Econ. Stat.* **2002**, *20*, 134–144. [CrossRef]

energies

MDPI

Article

Load Forecast Model Switching Scheme for Improved Robustness to Changes in Building Energy Consumption Patterns

Jaeyeong Yoo and Kyeon Hur *

School of Electrical and Electronic Engineering, Yonsei University, Seoul, 120-749, Korea; jy-yoo@yonsei.ac.kr
* Author to whom correspondence should be addressed; khur@yonsei.ac.kr; Tel.: +82-2-2123-5774.

Received: 4 January 2013; in revised form: 27 January 2013; Accepted: 26 February 2013;
Published: 5 March 2013

Abstract: This paper presents a new, accurate load forecasting technique robust to fluctuations due to unusual load behavioral changes in buildings, *i.e.*, the potential for small commercial buildings with heterogeneous stores. The proposed scheme is featured with two functional components: data classification by daily characteristics and automatic forecast model switching. The scheme extracts daily characteristics of the input load data and arranges the load data into weekday and weekend data. Forecasting is conducted based on a selected model among ARMAX (autoregressive moving average with exogenous variable) models with the processed input data. Kalman filtering is applied to estimate model parameters. The model-switching scheme monitors the accumulated error and substitutes a backup load model for the currently working model, when the accumulated error exceeds a threshold value, to reduce the increased bias error due to the change in the consumption pattern. This switching reinforces the limited performance of parameter estimation given a fixed structure and, thus, forecasting capability. The study results demonstrate that the proposed scheme is reasonably accurate and even robust to changes in the electricity use patterns. It should help improve the performance for building control systems for energy efficiency.

Keywords: load forecasting; data pattern classification; model-switching scheme (MSS); Kalman filtering; accumulated error; autoregressive moving average with exogenous variable (ARMAX)

1. Introduction

Buildings in every shape and size are envisioned to be a huge potential for the efficiency improvement of the power grid, especially because residential and commercial buildings are responsible for about 30- to 40-percent of primary energy consumption worldwide. A rich body of literature for developing high performance energy management systems (EMS) for buildings can be found to achieve significant energy savings [1–3]. Among the enabling technologies, load forecasting, in particular, short-term hourly load forecasting [4,5], should be the front-end application of the EMS, because it helps to better understand energy behavior and provides the baseline estimate of future real savings, especially under dynamic pricing. It helps analyze the load shape and variability and determine proper controls or demand response under a grid emergency, as well. The central aim of this paper is thus to develop an accurate and robust scheme for predicting the hourly power use of a building.

Several load forecasting techniques have been studied [6–8]. These studies can be categorized into three types: regression techniques [9,10], artificial neural network (ANN)-related methods [11–13] and time series approaches [14–16]. In regression techniques, linear representations are applied as the main forecasting function, where mathematical relationships among electrical load demand, weather and exogenous variables are intended. This technique has a test-feasibility and short handling of non-stationary temporal cases as an advantage and disadvantage. The ANN uses historical load and weather data to identify the load model and, therefore, has good approximation capabilities for a wide range of nonlinear models. However, the ANN often converges slowly in training mode and needs to manually determine the network structure and parameters.

In time series techniques, load demands are treated as time series signals. This technique predicts load demand using time series analysis. Among these approaches, stochastic time series techniques have the advantage of finding models with a minimum mean square forecasting error. However, the gradient search-based technique used by the basic stochastic time series (STS) approach is prone to finding local optimal points that build an insufficient forecasting model, because the forecasting error function of this approach possesses multiple minimum points. In summary, this approach is sufficient, but involves computational risk caused by numerical instability.

In the case of buildings, the increase of an accumulated error from load characteristic change, caused by consumption patterns and business and working hour changes, can occur. The Figure 1 shows that accumulated errors of the fixed model and our proposed model in load forecasting are increased dramatically when the load characteristic is changed. Figure 1 illustrates a case where the forecasting scheme using a fixed model fails to perform best, and thus, the accumulated error increases dramatically when the load characteristics are changed. It also includes a desired performance obtained through proposed model switching, as detailed in Section 3. Motivated by the observation that many of the previous forecasting methods using a fixed model do not perform well, as the above dynamic change may occur, we propose an accurate load forecasting technique that combines proven forecasting models with different model structures in order to improve the limited performance of existing techniques. In this research, the autoregressive moving average with exogenous variable (ARMAX) model with temperature as the exogenous variable is selected to represent the load behavior [13,17]. Our scheme has two core components: data pattern classification according to weekday and weekend characteristics and model switching in response to the significant change in the accumulated error. The improved robustness and resulting forecasting accuracy should help advance smart grid technologies, including energy management, security analysis, economic dispatch and power scheduling, by forecasting more accurate energy usage.

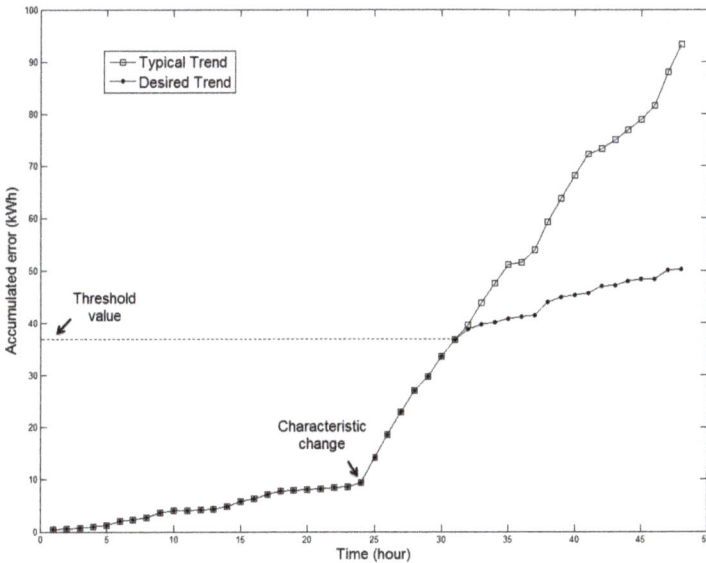

Figure 1. Trends of increasing accumulated error.

The remainder of this paper is organized as follows. Section 2 describes the ARMAX model and Kalman filtering as the forecast model and a tool for estimating parameters of the model. In Section 3, we propose the scheme for improving the accuracy and robustness by changing the model structure and reducing the accumulated error. Section 4 provides numerical study results using MATLAB simulations. Section 5 presents concluding remarks.

2. Background

The load pattern possesses nonlinear and dynamic characteristics, seasonal and diurnal variations and different weather conditions. The ARMAX model selection technique has been extensively studied, because it can describe the relationship between load and extra variables, including weather. In the ARMAX model, the load is denoted as a linear function that has an inaccessible white-noise input and an accessible exogenous input series. The characteristics of the load model are identified based on the variables in the function. The remaining problem becomes finding the proper values of variables in the model, namely parameter estimation.

By adequately constructing the load model and estimating parameters in the function, we can forecast the future load. Thus, an ARMAX-based load model and Kalman filtering for parameter estimation are introduced below as the processes of load forecasting.

2.1. ARMAX-Based Load Model

In this paper, the ARMAX model is used to define the relationship between load and temperature that is regarded as the exogenous variable influencing the load demand, because load demand in a building is susceptible to weather and the number of people in the building. The notation of the ARMAX model for this paper is represented as follows:

$$A(q)L(t) = B(q)u(t) + C(q)e(t)$$

$$(1)$$

where $L(t)$, $u(t)$ and $e(t)$ are the load, the exogenous variable and the white noise at time t, respectively; $A(q)$, $B(q)$ and $C(q)$ are the autoregressive (AR) part, the exogenous input part and the moving average (MA) part. Each part has a back-shift operator, q and an order parameter and can be represented as follows:

$$A(q) = 1 + a_1 q^{-1} + \cdots + a_n q^{-n}, B(q) = b_1 + b_2 q^{-1} + \cdots + b_m q^{-m+1}, C(q) = 1 + c_1 q^{-1} + \cdots + c_r q^{-r}$$

$$(2)$$

where $a_1, \cdots, a_n, b_1, \cdots, b_m$ and $c_1, \cdots, c_r$ are parameters of the autoregressive part, the exogenous input part and the moving average part; and n, m and r are the AR order, input order and MA order.

In determining the order number of each part, the sample autocorrelation function (ACF), the sample partial autocorrelation function (PACF) and the cross-correlation function (CCF) are employed [18]. In general, it is challenging to select appropriate model orders, and it is therefore essential to use a technique derived from experience.

In this paper, we modify the ARMAX model for our research. First, we divide the back-shift operator into day and time back-shift operators. Because we concentrate on forecasting energy consumption, and since time and date are important factors in load forecasting, we use the day and the 24 h factor as follows:

$$q^{-n} => d^{-j} t^{-k}$$

$$(3)$$

where d and t are the day and time back-shift operator, respectively; and j and k are their respective order. Each order has the following definition:

$$n = 24 * j + k$$

$$(4)$$

where j is a natural number; and k is a number from 0 to 23. For example, $n = 55$ means that j is 2 and k is 7. Thus, $q^{-55} = d^{-2}t^{-7}$, and it means two days and seven hours ago. Second, we define the load, the exogenous variable and the noise term. For the load term, we use the electrical consumption (kWh) of building in an hour. The exogenous variable is defined as outdoor temperature (°C). The last term (noise) is assumed to be white Gaussian noise.

2.2. Kalman Filter for the Parameter Estimation

This research briefly reviews the recursive discrete Kalman filter used for estimating parameters of the ARMAX model in line with the algorithm development. Details of the Kalman filtering approach to estimating parameters can be found in [19,20].

In order to define Kalman filtering, we have to consider the following discrete state equations:

$$x(k) = F(k)x(k-1) + v(k-1) \tag{5}$$

$$z(k) = H(k)x(k) + n(k) \tag{6}$$

where $x(k)$, $F(k)$, $z(k)$, $H(k)$, $v(k-1)$ and $n(k)$ are vectors of $n \times 1$ system states, the $n \times n$ dimensions of the state transition matrix, $m \times 1$ measurement vectors, the $m \times n$ output matrix, $n \times 1$ system error and $m \times 1$ measurement error, respectively. The noise vectors, $v(k-1)$ and $n(k)$, are drawn from white Gaussian noise that has a mean of zero and no time correlation, as shown below.

$$E[v(k)] = E[n(k)] = 0 \tag{7}$$

$$E\left[v(i)v^T(j)\right] = E\left[n(i)n^T(j)\right] = 0 \quad \text{for} \quad i \neq j \tag{8}$$

Q_1 and Q_2, covariance matrices, are defined as follows:

$$Q_1 = E\left[v(k)v^T(k)\right], Q_2 = E\left[n(k)n^T(k)\right] \tag{9}$$

Given the *a priori* estimate of the state vector, $\hat{x}(0) = \hat{x}_0$, and its error covariance matrix, $P(0) = P_0$, we set $k = 0$ and then apply the basic Kalman filter algorithm to estimate the next state by recursively computing the following equations:

$$K(k) = \left[F(k)P(k)H^T(k)\right]\left[H(k)P(k)H^T(k) + Q_2\right]^{-1} \tag{10}$$

$$\hat{x}(k+1) = F(k)\hat{x}(k) + K(k)[z(k) - H(k)\hat{x}(k)] \tag{11}$$

$$P(k+1) = [F(k) - K(k)H(k)]P(k)[F(k) - K(k)H(k)]^T + K(k)Q_2K^T(k) \tag{12}$$

where $K(k)$ is the Kalman gain. In this Kalman filter algorithm, it is important to choose an *a priori* estimate of the state $\hat{x}_0$ and its covariance error P_0, because an intelligent choice improves the accuracy and decreases the computational complexity of the algorithm. A few measurement vector samples can be considered as initial values for $\hat{x}_0$ and P_0 as follow:

$$\hat{x}_0 = \left[H^T Q_2^{-1} H\right]^{-1} H^T Q_2^{-1} z_0 \tag{13}$$

$$P_0 = \left[H^T Q_2^{-1} H\right]^{-1} \tag{14}$$

For our model, the discrete state equations in Equations (5) and (6) are defined for our forecasting model as follows:

1. The state transition matrix, $F(k)$, is a constant identity matrix;
2. The error covariance matrices, Q_1 and Q_2, are constant identity matrices;

3. The state vector, $x(k)$, has some parameters based on Equation (2);

4. The time-varying output matrix, $H(k)$, is derived from the load demand and temperature.

5. The observation value, $z(k)$, represents the load at time k. $z(k) = H(k)x(k)$ takes the following form, defined by Equations (1–3):

$$L(t) = \begin{bmatrix} -a_1 & \cdots & -a_{24*j_1+k_1} & b_1 & b_2 & \cdots & b_{24*j_2+k_2} \end{bmatrix}$$
$$\times \begin{bmatrix} L(t-1) & \cdots & L(t-24*j_1+k_1) & u(t) & u(t-1) & \cdots & u(t-(24*j_2+k_2)+1) \end{bmatrix}^T$$
$$+ \begin{bmatrix} 1 & c_1 & \cdots & c_{(t-24*j_3+k_3)} \end{bmatrix} \times \begin{bmatrix} e(t) & e(t-1) & \cdots & e(t-24*j_3+k_3) \end{bmatrix}^T \tag{15}$$

3. Enhanced Robustness of the Proposed Load Forecasting Scheme

Figure 2 shows the six steps of the proposed robust load forecasting. The first important step is data pattern classification. If the input data are well classified, we can determine a suitable model and reduce forecasting error. The second step, model selection, determines the structure of the forecasting model, model order and load and weather factor using the ARMAX model. The third step is model parameter estimation, during which optimal parameters of the forecasting model are estimated using the Kalman filter and database of past and present load and weather data. The third step leads to a forecast of the load at the next instant of time in the fourth step. The fifth and sixth steps are feedback processes.

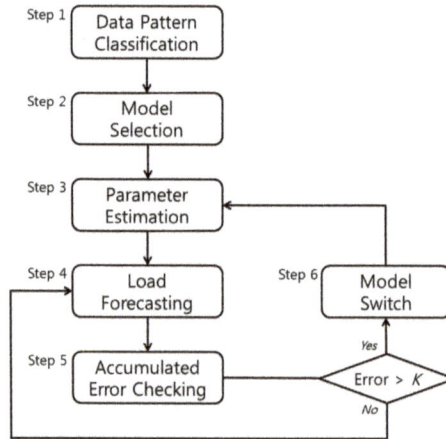

Figure 2. Steps for proposed load forecasting scheme.

In the fifth process, an accumulated error is obtained by summing up the absolute errors. If it exceeds the reasonable threshold value, K, the sixth step, model-switching scheme (MSS), is executed to replace the model structure. Through this process, we can respond to an increase of accumulated error and reduce total accumulated error. If it does not exceed the threshold value, the fourth step is executed. In this section, we describe data pattern classification for identifying the load model based on day characteristics and a model switching scheme for enhancing forecasting accuracy.

3.1. Data Pattern Classification for Selecting the Load Model

The main purpose of defining a forecasting model is to determine its order and the variables that have effects on the load. As mentioned in Section 2, we use the ARMAX model, which depends on load and temperature. Thus, we collected hourly load data for a building in March 2012, as well as

Energies **2013**, *6*, 1329–1343

temperature data, in order to reflect day characteristics of input data. Each data point is categorized into two databases, one for the weekday and the other for the weekend.

Figures 3 and 4 show the load shape over 24 hours of weekdays and weekends in the building. These data were collected by Korea Energy Management Corporation (KEMCO) during March 2012 in Korea. As shown in Figures 3 and 4, the graph shape differs for weekdays and weekends. Because people usually work on weekdays, the load characteristic of weekdays differs from that of the weekend: this research does not consider the holiday case on account of its irregularity. Temperature is also an influential factor. Based on the tendencies of loads due to weather and time, the input data sequences are divided as in Figure 5.

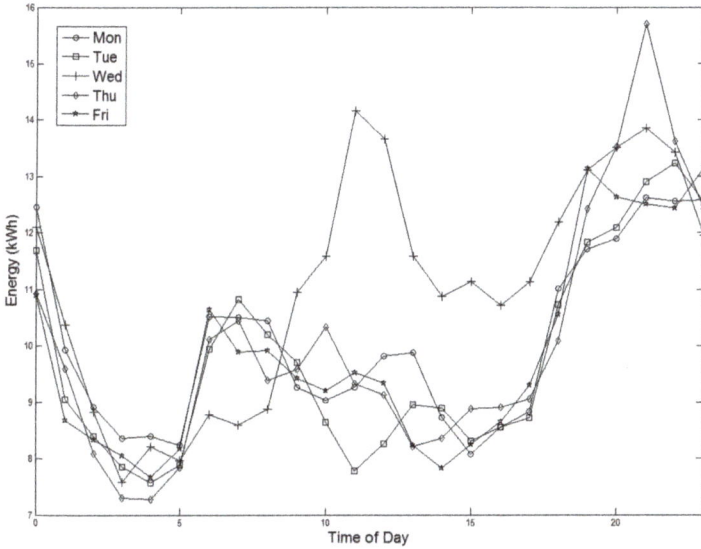

Figure 3. Sample of hourly electrical demand during the weekday.

Energies **2013**, *6*, 1329–1343

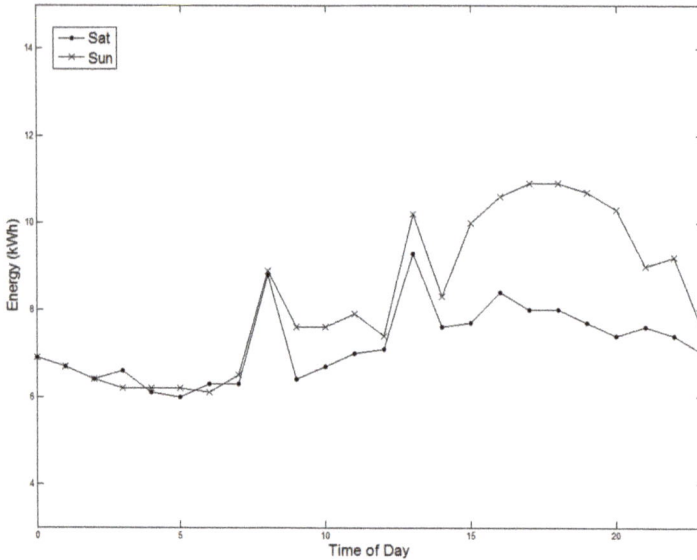

Figure 4. Sample of hourly electrical demand during the weekend.

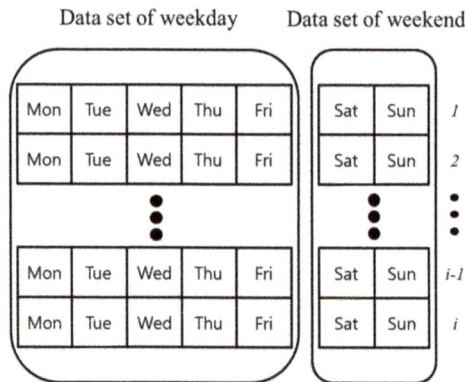

Figure 5. Sample of hourly electrical demand for *i* weeks.

Thus, in the weekday case, this model uses data, including days from Monday to Friday, such as $\{mon_1, tue_1, wed_1, \cdots, thu_i, fri_i\}$. Similarly, the weekend case uses data composed only of Saturday and Sunday: $\{sat_1, sun_1, \cdots, sat_i, sun_i\}$. The value i of the data sequence represents the ith week data. Also, each datum has a 24-hourly load datum. In this data pattern classification, we reflect the characteristic of two data by using a general load model. By using Equations (1) and (2), this general model can be expressed as:

$$
\begin{aligned}
L(t) = - & (a_1 L(t-1) + \cdots + a_n L(t - (24 * j_l + k_l))) \\
+ & (b_1 T(t) + b_2 T(t-1) + \cdots + b_m T(t - (24 * j_u + k_u) + 1)) \\
+ & (e(t) + c_1 e(t-1) + \cdots + c_r e(t - (24 * j_e + k_e)))
\end{aligned}
\tag{16}
$$

98

where at any instant, t, $L(t)$ is the load; $T(t)$ is the temperature and $e(t)$ is the noise; $L(t - (24 * j + k))$ is the previous load at time; $24 * j + k$, where j is a day and k is a time; Similarly, $T(t - (24 * j + k) + 1)$ and $e(t - (24 * j + k))$ are the previous temperature and noise at time, $24 * j + k$, respectively. This basic load model illustrates the characteristics of weekday and weekend cases by using the categorized input data. In this paper, we define the order of the load model by the following equation:

$$L(t) = - a_1 L(t - 1) + a_2 L(t - (24 * 7)) + a_3 L(t - (24 * 7 + 1))$$
$$+ b_1 T(t - 1) + b_2 T(t - (24 * 7)) + b_3 T(t - (24 * 7 + 1)) + e(t) \tag{17}$$

Each term of the equation is labeled as follows: $L(t)$ is the forecasted load at the next step; $L(t - 1)$ is the load value one hour before on the same day; $L(t - (24 * 7))$ is the load during the same hour 7 days prior; $L(t - (24 * 7 + 1))$ is the load one hour before the hour of 7 days prior; $T(t - 1)$, $T(t - (24 * 7))$ and $T(t - (24 * 7 + 1))$ follow the same rule of load term. The last term, $e(t)$, represents total noise.

3.2. Division of Input Data Sequences

Most previous studies of load forecasting only dealt with estimating parameters of the load structure. Hence, this aspect tends to face increasing accumulated error caused by changes of input data characteristics. To overcome this problem, we substitute a candidate model for the current model when accumulated error exceeds a threshold value in our model. Therefore, candidate models reduce accumulated error as a back-up model. The candidate load models used in this paper and those characters are shown in Table 1. Finding the proper threshold value is important for defining scheme characteristics. Too small of a threshold may cause unnecessary model switching, but too large of a value may not provide the desired robustness to the behavioral change in load in time. The determination of the optimal threshold value should be different for different cases and may require considerable experience. However, it may be advised to set a value with some margin, calculated when reasonably accurate forecasting performance is observed during the initial calibration period. In this research, this value is assumed to be 250 kWh. The proposed scheme also provides two operation modes in order to help determine the appropriate model. The first is an initial mode and the second is an executing mode. Details of these modes are presented in the following subsections.

Table 1. Candidate load models.

Name	Model	Note
Candidate 1	$L(t) = a_1 L(t - 1) + b_1 T(t - 1) + e(t)$	Use data, an hour ago
Candidate 2	$L(t) = a_1 L(t - (24 * 7)) + b_1 T(t - (24 * 7)) + e(t)$	Use data, a week ago
Candidate 3	$L(t) = a_1 L(t - 1) + a_2 L(t - 2) + b_1 T(t - 1) +$ $b_2 T(t - 2) + e(t)$	Use data, an hour and two hours ago

3.2.1. Initial Mode

The initial mode of MSS aims to provide a new structure quickly. When accumulated error exceeds a threshold value, there are not enough data to select the best candidate model. Because it needs two weeks at least in order to obtain sufficient data, MSS activates training mode to reduce accumulated error until the system collects sufficient load data. As shown in Figure 6, when this mode activates, our system generates estimated load data based on past data and chooses a model randomly or empirically among a set of models. Then, the basic model for load forecasting is substituted by the structure that was selected from the candidate models. Our scheme then forecasts the next time load using this model. Although there is a possibility that the accumulated error of the selected model again reaches the critical value, the initial mode plays a role in the lack of load data.

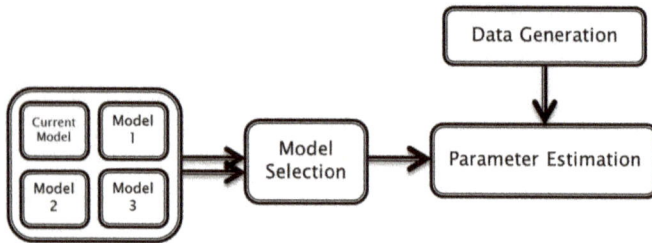

Figure 6. Process of the initial mode.

3.2.2. Executing Mode

While the purpose of the initial mode is to provide a model quickly in the case of a lack of load data, the executing mode aims to select the most accurate of the candidate structures by using sufficient data. When this mode activates, the system starts collecting sufficient load data to estimate the model coefficient for two weeks. This mode then simulates all candidates by using collected data, as shown in Figure 7. Then, our system selects the candidate that has a minimum error. Finally, the model with the least accumulated error during this period replaces the current forecasting model. Though it takes a longer time when we have more models, this can provide the best one.

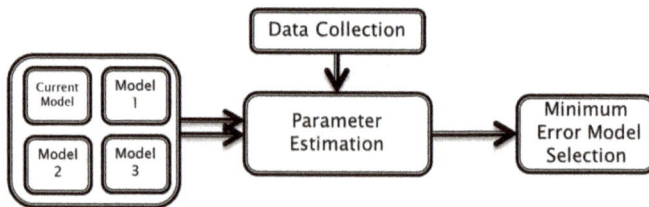

Figure 7. Process of accurate mode.

4. Simulation Studies

Simulation studies are conduced to demonstrate the efficacy of the proposed load forecasting scheme with actual hourly load data as the basic input data. By using our basic load model, the accumulated and average errors of data pattern classification are evaluated with reference to errors without the proposed classification. The proposed scheme for enhancing the model robustness is also implemented and evaluated. Simulations for two modes are carried out to evaluate performance for the same scenarios. The scenario includes a changing load characteristic so that accumulated error increases. The weather data is obtained from the Korea Meteorological Administration [21].

4.1. Data Pattern Classification

To forecast the load, initial parameters, such as x_0, a set of a, b, c and P_0 in Equations (13–15) must be defined. In this simulation, we set those parameters arbitrarily. Using these parameters, load demand and temperature obtained from measuring the Kalman filter is applied to estimate coefficients of the basic load model for each data pattern classification case (Table 2).

The actual daily load for the week is shown in Figure 8. By using the above values, the forecasted loads with data pattern classification are presented in Figure 9. In particular, Figure 9 illustrates weekday and weekend cases simultaneously. Table 3 compares the performance of the models with and without data pattern classification in terms of max, min and average errors. It clearly indicates that data classification helps prevent interference between the weekday load and weekend load and improves the forecasting accuracy.

Table 2. Estimated coefficients for the load model.

	a_1	a_2	a_3	b_1	b_2	b_3	c_1
Weekday case	0.6626	−0.2632	0.5300	−0.0090	0.1803	−0.1357	0.0223
Weekend case	0.1136	0.1548	0.6464	−0.0218	0.0696	0.0105	−0.0207
Whole-week case	0.6052	−0.1872	0.5173	0.0090	0.1932	−0.1642	0.0078

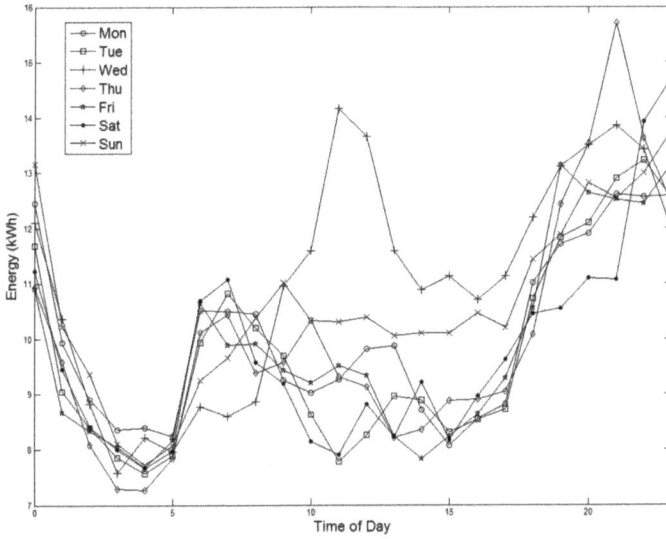

Figure 8. Actual load demand.

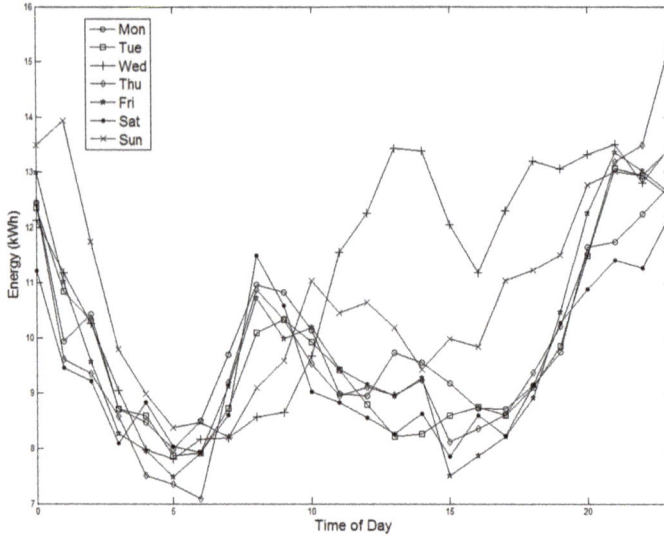

Figure 9. Forecasted load demand with data pattern classification.

Table 3. Error performance for data pattern classification.

	Max error	Min error	Average error	Accumulated error
Included	0.9705	0.0169	0.3914	9.3928
Not included	1.3230	0.0430	0.5752	13.8044
Difference	0.3525	0.0261	0.1838	4.4116

4.2. Performance of the Model Switching Scheme

This subsection demonstrates the performance of the MSS. It is important to note that forecasting error is unavoidable and the accumulated error tends to increase continuously. Thus, our algorithm initializes the accumulated error to an initial value every regular checking period, e.g., one week in this paper. The threshold value of accumulated error is empirically chosen to be 250 kWh. Candidate forecasting models are selected, as shown in Table 1. We then investigate numerical simulations using new weekly data in Figure 10 to address the load characteristic change.

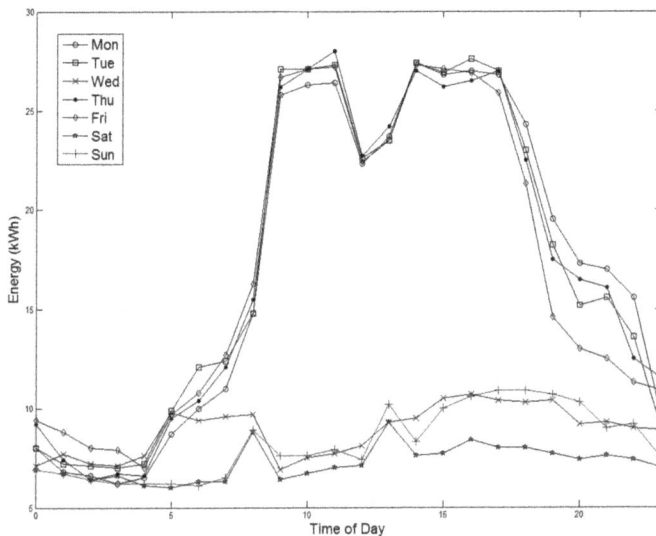

Figure 10. Weekly load as new incoming data.

When the characteristics of input data change, the accumulated error of the load model increases, as shown in Figure 11. This figure shows that accumulated error is sharply increased after characteristic change and exceeds a threshold value. Because the accumulated error exceeds the threshold value, MSS starts one of the modes to substitute for the basic load model.

In the case of the initial mode, there are not sufficient load data. Thus, the system starts generating estimated load data and chooses the first candidate randomly. In the executing mode, our system has two weeks of data. Then, this mode simulates every candidate model using those data. Figure 12 indicates the improved performance of the executing mode in the proposed scheme. In this figure, the gray area means the difference between two modes. The consequent accumulated errors for initial and executing modes are 186.78 kWh and 104.64 kWh, respectively. As you can see, executing mode is superior to the initial mode in reducing the error. However, that mode has a disadvantage, because it cannot control accumulated error for two weeks, because the current model must be used until data collection is finished. The initial mode randomly chooses one of the candidates and uses estimated load data in order to estimate the coefficient of the load model. Even though the initial mode has lower performance than the executing mode, it is a significant operation mode, because it is able to decrease accumulated error until the system collects enough load data.

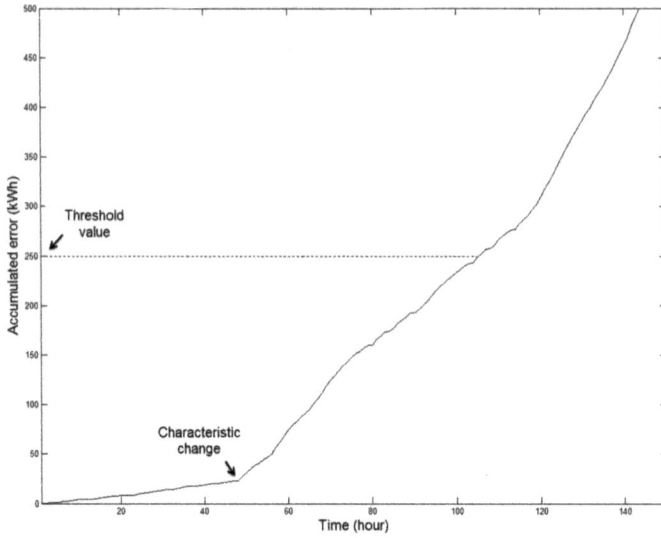

Figure 11. Trend of the accumulated error.

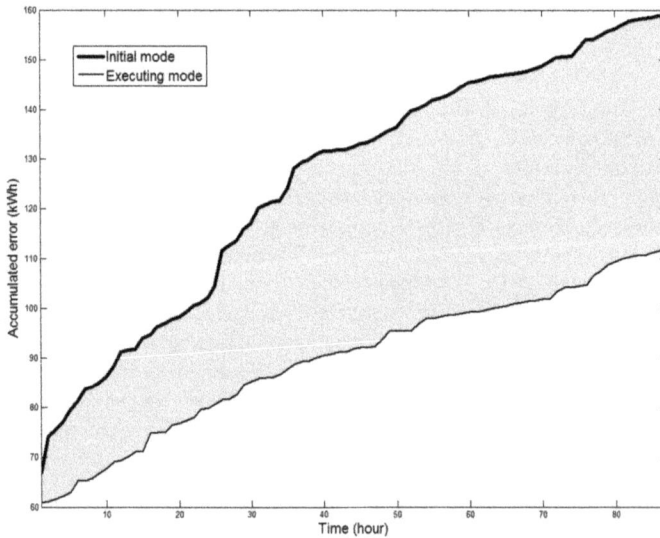

Figure 12. Comparison between initial and executing modes.

5. Conclusions

To improve the accuracy and robustness of the load forecasting, this paper has presented a new load forecasting strategy incorporating data pattern classification and automatic model switching and demonstrated the enhanced performance through case studies using the real energy consumption data. Specifically, this research improved accuracy by processing the input load data in terms of daily characteristics and reinforced the robustness against the structural bias error due to any change in electricity consumption pattern in the building by allowing the forecast model change. The proposed

Energies **2013**, *6*, 1329–1343

strategy has adopted ARMAX models with the Kalman filter for estimating their parameters, because they are easily implementable on top of their proven performance in time series analysis. However, it is worth noting that the forecasting model is not limited to the ARMAX models, as this study has investigated. Several candidate models with complementary structures should work with flexible and scalable architecture and interfaces, allowing for seamless transition from one model to the other in order to improve the accuracy and robustness. Accuracy and robustness against any uncertainty this research provides should help understand the energy behavior of buildings and enable the potential savings through proper load controls and demand responses, which should contribute to the cost-effective operation and stabilization of an entire generation and distribution systems, as well.

Acknowledgments: This work was supported in part by the Basic Science Research Program through the National Research Foundation of Korea (NRF), funded by the Ministry of Education Science and Technology (No. 20110014440). This work was supported in part by the Human Resource Development of the Korea Institute of Energy Technology Evaluation and Planning (KETEP) grant funded by the Korean government Ministry of Knowledge Economy (No.20124030200040).

References

1. Clarkea, J.A.; Cockroftb, J.; Connera, S.; Handa, J.W.; Kellya, N.J.; Mooreb, R.; OBriena, T.; Strachan, P. Simulation-assisted control in building energy management systems. *Energy Build.* **2002**, *34*, 933–940. [CrossRef]

2. Kolokotsaa, D.; Pouliezosb, A.; Stavrakakisc, G.; Lazosc, C. Predictive control techniques for energy and indoor environmental quality management in buildings. *Energy Build.* **2009**, *44*, 1850–1863. [CrossRef]

3. Tyler, J.; Auslander, D.M. Improved Methods to Load Prediction in Commercial Buildings. In Proceedings of the 2012 ACEEE Summer Study on Energy Efficiency in Buildings, Pacific Grove, CA, USA, 12–17 August 2012; pp. 173–184.

4. Vandaele, W. *Applied Time Series and Box-Jenkins Model*; Academic Press: New York, NY, USA, 1983.

5. Gross, G.; Galiana, F.D. Short term load forecasting. *Proc. IEEE* **1987**, *75*, 1588–1573. [CrossRef]

6. Kim, K.H.; Park, J.K.; Hwang, K.J.; Kim, S.H. Implementation of hybrid short-term load forecasting system using artificial neural networks and fuzzy expert systems. *IEEE Trans. Power Syst.* **1995**, *10*, 1534–1539.

7. Srinivasan, D.; Swee, T.S.; Cheng, C.S.; Chan, E.K. Parallel neural network-fuzzy expert system strategy for short-term load forecasting: System implementation and performance evaluation. *IEEE Trans. Power Syst.* **1999**, *6*, 1100–1106. [CrossRef]

8. Park, D.C.; El-Sharkawi, M.A.; Marks II, R.J.; Atlas, L.E.; Damborg, M.J. Electric load forecasting using an artificial neural network. *IEEE Trans. Power Syst.* **1991**, *6*, 442–449. [CrossRef]

9. Mori, H.; Kosemura, K. Optimal Regression Tree Based Rule Discovery for Short-Term Load Forecasting. In Proceedings of the IEEE Power Engineering Society Winter Meeting, Columbus, OH, USA, 28 January–1 February 2001; pp. 421–426.

10. Rahman, S.; Hazim, O. A generalized knowledge-based short-term load-forecasting technique. *IEEE Trans. Power Syst.* **1993**, *7*, 508–514. [CrossRef]

11. Lee, K.Y.; Cha, Y.T.; Park, J.H. Short-term load forecasting using an artificial neural network. *IEEE Trans. Power Syst.* **2005**, *7*, 2499–2504. [CrossRef]

12. Hsu, H.H.; Ho, K.L. Fuzzy expert systems: An application to shortterm load forecasting. *Proc. Inst. Elect. Eng. Gen. Transm. Distrib.* **1992**, *139*, 471–477. [CrossRef]

13. Yang, H.T.; Huang, C.M.; Huang, C.L. Identification of ARMAX model for short term load forecasting: An evolutionary programming approach. *IEEE Trans. Power Syst.* **1996**, *11*, 403–408. [CrossRef]

14. Ruzic, S.; Vuckovic, A.; Nikolic, N. Weather sensitive method for short term load forecasting in electric power utility of Serbia. *IEEE Trans. Power Syst.* **2003**, *18*, 1581–1586. [CrossRef]

15. Herter, K. Residential implementation of critical-peak pricing of electricity. *Energy Policy* **2007**, *35*, 2121–2130. [CrossRef]

16. Hagan, M.T.; Behr, S.M. The time series approach to short-term load forecasting. *IEEE Trans. Power App. Syst.* **1987**, *PAS-2*, 785–791. [CrossRef]

17. Wang, B.; Tai, N.L.; Zhai, H.Q.; Ye, J.; Zhu, J.D.; Qi, L.B. A new ARMAX model based on evolutionary algorithm and particle swarm optimization for short-term load forecasting. *Electr. Power Syst. Res.* **2008**, *78*, 1679–1685. [CrossRef]

18. Box, G.E.P.; Jenkins, G.M. *Time Series AnalysisForecasting and Control*; Holden-Day: San Francisco, CA, USA, 1976.

19. Brown, R.G.; Hesterburg, T. *Introduction to Random Signal Analysis and Kalman Filtering*; Wiley: New York, NY, USA, 1983.

20. Ristic, B.; Arulampalam, S.; Gordon, N. *Beyond the Kalman Filter*; Artech House: Norwood, MA, USA, 2004.

21. Korea Meteorological Administration. Available online: http://web.kma.go.kr/eng (accessed on 26 February 2013).

energies

MDPI

Article

A New Two-Stage Approach to Short Term Electrical Load Forecasting

Miloš Božić [1,*], **Miloš Stojanović** [2], **Zoran Stajić** [1] and **Dragan Tasić** [1]

[1] Faculty of Electronic Engineering, University of Niš, Aleksandra Medvedeva 14, Niš 18000, Serbia; zoran.stajic@alfatec.rs (Z.S.); Dragan.Tasic@elfak.ni.ac.rs (D.T.)

[2] School of Higher Technical Professional Education, Aleksandra Medvedeva 20, Niš 18000, Serbia; milosstojanovic10380@gmail.com

* Author to whom correspondence should be addressed; milos1bozic@gmail.com; Tel.: +381-69-222-3343; Fax: +381-18-293-921.

Received: 28 December 2012; in revised form: 11 March 2013; Accepted: 1 April 2013; Published: 18 April 2013

Abstract: In the deregulated energy market, the accuracy of load forecasting has a significant effect on the planning and operational decision making of utility companies. Electric load is a random non-stationary process influenced by a number of factors which make it difficult to model. To achieve better forecasting accuracy, a wide variety of models have been proposed. These models are based on different mathematical methods and offer different features. This paper presents a new two-stage approach for short-term electrical load forecasting based on least-squares support vector machines. With the aim of improving forecasting accuracy, one more feature was added to the model feature set, the next day average load demand. As this feature is unknown for one day ahead, in the first stage, forecasting of the next day average load demand is done and then used in the model in the second stage for next day hourly load forecasting. The effectiveness of the presented model is shown on the real data of the ISO New England electricity market. The obtained results confirm the validity advantage of the proposed approach.

Keywords: short-term load forecasting; least-squares support vector machines; average daily load; two-stage approach

1. Introduction

With the deregulation of the energy market and the promotion of the smart grid concept, load forecasting has gained even more significance. Generation capacity scheduling, coordination of hydro-thermal systems, system security analysis, energy transaction planning, load flow analysis and so on are all tasks which rely on accurate short-term load forecasting (STLF) [1]. On the other hand, electric load is a random non-stationary process which is influenced by a number of factors, including: economic factors, time, day, season, weather and random effects, all of which leads to load forecasting being a challenging subject of inquiry.

During the past few decades, a wide variety of models have been proposed for the improvement of STLF accuracy. Conventional methods include: linear regression methods [2], exponential smoothing [3] and Box–Jenkins ARIMA approaches [4] which are linear models which cannot properly represent the complex nonlinear relationships between loads and their various influential factors. Artificial intelligence-based techniques are employed because of the good approximation capability for non-linear functions. These methods include: Kalman filters [5], fuzzy logic [6,7], knowledge-based expert system models [8], artificial neural network (ANN) models [9,10] and support vector machines (SVMs) [11,12]. No single model has performed well in STLF and hybrid approaches are being proposed to take advantage of the unique strength of each method. An adaptive two-stage hybrid network with a self-organized map and support vector machines is presented in [13]. A hybrid

Energies **2013**, *6*, 2130–2148; doi:10.3390/en6042130

107

method composed of a wavelet transform, neural network and evolutionary algorithm is proposed in [14]. A combined model based on the seasonal ARIMA forecasting model, the seasonal exponential smoothing model and weighted support vector machines is presented in [15] with the aim of effectively accounting for the seasonality and nonlinearity shown in the electric load. Another seasonal model which combines the seasonal recurrent support vector regression with a chaotic artificial bee colony algorithm is proposed in [16] to determine the appropriate values of three parameters of SVRs.

In spite of all the performed research in the area of STLF, more accurate and robust load forecast methods are still required. One can also highlight some interesting works in this area, especially in recent years. A combined aggregative STLF method for smart grids which obtain a global forecasting by summing up the forecasts on the compounding individual loads is introduced in [17], with three new approaches proposed: bottom-up, top-down and regressive aggregation. A new singular value decomposition based exponential smoothing method is presented in [18], where it is applied to the intraweek cycle, which leads to a simpler and potentially more efficient model formulation. The new method is similar to the Holt-Winters exponential smoothing method, but both were outperformed by the unrestricted form of intraday cycle exponential smoothing. A combined forecast model constructed as the simple average of the weather-based method, the Holt-Winters exponential smoothing and proposed method, obtained the best results at all horizons. Also, these univariate methods outperformed a weather-based method up to about five hours ahead. In [19] an integrated approach which combines a self-organizing fuzzy neural network method with a bilevel optimization method is proposed for STLF. The proposed approach uses self-organizing fuzzy neural network advantage to automatically determine both the model structure and parameters, and bilevel optimization method advantage to automatically select the best pre-training parameters to ensure that the best fuzzy neural networks are identified. In [20], the comparison between the frequently used radial basis function network in STLF and the modified radial basis function network with a genetic algorithm for weight estimation and a nonsymmetrical penalty function with different penalties for over-forecasting and under-forecasting is presented. The obtained results show the efficiency of the proposed method with the new forecasting metric which is the extension of the conventional sum of the squared error metric. Two methodologies for bus load forecasting, *i.e.*, multimodal load forecasting are proposed in [21], where one individually forecasts the local loads while the second forecasts the global load and then individually forecasts the load participation factors to estimate the local loads. In both methodologies a modified general regression neural network with automatic feature selection to reduce the number of inputs of the artificial neural networks is used.

In order to improve forecasting accuracy, in this paper emphasis is placed on model features in the context of machine learning models. It is well known that the balance between the size of the feature set and the quality of the chosen features is important, regardless of which method is used for modeling. A small feature set cannot provide enough information about the load and, on the other hand, too many features do not necessarily provide more information, but may bring noise to the model. The selection of appropriate model features which carry the right information about load behavior is one of the most important tasks. An analysis of what kind of information should be included in the model for mid-term load forecasting was done in [11] and a winning model feature set consists of calendar weekday features and time-series past load demand features. The approach in [22], in addition to the weekday calendar features, proposed using the hour of the day feature in STLF problems, and also suggested the use of temperature as the most important weather variable because of the strong correlation between temperature and load. Other weather variables (wind velocity and cloud cover) are also analyzed but in the end are neglected. The final feature set consists of an hour indicator, day indicator and estimated temperature at the hours k, $k-1$ and $k-2$, without using time-series past load. As load time series indicated a clear daily and weekly seasonality, in [23] the effects of the days of the week and special days, such as holidays, are included in the model. To model these effects, several features are introduced besides weekday features such as holidays, working days after or before a holiday, work only during the mornings or only during the afternoons, the Saturday after a holiday, special holidays and so on. Also, in order to choose the appropriate feature subset which best describes the load, in some papers the choice of features is not done manually, and it is common to use some of the algorithms for feature selection. In [24], ant colony optimization is applied to yield optimal feature subsets. The initial feature set is composed of 38 features which are selected to describe hourly and weekly load behavior and the correlation with weather variables. Some included features are the maximum, minimum and average temperatures during the last seven days, six temperature points on the forecasted day, forecasted day rainfall, wind speed, humidity, cloud cover, month, season, week, whether the day is a holiday or not, whether the day is a weekend or not and so on. At the end of the feature selection, 21 features were dropped from the initial set. The features have been selected by using a cross-correlation analysis in [25]. The feature set is composed of the previous hour load, the load of the previous day, load of the previous week and the load from two weeks ago.

It may be noted that the list of used features is wide and varies from work to work but they all have the same goal, to improve the model and achieve the best forecast accuracy. With the same aim, in this paper a new approach to STLF is proposed. An additional feature, next day average load demand, is appended to the STLF model feature set. As this feature is unknown for the next day, in the first stage, the forecasting of the average daily load is carried out. Then, in the second stage, the forecasted average daily load is incorporated into the STLF model and the forecasting of the hourly load for the next day is carried out. It is important to emphasize here that the proposed approach is distinguished from others by the use of the average load in the model, such as for example the Box-Jenkins approach, in terms of using it in the context of the machine learning model, more concretely the LS-SVM. In this way this feature has direct influence in the training phase of the model formation. The results obtained from experiments on the real electricity market data indicate the validity and advantage of this approach.

The rest of the paper is organized as follows: Section 2 presents the basics of least-squares support vector machines (LS-SVM) used in the regression. Next, Section 3 shows electrical load data features and presents the proposed STLF approach. Section 4 includes a variety of experiments to verify the proposed approach. Finally Section 5 outlines the conclusions.

2. Least Squares Support Vector Machines Model

The brief basic concepts of LS-SVMs are introduced. SVMs were proposed by Vapnik in [26], and are widely used for load forecasting, in addition to ANNs which also show a good approximation capability for non-linear functions. However, SVMs are based on the structural risk minimization principle in order to minimize the upper limit of the estimation error, rather than the empirical risk minimization which minimizes the training error used by ANNs. Consequently, by solving the quadric programming (QP) optimization problem, SVMs always manage to achieve the global optimum solution, instead of possibly stocking the local optimum like ANNs models. This approach, by using nonlinear kernels, leads to a very good generalization performance and sparse solutions. LS-SVMs, defined in [27], as reformulations of standard SVMs instead of solving the QP problem, which is complex to compute, obtain a solution from a set of linear equations. Therefore, LS-SVMs have a significantly shorter computing time and they are easier to optimize.

Let us consider a given training set $\{x_k, y_k\}$, $k = 1, \dots, n$ with inputs $x_k \in R^p$ and outputs $y_k \in R$. The following regression model can be built by using a non-linear mapping function $\phi(\cdot): R^p \to R^{p_h}$ which maps the input space into a high-dimensional feature space and constructs a linear regression in it. The regression model in primal weight space is expressed as follows:

$$y(x) = \omega^T \phi(x) + b \tag{1}$$

where ω represents the weight vector and b is a bias term.

LS-SVM formulates the optimization problem in primal space presented as follows:

$$\min J_p(\omega, e) = \frac{1}{2}\omega^T \omega + \frac{1}{2}\gamma \sum_{k=1}^{n} e_k^2$$

(2)

subject to equality constrains expressed as follows:

$$y_k = w^T f(x_k) + b + e_k, k = 1, \ldots, n$$

(3)

where e_k represents error variables; γ is a regularization parameter which gives the relative weight to errors and should be optimized by the user.

In order to solve the optimization problem defined with Equations (2) and (3), it is necessary to construct a dual problem using the Lagrange function. Once the mathematical calculations were carried out, described in detail in [27], the following linear system was obtained:

$$\begin{bmatrix} 0 & 1_v^T \\ 1_v & \Omega + I\gamma^{-1} \end{bmatrix}\begin{bmatrix} b \\ \alpha \end{bmatrix} = \begin{bmatrix} 0 \\ y \end{bmatrix}$$

(4)

In Equation (4), $y = [y_1, \ldots, y_n]^T$, $1_v = [1, \ldots, 1]^T$, $\alpha = [\alpha_1, \ldots, \alpha_n]^T$, there are Lagrange multipliers, I is an identity matrix and $\Omega_{kl} = \phi(x_k)^T \phi(x_l) = K(x_k, x_l), k, l = 1, \ldots, n$ denotes the kernel matrix.

Once the system defined in Equation (4) is solved, the solutions for α and b are obtained. It is shown in [27] that usually all Lagrange multipliers are non-zero, which means that all training data participate in the solution, *i.e.*, every data point represents a support vector. Compared with SVM, the LS-SVM solution is not sparse.

The resulting LS-SVM model for function estimation in dual form is defined as follows:

$$y(x) = \sum_{k=1}^{n} \alpha_k K(x, x_k) + b$$

(5)

The dot product $K(x, x_k) = \phi(x)^T \phi(x_k)$ is known as a kernel function. Kernel functions that satisfy Mercer's condition enable computation of the dot product in a high-dimensional feature space by using data inputs from the original space, without explicitly computing $\varphi(x)$.

A commonly used kernel function in non-linear regression problems, one that is employed in this study, is a radial basis function represented as follows:

$$k(x, x_k) = e^{\frac{\|x - x_k\|^2}{\sigma^2}}$$

(6)

where the kernel parameter σ^2 denotes the squared variance of the Gaussian function.

When choosing the RBF kernel function with the LS-SVM, the optimal parameter combination (γ, σ) should be established, where γ denotes the regularization parameter and σ is a kernel parameter. It can be noticed that only two additional parameters (γ, σ) need to be optimized, instead of three (γ, σ, ε) as in SVM. Parameter selection is the most significant part during the formation of the LS-SVM regression model, because it has a significant effect on the performance, both in terms of accuracy and computing time. Accordingly, for this purpose, a grid search algorithm in combination with k-fold cross validation was used in this study.

3. Model Formation

3.1. Features of Electric Load

The electric load is a random non-stationary process influenced by a number of factors which makes it difficult to model. Choosing appropriate input features to build the model is an important task in load forecasting. There is no general approach to conduct this problem, but load curve analyses and statistical analyses can be helpful for choosing key features to build a good load forecasting model.

The real-life STLF test case is considered in this paper to evaluate the performance of the proposed forecast approach. This STLF test case is related to the ISO New England power system, which is an electricity market in the U.S. The employed data for the load in this test case are publicly available data obtained from a website [28]. Figure 1 shows the power load curves for four months, which are typical representatives of each quarter of the year.

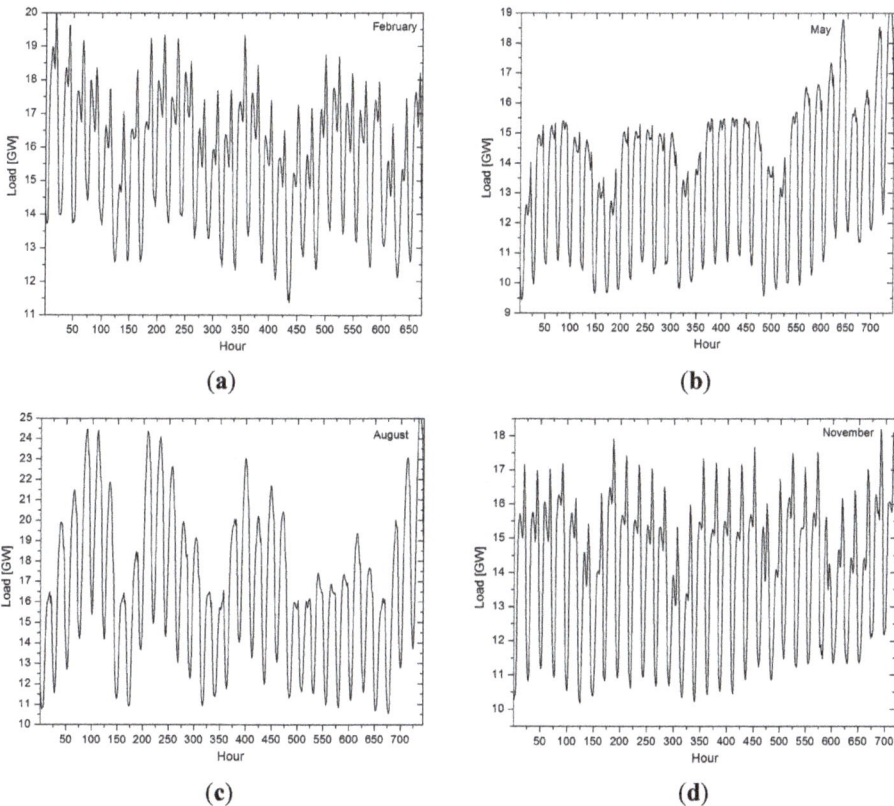

Figure 1. Power load curves. (**a**) February 2011; (**b**) May 2011; (**c**) August 2010; (**d**) November 2010.

In Figure 2, the hourly load during the week is presented for four weeks in February, May, August and November. It is obvious that the daily load on work days is greater than the load on weekends. The reasons for this are people's behavior during the week, and this pattern is periodically repeated each week. All this imposes using the day of the week for the features in the model.

Figure 3 shows hourly load during the day for each day in one week in February, May, August and November. This curve is influenced and shaped by people's daily habits. The load changes from hour to hour during the day, indirectly following consumer behavior. This brings one more important variable to the feature set, and that is the hour of the day. Also, it can be noticed that the curves have a similar shape but different magnitude from day to day in the week. This also confirms the validity of using the day of the week for the model feature with the aim of mapping this property. However, from Figures 2 and 3 it can be observed that the daily load curve is different for the four given months. This difference is reflected not only in load magnitude but also in the shape of the load curves.

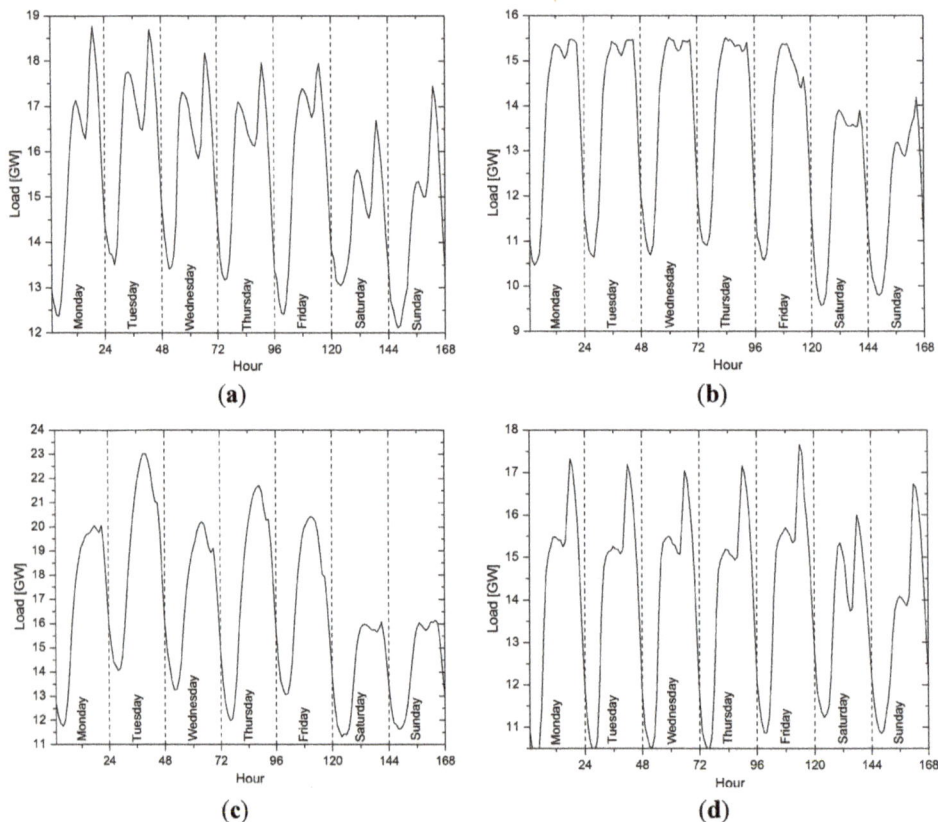

Figure 2. Hourly load during the week in (**a**) February 21–27; (**b**) May 16–22; (**c**) August 16–22; (**d**) November 15–21.

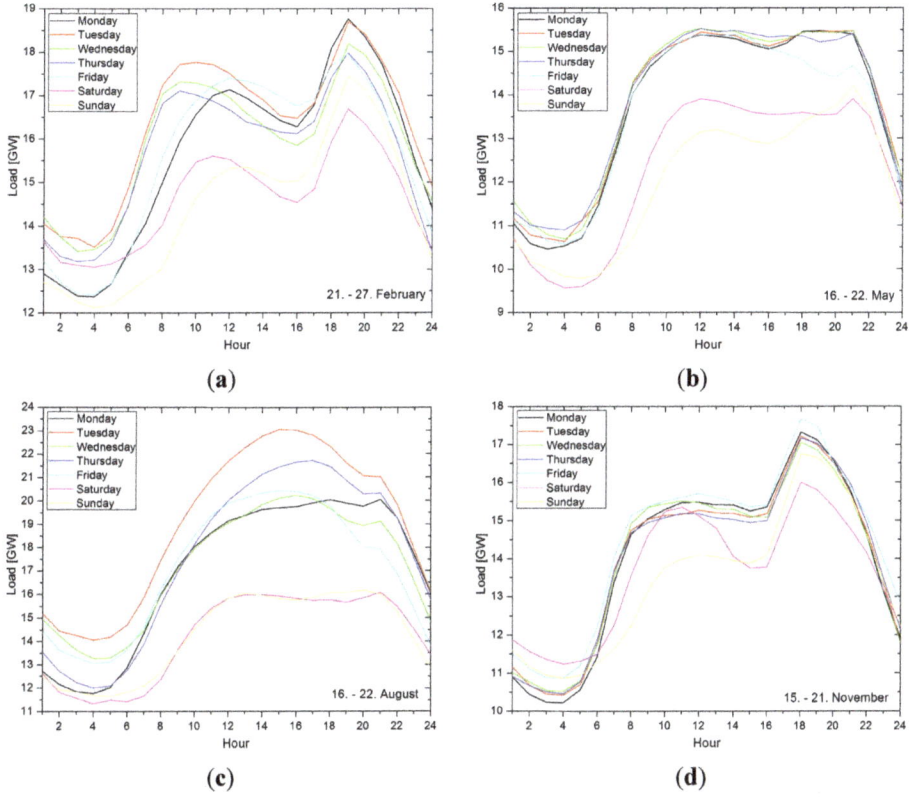

Figure 3. Hourly load during the day. (**a**) February 21–27; (**b**) May 16–22; (**c**) August 16–22; (**d**) November 15–21.

In Figure 4, the average daily loads in February, May, August and November are presented. The start of the week (Monday) is marked with dashed vertical lines.

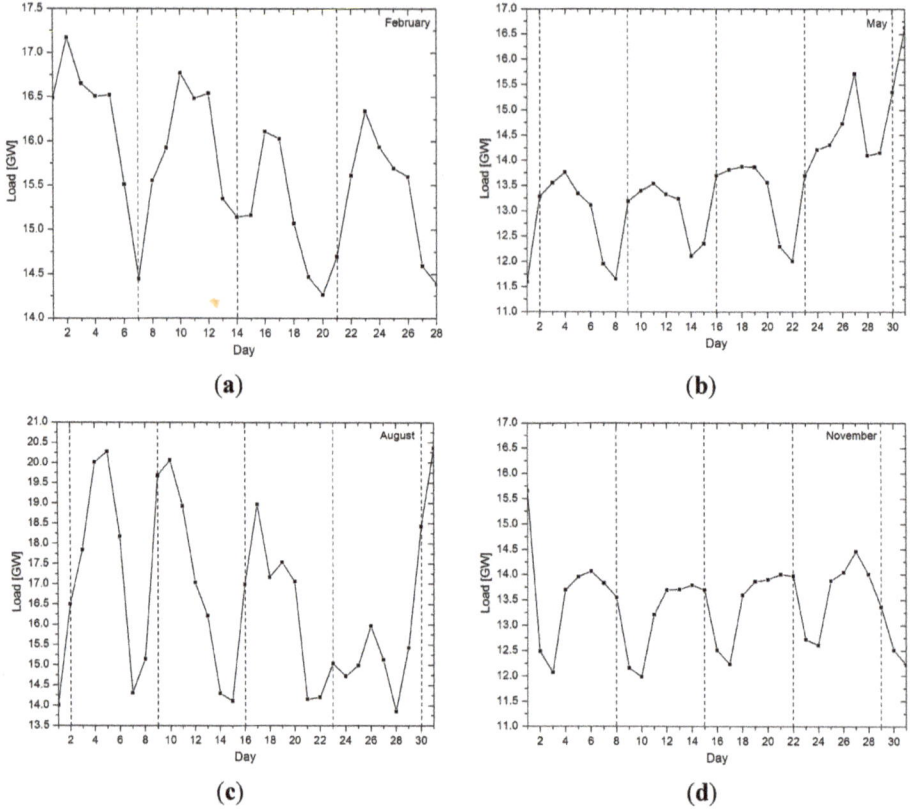

Figure 4. Average daily load in (**a**) February 2011; (**b**) May 2011; (**c**) August 2010; (**d**) November 2010.

It is clear that the average daily load on the weekends is smaller than on week days. It also can be seen that power consumption on Tuesday and Wednesday is much greater than on the other days.

3.2. The Proposed Approach

As previously described, electric load is a nonlinear, time variant and multi-variable function. It is very difficult to capture the correct mapping function of such a signal in all the time spans. To solve this problem, a new two-stage STLF approach based on least squares support vector machines with the architecture shown in Figure 5 is proposed in this paper. Beside the Figure 5 which shows graphical representation of the proposed approach, a step-by-step procedure of two-stage LS-SVM model training and forecasting is given in algorithm 1.

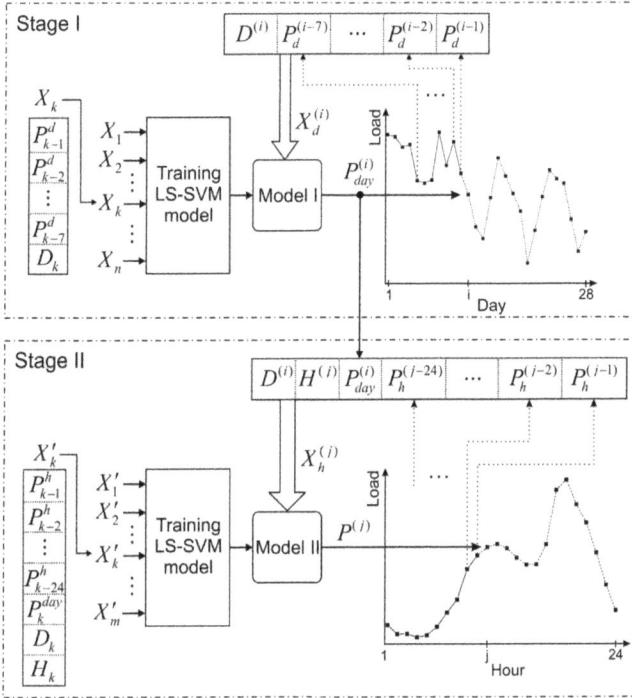

Figure 5. The proposed two-stage model architecture.

In the first prediction stage, Stage I, forecasting of the next day average load is done. This is performed by Model I, whose inputs consist in total of $t + s$ features, where t is the number of past average daily load time-series features and s is the number of non-time series features. The past average daily load time horizon is set to $t = 7$, *i.e.*, the model uses the last seven average daily loads from the prediction moment (P^d_{k-i}, $i = 1, \ldots, 7$). To map the weekly load behavior, the day of the week feature (D_k, $D_k \in \{1, 2, \ldots, 7\}$ where 1 corresponds to Monday, 2 to Tuesday and so on) is included in the feature set and this feature is the only non-time series feature, *i.e.*, $s = 1$.

Algorithm 1. The two-stage LS-SVM model training and forecasting procedure

1. **Stage I**

   1.1. Model I training set formation using daily average load data for the past three years. This training set contains 1095 vectors in total and each vector is composed of features from seven past average daily loads and the current day of the week indicator. Normalize all of the features in the [0–1] range by using min-max normalization,

   1.2. Based on this training set and grid-search algorithm with a k-fold cross validation procedure ($k = 10$), obtain the optimal parameters γ and σ for the LS-SVM Model I,

   1.3. Using Equations (5) and (6) and the previously optimized parameters γ and σ train the LS-SVM forecasting Model I,

   1.4. In order to predict the average load for one step ahead, *i.e.*, for the next day, seven past average daily loads and the next day of the week indicator form the input test vector for model I,

1.5. At the end of stage I, the average load for the next day is obtained and passed on to Stage II.

2. **Stage II**

2.1. Model II training set formation using hourly load data for the corresponding months from three previous years. This training set contains 2016 vectors in total and each vector is composed of features from 24 past hourly loads, the current day of the *week* indicator, the current hour of the day indicator and the current average daily load. Normalize all of the features in the [0–1] range by using min-max normalization,

2.2. Based on this training set and grid—search algorithm with a *k*-fold cross validations procedure (*k* = 10) obtain the optimal parameters γ and σ for the LS-SVM model II,

2.3. Using expressions (5) and (6) and the previously optimized parameters γ and σ train the LS-SVM forecasting Model II,

2.4. Now, the input test vector is formed from the 24 past hourly loads, the next day of the week indicator, the next hour of the day indicator and the average load for next day, obtained from Model I in Stage I,

2.5. Employ model II with the test vector for the prediction of the hourly load for one step ahead, *i.e.*, for the next hour,

2.6. Update the test vector, first shift the 24 past hourly loads one place to the left and then add the prediction for the past hour in last place, then, update the hour of the day indicator (the day of the week indicator and the daily average load remains the same),

2.7. Go to Step 2.5. until the prediction of the hourly loads for the 24 steps ahead are obtained,

2.8. At the end of stage II, the hourly load for next day is obtained.

When the structure of the inputs is defined, the training set which contains an *n* number of inputs is formed. For Model I training, the total number of inputs is set to *n* = 1095, *i.e.*, the training set contains inputs for the previous three years before the prediction moment. As the experiment results will show, this value is sufficient to catch the evolving nature of the average load pattern. After establishing the training set, the training of the LS-SVM forecasting of Model I is performed. In order to have an optimal training of the model, the data set has to be normalized before training. This prevents the dominance of any features in the output value and provides faster convergence and better accuracy of the learning process. Accordingly, all of the features are normalized within the range [0–1]. After that, the optimal (γ, σ) pair is determined on a training set using a grid search with *k*-fold cross validations, as mentioned in Section 2.

The training set is randomly subdivided into *k* disjoint subsets of approximately equal size and the LS-SVM model is built *k* times with the current pair (γ, σ). Each time, one of the *k* subsets is used as the test set and the other *k-1* subsets are put together to form a training set. After *k* iterations, the average model error is calculated for the current pair (γ, σ). The entire process is repeated with an update of the parameters (γ, σ) until the given stopping criterion (*e.g.*, Mean Squared Error) is reached. The parameters (γ, σ) are updated exponentially in the given range using predefined equidistant steps, according to the grid-search procedure. After obtaining the optimal (γ, σ) combination, values for α and *b* are obtained from Equation (4), and the LS-SVM Model I is formed according to Equations (5) and (6). The test vector is constructed in regard to the previously defined feature set structure and Model I is then employed for the prediction of the average load for one step ahead, *i.e.*, for the next day. When the next day average load is obtained, it is passed to Stage II where the forecasting of the next day hourly loads is done. The Model II feature set is also composed of both time-series and non time-series types of features. The past hourly load time horizon used for this model is *t* = 24, *i.e.*, the model uses the last 24 hour loads from the prediction moment (P_{k-i}^h, *i* = 1, ... , 24). In addition to these time-series parts, the model feature set contains three non time-series features (*s* = 3): the hour of the day H_k, $H_k \in \{1, 2, ..., 24\}$, the day of the week D_k, $D_k \in \{1, 2, ..., 7\}$ and average daily load P_k^{day}. In the training phase of Model II, the last mention feature is obtained as an average from the history of hourly loads, and therefore has an exact value, while in the prediction phase this value is obtained in Model I, and therefore represents a predicted value. After defining the structure of training inputs, the Model II training set is formed from approximately $m = 2016$ inputs, *i.e.*, the training set contains hourly inputs from three months in the past three years, e.g. if the hourly loads for each day in February 2012 need to be predicted, the training set consist of the inputs from February 2009, 2010 and 2011. This is not necessary but it is shown in [11] that the training set calendar congruence with the predicted period produces better forecasting accuracy and reduces the time needed for model formation. After establishing the training set, training of the LS-SVM forecasting Model II is performed in the same manner as the training of Model I. After Model II is trained, it is then committed with the test vector which is formed in regard with the previously defined feature set structure, and the prediction of load for one step ahead, *i.e.*, for the next hour is done. After that it is necessary to update the test vector for the next prediction step, *i.e.*, for the next hour. The update is needed because the exact values of the load for the past 24 hours are available only for the first prediction step. After that, for the next predictions, the predicted values from the previous steps are used instead of the exact ones, which are unknown at that moment. Accordingly, the test vector is first shifted left for one place, the hour feature is updated (the day and average load features remain for the current day) and prediction from the previous step is placed in the final position. The whole process is repeated 24 times and in the end, hourly predictions for the next day will be obtained.

4. Experimental Results

For the evaluation of the proposed STLF approach, the forecasting of hourly loads for four typical month representative of each quarter of the year was done for each day. The results are obtained for August 2011, November 2011, February 2012 and May 2012. This implies that the results from the Stage I forecasting model for the prediction of the next day average load, must first be obtained. Also, the evaluation of these results is important, because they directly influence final STLF accuracy and provide insight into the extent of this dependence, and that is a useful indicator of new feature contributions to STLF accuracy.

The prediction quality is evaluated using the Mean Absolute Percentage Error (MAPE), Maximum Error (ME) and Absolute Percent Error (APE) as follows, respectively:

$$MAPE[\%] = 100 \frac{1}{n} \sum_{i=1}^{n} \left| \frac{P_i - \hat{P}_i}{P_i} \right|,$$

(7)

$$ME = \max_i \left| P_i - \hat{P}_i \right|,$$

(8)

$$APE[\%] = \left| \frac{P_i - \hat{P}_i}{P_i} \right| \cdot 100,$$

(9)

where P_i and $\hat{P}$ are the real and the predicted value of the load demand in the ith hour and n is the number of hours.

Real and predicted average daily loads are shown in Figure 6 for August, November, February and May respectively. In the same Figure, daily APEs are given to illustrate the deviation in the prediction of next day average load.

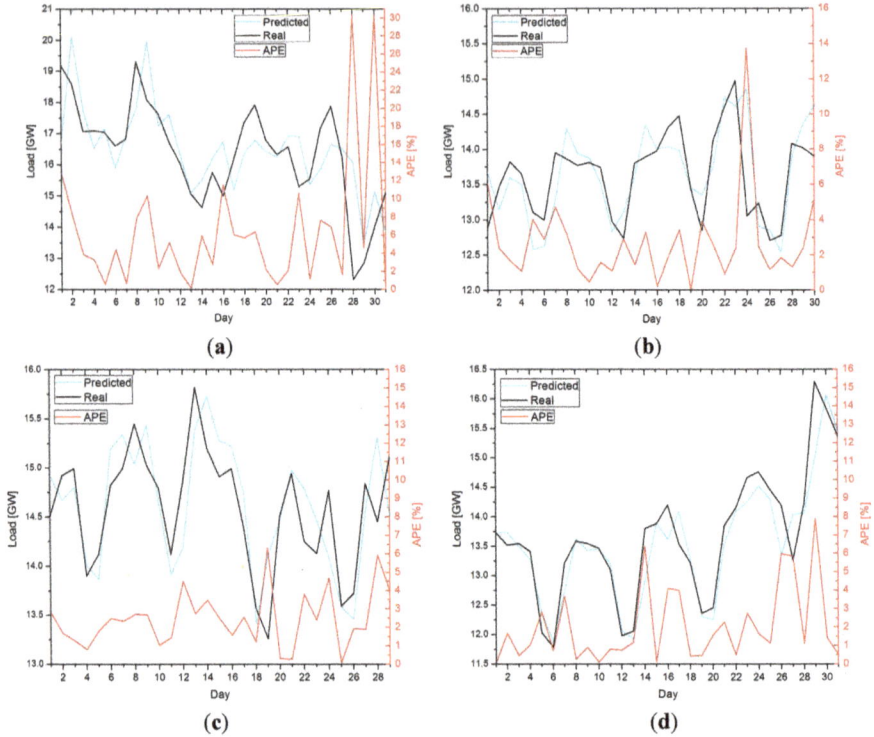

Figure 6. Real and predicted average daily load with APE. (**a**) August 2011; (**b**) November 2011; (**c**) February 2012; (**d**) May 2012.

In Table 1, minimum, average and maximum APE values of entire test sets are shown to also give an indication of the range of APE values in addition to the graphic representations. The first column indicates the test month set, while the second to forth indicate minimum, average and maximum monthly APE values. These APE values fall within scope of interest not because the development and evaluation of the next day average load forecasting model was carried out here, but because we are interested in how the proposed STLF model will behave using the predicted next day average load values in that range.

Table 1. Daily APE for next day average load prediction.

Set	APE		
	Minimum	Average	Maximum
August	0.14	6.12	30.47
November	0.06	2.72	13.73
February	0.08	2.52	6.32
May	0.05	2.02	7.88

Figure 6 and Table 1 give as a sense of the range of the forecasted average load APE for each day in test sets. Thus the days that do not have a satisfactory average load forecasting accuracy can be identified with the aim of monitoring the results of hourly load forecasting on these days. It is of interest because the forecasted average load at stage I is used as input at stage II, where hourly load forecasting is done, as stated above.

To examine the STLF model behavior when it uses the next day average load feature with different APE values, three sets for two test month of next day average loads were artificially generated using the reverse process of calculating APEs with respect to APE values of 2.5, 5 and 7.5%. This resembles a prediction of next day average loads, where the obtained values are in the range of 2.5, 5 and 7.5 of the APE for each day in the test set. When these artificially generated values are collected, they are used as a feature in the input vector for Model II and the forecasting of the next day hourly loads are carried out. In Table 2, the STLF results obtained using artificially generated values for next day average loads are shown. The first column indicates the test month set and the second, the artificially generated value in the input vector, where I2.5 means an artificial next day average load with 2.5 APE, I5 with 5 and I7.5 with 7.5 APE. The remaining columns contain values for minimum, average and maximum monthly values of MAPE and ME. From this table it can be observed that the MAPE and ME values, regardless of whether they are minimum, average or maximum values, increase with the rise in the APE of the next day average load artificially generated values used in the input vector. Thus, it can be noted that the accuracy of the proposed STLF model will increase with an increase in the next day average load forecasting model accuracy, *i.e.*, if the next day average load predicted value is closer to the real value, then the STLF model will also give accurate predictions.

Table 2. Average, max and min daily MAPEs and MEs, obtained with artificial inputs during Stage II.

Set	Input	MAPE			ME		
		Minimum	Average	Maximum	Minimum	Average	Maximum
February	I2.5	2.43	2.92	4.38	0.58	0.96	2.05
	I5	4.48	5.33	7.04	0.92	1.55	2.14
	I7.5	6.57	7.43	8.58	1.49	2.1	2.91
May	I2.5	2.26	3.16	5.71	0.63	0.98	1.6
	I5	4.14	5.12	6.14	0.88	1.44	2.26
	I7.5	5.81	7.32	9.77	1.18	1.99	2.81

To give a graphic representation of the STLF accuracy of the proposed approach, from its obtained results for test sets, daily MAPEs are calculated and shown in Figure 7. In this figure, five curves for each test month can be seen, each corresponding to the LSSVM-I, LSSVM-TSTL, LSSVM-TS, DS-ARIMA and DS-EST model respectively. The LSSVM-I (least square support vector machines initial) model curves represent daily MAPEs for initial model forecasting, *i.e.*, a model whose feature set consists of 26 features: days of the week, hours of the day and 24 past load time-series features. In addition to the features in the LSSVM-I model, models LSSVM-TSTL (least square support vector machines two-stage true average load) and LSSVM-TS (least square support vector machines two-stage) have one more feature, the next day average daily load. Although the LSSVM-TSTL and LSSVM-TS models share the same model structure, they have different inputs in the prediction step. The LSSVM-TSTL model in the input vector for next day average load feature uses exact values, which cannot be used in the real scenario because this value is not known for the step forward, while the LSSVM-TS model uses previously predicted values from Stage I. In addition, due to the verification of performance of a proposed method, the double seasonal ARIMA model (DS-ARIMA) proposed by Taylor *et al.* [29] and the double seasonal exponential smoothing model (DS-EST) proposed by Taylor [30], are also involved in the comparison.

Bearing in mind the obtained results for average daily load in Figure 6, the days characterized by higher MAPEs can be recognized. This refers to the days when the MAPEs are at least twice the values of the average daily MAPEs for a given month. As can be seen in Figure 7, on these days daily MAPEs for the proposed model LSSVM-TS are higher compared to the model LSSVM-TSTL which uses a true next day average load, *i.e.*, prediction accuracy is reduced as a result of inaccurate next day average load forecasting at stage I. This behavior is especially pronounced in several days in each test month, so for example on days 1, 9, 16, 23, 28 in August, 1, 7, 24, 30 in November, 1, 6, 12, 19, 22, 24, 28 in

February and 5, 7, 14, 16, 17, 26, 27, 29 in May. On these days the difference in MAPEs is significantly expressed compared to the LSSVM-TSTL model, but on the other hand on days when the predicted average daily load is nearly equal to the real average daily load, there was a significant improvement in the forecasting accuracy at stage II. This does not mean that the on previously mentioned days with a slightly larger MAPE at stage I there was no improvement compared to the initial LSSVM-I model, which does not use next day average load in the feature set. Also, it should be noted that there are days for the proposed LSSVM-TS model with obtained MAPEs greater than those of the initial LSSVM-I model. These are for example the following days: 1, 7, 17 in August, 15 in November, 5, 6, 19, 24 in February and 5, 6, 7, 14 in May. The reason for this is that on these days the inaccurate next day average load was used in stage II, *i.e.*, as can be seen in Figure 7 on these days in the LSSVM-TSTL model with real next day average load gain, better MAPEs were determined compared to the proposed LSSVM-TS model, but also compared to the initial LSSVM-I model. This is not entirely true for days 7 in August, 6 in February and 6 in May where the initial LSSVM-I model obtained better MAPEs than the LSSVM-TSTL model. That can be expected in some situation when the hourly load curve is not strongly correlated with the daily average load, which then gives faulty information to the model.

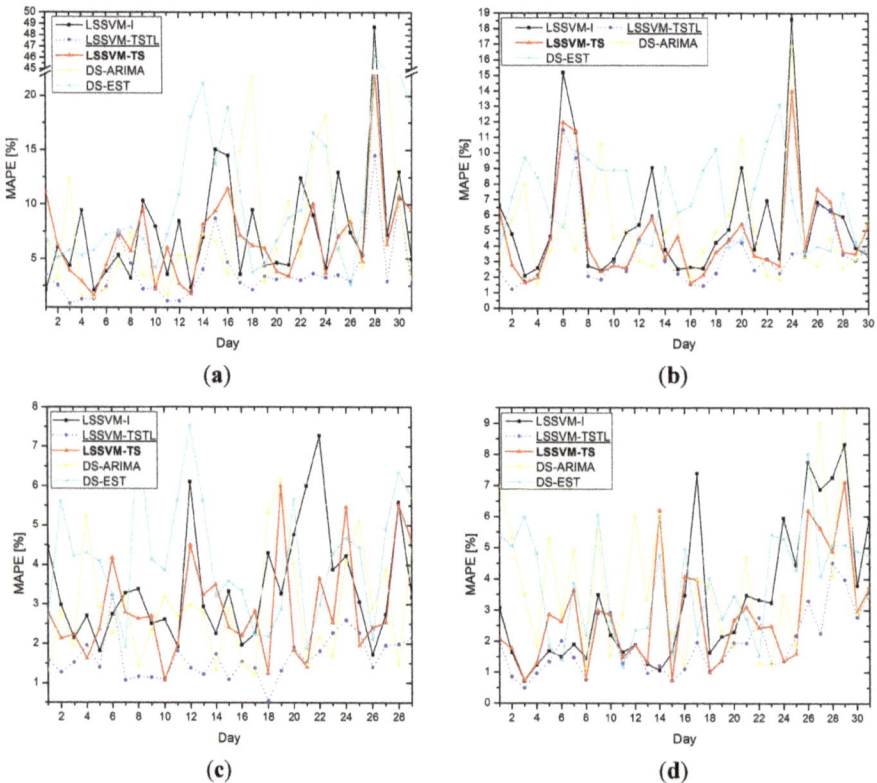

Figure 7. Daily MAPEs for all of STLF models. (**a**) August; (**b**) November; (**c**) February; (**d**) May.

Table 3 shows the minimum, average and maximum values of MAPEs and MEs in the third to the fifth, *i.e.*, in the sixth to the eighth column, respectively, where the first column indicates the test set and the second column indicates the model. Table 3 provides a general overview of the behavior of the proposed LSSVM-TS model compared to not only the initial LSSVM-I model and LSSVM-TSTL model, but also compared to the DS-ARIMA and DS-EST models which take into account the time series trend and seasonality. The proposed LSSVM-TS model has smaller MAPE values than the LSSVM-I, DS-ARIMA and DS-EST models for all the test months. It should be noted that in Figure 7 there are days when the DS-ARIMA and DS-EST models gain better accuracy than the proposed LSSVM-TS model but on a monthly average the LSSVM-TS model is superior. The reasons why the proposed LSSVM-TS model has obtained smaller MAPEs can be found in several facts: the nonlinear mapping capabilities and structural risk minimization of LS-SVM model itself, the recurrent mechanism with superior capability to capture more data pattern information from the past load data and the indirect trend adjustment with an introduction of average daily load in the feature set. However, the proposed model prediction accuracy can be distorted because of these aforementioned facts, due to the using inaccurate prediction of the next day average load at Stage II.

Table 3. Average, max and min daily MAPEs and MEs.

Set	Model	MAPE (%)			ME (GW)		
		Min.	Avr.	Max.	Min.	Avr.	Max.
August	LSSVM-I	2.1	8.31	48.73	0.7	2.74	10.92
	LSSVM-TSTL	0.85	3.73	17.47	0.4	1.29	3.95
	LSSVM-TS	1.55	7.09	32.06	0.63	2.29	7.99
	DS-ARIMA	1.38	8.44	30.1	0.59	2.17	6.6
	DS-EST	2.55	12.22	46.14	1.06	3.22	10.23
November	LSSVM-I	2.09	5.56	18.62	0.62	1.64	5.41
	LSSVM-TSTL	1.2	3.67	11.46	0.45	1.17	3.42
	LSSVM-TS	1.59	4.69	13.96	0.44	1.5	4.25
	DS-ARIMA	1.5	4.94	16.83	0.51	1.34	4.52
	DS-EST	3.42	6.95	13.11	1.06	1.81	2.69
February	LSSVM-I	1.73	3.42	7.28	0.51	0.98	2.11
	LSSVM-TSTL	0.53	1.63	3.22	0.23	0.64	1.63
	LSSVM-TS	1.07	2.9	6	0.39	0.94	1.97
	DS-ARIMA	1.22	2.97	6.15	0.39	1	2.01
	DS-EST	1.87	4.16	7.53	0.59	1.31	2.32
May	LSSVM-I	0.71	3.35	8.33	0.26	1.01	2.93
	LSSVM-TSTL	0.48	1.89	4.51	0.24	0.67	1.83
	LSSVM-TS	0.71	2.82	7.1	0.21	0.85	2.24
	DS-ARIMA	1.22	3.71	9.44	0.12	0.96	1.74
	DS-EST	1.15	3.86	8.02	0.47	1.21	2.63

5. Conclusions

Electric load forecasting is a complex problem and electric load data present nonlinear data patterns caused by influencing factors. In order to overcome this, one approach for improving short-term load forecasting is presented in this paper. The proposed approach is based on two LS-SVM prediction models, in two stages, where the first stage introduces a new feature, average daily load, into the second stage. The introduction of the average load into the feature set for the next day hourly load forecasting model is done with aim to examine its potential in the electric STLF. Moreover, this paper studied and revealed the influence of a new type of feature on STLF accuracy, besides the widely used calendar, climate and time-series features, and provided an efficient method for forecasting it.

Three other alternative models, LSSVM-I, DS-ARIMA and DS-EST models are used to compare the forecasting performance. The experiment results indicate that the proposed LSSVM-TS model has significant improvements among other alternatives in terms of forecasting accuracy. Furthermore, it has been shown that the quality of the proposed LSSVM-TS model directly depends on the quality of the next day average load predictions. As the experiment results have shown, by generating artificial average load samples, the accuracy of forecasting at stage II increases with an increase in the forecasting accuracy in stage I. Also, despite the usage of predicted or true value for next day average load, *i.e.*, LSSVM-TS or LSSVM-TSTL models, in both cases the generated STLF models generally performed better than the initial LSSVM-I model. Of course, usage of the exact next day average load in the STLF model input obtained the best forecasting results. However, this value is unknown and attempts should be made to obtain a value as close to the true value as possible, which would improve STLF accuracy.

Although the results are promising, further work could consider the development of a more advanced model for the prediction of average daily load for one day ahead in order to make it more accurate and thus improve STLF accuracy even more.

Acknowledgments: This work was supported by the Ministry of Science and Technological Development, Republic of Serbia (Project number: III 44006). We would like to extend our thanks to all of the anonymous reviewers for all their helpful comments and suggestions, which have helped to improve the quality of this paper.

References

1. Soliman, S.A.; Alkandari, A.M. *Electrical Load Forecasting: Modeling and Model Construction*; Butterworth-Heinemann: Burlington, MA, USA, 2010.
2. Papalexopoulos, A.D.; Hesterberg, T.C. A regression-based approach to short-term system load forecasting. *IEEE Trans. Power Syst.* **1990**, *5*, 1535–1547.
3. Christiaanse, W.R. Short-term load forecasting using general exponential smoothing. *IEEE Trans. Power Appar. Syst.* **1971**, *PAS-90*, 900–911.
4. Vähäkyla, P.; Hakonen, E.; Léman, P. Short-term forecasting of grid load using Box-Jenkins techniques. *Int. J. Electr. Power Energy Syst.* **1980**, *2*, 29–34.
5. Irisarri, G.D.; Widergren, S.E.; Yehsakul, P.D. On-line load forecasting for energy control center application. *IEEE Power Eng. Rev.* **1982**, *PAS-101*, 71–78.
6. Mori, H.; Kobayashi, H. Optimal fuzzy inference for short-term load forecasting. *IEEE Trans. Power Syst.* **1996**, *11*, 390–396.
7. Ranaweera, D.K.; Hubele, N.F.; Karady, G.G. Fuzzy logic for short term load forecasting. *Int. J. Electr. Power Energy Syst.* **1996**, *18*, 215–222.
8. Rahman, S.; Bhatnagar, R. An expert system based algorithm for short term load forecast. *IEEE Trans. Power Syst.* **1988**, *3*, 392–399.
9. Dillon, T.S.; Sestito, S.; Leung, S. Short term load forecasting using an adaptive neural network. *Int. J. Electr. Power Energy Syst.* **1991**, *13*, 186–192.
10. Hippert, H.S.; Pedreira, C.E.; Souza, R.C. Neural networks for short-term load forecasting: A review and evaluation. *IEEE Trans. Power Syst.* **2001**, *16*, 44–55.
11. Chen, B.-J.; Chang, M.-W.; Lin, C.-J. Load forecasting using support vector Machines: A study on EUNITE competition 2001. *IEEE Trans. Power Syst.* **2004**, *19*, 1821–1830.
12. Hong, W.-C. Electric load forecasting by support vector model. *Appl. Math. Model.* **2009**, *33*, 2444–2454.
13. Fan, S.; Chen, L. Short-term load forecasting based on an adaptive hybrid method. *IEEE Trans. Power Syst.* **2006**, *21*, 392–401.
14. Amjady, N.; Keynia, F. Short-term load forecasting of power systems by combination of wavelet transform and neuro-evolutionary algorithm. *Energy* **2009**, *34*, 46–57.
15. Wang, J.; Zhu, S.; Zhang, W.; Lu, H. Combined modeling for electric load forecasting with adaptive particle swarm optimization. *Energy* **2010**, *35*, 1671–1678.
16. Hong, W.-C. Electric load forecasting by seasonal recurrent SVR (support vector regression) with chaotic artificial bee colony algorithm. *Energy* **2011**, *36*, 5568–5578.
17. Borges, C.; Penya, Y.; Fernandez, I. Evaluating combined load forecasting in large power systems and smart grids. *IEEE Trans. Ind. Inf.* **2013**. [CrossRef]
18. Taylor, J.W. Short-term load forecasting with exponentially weighted methods. *IEEE Trans. Power Syst.* **2012**, *27*, 458–464.
19. Mao, H.; Zeng, X.-J.; Leng, G.; Zhai, Y.-J.; Keane, J.A. Short-term and midterm load forecasting using a bilevel optimization model. *IEEE Trans. Power Syst.* **2009**, *24*, 1080–1090.

20. Kebriaei, H.; Araabi, B.N.; Rahimi-Kian, A. Short-term load forecasting with a new nonsymmetric penalty function. *IEEE Trans. Power Syst.* **2011**, *26*, 1817–1825.

21. Nose-Filho, K.; Lotufo, A.D.P.; Minussi, C.R. Short-term multinodal load forecasting using a modified general regression neural network. *IEEE Trans. Power Deliv.* **2011**, *26*, 2862–2869.

22. Kandil, N.; Wamkeue, R.; Saad, M.; Georges, S. An efficient approach for short term load forecasting using artificial neural networks. *Int. J. Electr. Power Energy Syst.* **2006**, *28*, 525–530.

23. Soares, L.J.; Medeiros, M.C. Modeling and forecasting short-term electricity load: A comparison of methods with an application to Brazilian data. *Int. J. Forecast.* **2008**, *24*, 630–644.

24. Niu, D.; Wang, Y.; Wu, D.D. Power load forecasting using support vector machine and ant colony optimization. *Expert Syst. Appl.* **2010**, *37*, 2531–2539.

25. Kelo, S.M.; Dudul, S.V. Short-term Maharashtra state electrical power load prediction with special emphasis on seasonal changes using a novel focused time lagged recurrent neural network based on time delay neural network model. *Expert Syst. Appl.* **2011**, *38*, 1554–1564.

26. Cortes, C.; Vapnik, V. Support-vector networks. *Mach. Learn* **1995**, *20*, 273–297.

27. Suykens, J.A.K.; Gestel, T.V.; Brabanter, J.D.; Moor, B.D.; Vandewalle, J. *Least Squares Support Vector Machines*; World Scientific Publishing Company: Singapore, 2002.

28. ISO New England Historical Data. Available online: http://www.iso-ne.com/markets/hst_rpts/hstRpts. do?category=Hourly (accessed on 12 April 2013).

29. Taylor, J.W.; de Menezes, L.M.; McSharry, P.E. A comparison of univariate methods for forecasting electricity demand up to a day ahead. *Int. J. Forecast.* **2006**, *22*, 1–16.

30. Taylor, J.W. Short-term electricity demand forecasting using double seasonal exponential smoothing. *J. Oper. Res. Soc.* **2003**, *54*, 799–805.

Article

Short-Term Load Forecasting for Microgrids Based on Artificial Neural Networks

Luis Hernandez [1], Carlos Baladrón [2], Javier M. Aguiar [2,*], Belén Carro [2], Antonio J. Sanchez-Esguevillas [2] and Jaime Lloret [3,*]

[1] Centro de Investigaciones Energéticas, Medioambientales y Tecnológicas (CIEMAT), Autovía de Navarra A15, salida 56, Lubia 42290, Soria, Spain; luis.hernandez@ciemat.es

[2] Universidad de Valladolid, Escuela Técnica Superior de Ingenieros de Telecomunicación, Campus Miguel Delibes, Paseo de Belén 15, Valladolid 47011, Spain; cbalzor@ribera.tel.uva.es (C.B.); belcar@tel.uva.es (B.C.); antsan@tel.uva.es (A.J.S.-E.)

[3] Universidad Politécnica de Valencia, Departamento de Comunicaciones, Camino Vera s/n. 46022, Valencia, Spain

* Authors to whom correspondence should be addressed;
javagu@tel.uva.es (J.M.A.); jlloret@dcom.upv.es (J.L.); Tel.: +34-983423704 (J.M.A.); +34-609549043 (J.M.A.); Fax: +34-983423667 (J.M.A.).

Received: 28 November 2012; in revised form: 18 February 2013; Accepted: 20 February 2013; Published: 5 March 2013

Abstract: Electricity is indispensable and of strategic importance to national economies. Consequently, electric utilities make an effort to balance power generation and demand in order to offer a good service at a competitive price. For this purpose, these utilities need electric load forecasts to be as accurate as possible. However, electric load depends on many factors (day of the week, month of the year, *etc.*), which makes load forecasting quite a complex process requiring something other than statistical methods. This study presents an electric load forecast architectural model based on an *Artificial Neural Network* (*ANN*) that performs *Short-Term Load Forecasting* (*STLF*). In this study, we present the excellent results obtained, and highlight the simplicity of the proposed model. Load forecasting was performed in a geographic location of the size of a potential *microgrid*, as *microgrids* appear to be the future of electric power supply.

Keywords: artificial neural network; distributed intelligence; short-term load forecasting; smart grid; microgrid; multilayer perceptron

1. Introduction

One of the most remarkable characteristics of the traditional energy production and distribution system is that most power is generated at large plants located far from the end-use points. This causes losses during transport and hinders the possibility of decentralizing power generation, resulting in a high dependence on large generation plants. In recent times, a conceptual change has been proposed so as to make the current supply system more sustainable in economic and environmental terms, as reflected for instance in the Lisbon Treaty [1].

According to these new concepts, and in order to increase sustainability and optimize resource consumption, electric utilities are constantly trying to adjust power supply to the demand. Taking into account that it is extremely difficult to store energy at a large scale, power generation has to be adjusted to demand in real time. Accordingly, it is important that electric load forecasting be as accurate as possible.

However, electric power demand depends on many factors, as the day of the week, the month of the year, *etc.*, which makes electric load forecasting quite a complex process that involves more than only statistical methods. In recent years, electric load forecasting is being performed using several prediction algorithms, and among them, *Artificial Neural Networks* (*ANNs*) are one of the most popular options due to their ability to automatically learn from experience and adapt themselves [2].

On the other hand, the need for achieving a balance between electric power generation and demand has added to the emergence of smaller electric power generation and demand environments called *microgrids*, in which adaptation of production to load can be performed much more dynamically due to their distributed smaller elements and the geographical proximity of all elements (which in addition helps reduce transport loses). The load curve for a *microgrid* disaggregates electric power consumption data, making traditional methods (designed for nation- or region-wide forecasting) unsuitable for its direct application because of two main reasons. In *microgrids*, not only the aggregated consumption figure is several times smaller than in region-wide areas, but the load curve presents a much higher variability and does not always conform to the same shape. Some examples of typical load curves for different environments are presented in Figure 1 in order to illustrate the differences. It is easy to realize that the typical load curve is noisier and presents abrupter changes as the environment is more disaggregated.

This paper presents an *ANN*-based architectural model for *Short-Term Load Forecasting* (*STLF*) in small *microgrid* scenarios. After this introduction, Section 2 briefly presents the global concept of *Smart Grid* (*SG*) and *microgrid* (which represents an evolution of traditional grids into more localized power generation systems) and new distributed-intelligence technologies, which are expected to be incorporated into different components of the grid. Section 3 reviews the state of the art of the application of *ANNs* in load forecasting. Section 4 describes a new proposal for an *ANN*-based architectural model for *STLF* in *microgrid* environments. Section 5 presents the validation of the model with real world data. Section 6 analyzes the results obtained and, finally, Section 7 summarizes the conclusions of this study.

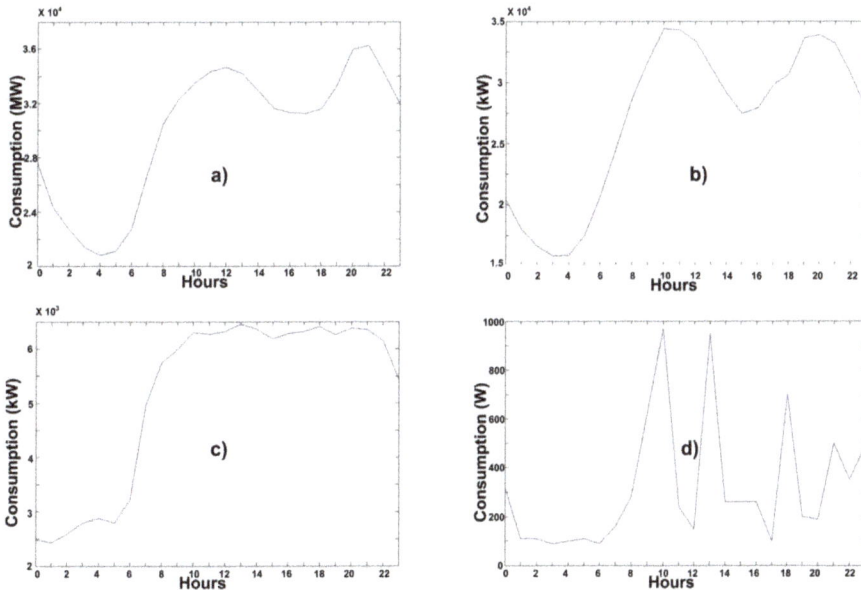

Figure 1. (**a**) Country; (**b**) substation (potential *microgrid*); (**c**) industrial park (potential *microgrid*); (**d**) home consumption.

2. Evolution of Power Supply Systems: Smart Grids and *Microgrids*

2.1. Smart Grids and Microgrids

In recent years, national administrations and international institutions are adopting strategic plans to accelerate the development and deployment of low carbon technologies, putting in place several initiatives to concentrate, promote and reinforce efforts aimed at reducing carbon emissions in Europe. Some examples are the European *Strategic Energy Technology Plan (SET-PLAN)* [3], the Spanish platform *FutuRed* [4], or the European technological platform *Smartgrids* [5]. Public and private efforts are leading to a transition from the traditional grid to new electric power supply models based on *SG*. The term *SG* is used to describe a *"smart"* electric power supply system that uses *Information and Communications Technologies (ICT)* to optimize electric power generation and distribution, and achieve a balance between electric power generation and demand. *SGs* are based on the usage of *Smart Meters (SM)* to retrieve real time data from users and elements of the grid and the application of intelligent algorithms to adapt the behavior of the nodes so as to improve the performance of the network at various levels.

On the other hand, a *microgrid* is a localized physical space consisting of distributed power generation, storage and consumption. According to the *Consortium for Electric Reliability Technology Solutions (CERTS)*, a *microgrid* is an *"aggregation of loads and micro-power units jointly operating as a single system to provide both electric power and heat, includes power units, energy storage and interconnected loads that can operate both connected to the bulk power system and in isolation from the grid in case disturbances may arise"*. Therefore, *microgrids* have the potential to become autonomous and independent energy systems capable, while they are still connected to the global network to allow higher level interactions.

When *ICTs* are incorporated into a *microgrid*, it becomes a *SG* of a specific size. In this case, in order to adjust electric power production of its generation elements, disaggregated load forecasting is required within the *microgrid*.

2.2. New Distributed Intelligence Elements in the Grid

The imminent deployment of *SM* at end-points will enable utilities to accurately identify demand patterns. *Microgrid* operators will get more reliable values from disaggregated profiles, which will enable them, for example, to perform more reliable *Demand Response (DR)* and make more accurate aggregated forecasts based on disaggregated data.

The new concept and physical distribution of *SG* and *microgrids* will require the deployment of *Distributed Intelligence (DI)* in traditional sites where to date there were no distributed electronics. This intelligence will control the behavior of the different smart elements of the grid. Figure 2 shows a hypothetical *microgrid* including *DI*, *Distributed Generation (DG)*, end-point and storage elements. One of the most important inputs to this *DI* scenario is the disaggregated load forecasting, which allows smart elements in the grid to react in advance to the demand. *Microgrids* use techniques with Multi-Agent Systems for island mode operation [6,7] and for strategic control [8]. Similarly, [9] present a new nonintrusive energy monitoring method using *ANN*.

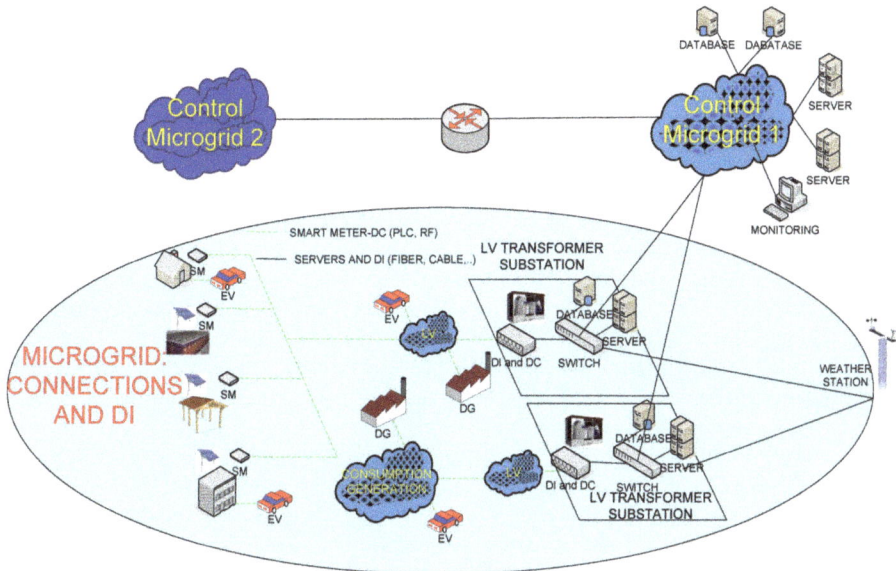

Figure 2. *Microgrid* example.

3. Artificial Neural Networks for Electric Power Load Forecasting in *Microgrids*

3.1. Background

Load forecasting is a challenging task, as there are a large number of influential relevant variables that must be considered, and several strategies have been used to deal with this complex problem. Forecasting models can be classified according to the factors considered as time series models (univariate) and causal models. The former methods model energy load on the basis of past data [10–14], while the latter model electric load on the basis of exogenous and social factors [15–22]. Intelligence-based forecasting techniques have also been employed as those based on expert systems [23,24], fuzzy inference [25] and fuzzy-neural [26,27].

However, one of the most popular methods for load forecasting are *ANNs*, in all their different flavors. There are *ANNs* based on the *MultiLayer Perceptron* (*MLP*) developed by Rumelhart [28]; others employ *Radial Basis Functions Networks* (RBF), proposed by Bromhead and Lowe [29]; recurrent networks, such as those proposed by Elman [30,31], and other models are based on *Self-Organizing Maps* (*SOM*), which were introduced by Kohonen [32]. Cascade combinations of some of the models above and others have also been employed for a wide array of tasks related to data analysis, prediction, estimation, *etc.* [33,34].

In the work reported by Park *et al.* [35], an *ANN* system with one output neuron is employed for hourly, total and peak load forecast. Ho *et al.* [36] perform a peak load forecast 24 h ahead; the same forecast is used by Ho *et al.* [23], as input to an expert system that performs 24-hour ahead load forecasting. *ANNs* with one output can be repeatedly used to forecast load curves, as in [37,38] or by using a 24-hour parallel system, as shown McMenamin *et al.* [39]. Lee *et al.* [40] present a day divided into three periods having one *ANN* forecasting the load for each period. Lu *et al.* [41] conducted an experiment with three *ANN* models of two utilities, and conclude that systems are dependent and must be adjusted to each of the utilities. With Papalexopoulos *et al.* [42], temperature is represented by non-linear functions, which are used as input, and suggests a set of measures to improve load performance in public holidays. Barkitzis *et al.* [43] present an improved model that considers public holidays.

127

Some publications present systems where a set of *ANNs* work together to compute a forecasting. Alfuhaid *et al.* [44] use a small *ANN* that pre-processes a data set and produces peak, valley and total load forecasts; these forecasts, in combination with other data, are used as input to a larger *ANN* to obtain next-day load forecast. Lamedica *et al.* [45] present 12 *ANNs*—one for each month of the year—where load curves are classified using Kohonen's Self-Organized Map.

Artificial intelligence techniques as fuzzy logic have been combined with *ANNs*. With Srinivasan *et al.* [46], quantitative and qualitative data are presented to a "front-end processor", which assigns four fuzzy numbers measuring the expected load change to each of the four periods of the day. Each number together with temperature data are presented to the *ANN*, which produces a load forecasting. In the work of Kim *et al.* [47], an *ANN* produces a provisional load forecast, then, a fuzzy expert system is used to modify the provisional load forecast on the basis of temperature data and day type (workday/holiday). Daneshdoost *et al.* [48] classify data into 48 fuzzy subsets by temperature and humidity, then each subset is modeled by its own *ANN*. Senjyu *et al.* [49] present a hybrid correction method where fuzzy logic, based on "similar days", corrects the neural network output to obtain next-day load forecast.

Basically, [35–39,49,50] present peak load or aggregated daily predictions, which is a very useful parameter for instance for plant operations planning, but not detailed enough to perform other precise activities such as *DR*. For these, more detailed approaches calculating several predictions a day are required, in order to identify the nuances of the predicted load.

References [51–53] describe complex models capable of hourly prediction 24-hour in advance, but they use between 40 and 50 input variables and a hidden layer with a number of neurons ranging from 24 to 50. A similar case is presented in [44], where 30-minute predictions are provided 24 h in advance, but using more than 50 input variables. These works are prone to the curse of dimensionality effect as reported in [54]: the number of training patterns required to properly train the network increases exponentially with the dimension of the input space. This means that the high dimensional input of these solutions will take more measures to be properly trained, and as such, when installed in a new environment, a solution with a smaller number of inputs will start to output better results sooner.

3.2. Geographical Area in Load Forecasting

There is a variety of experiments reported which apply load forecasting methods to very different geographical areas: nations, regions and big metropolitan areas. In the work by Hsu *et al.* [50], peak and valley loads are forecasted for the city of Taiwan, which presents 5500–9000 MW loads. Taylor *et al.* [55] present load forecasts for England and Wales, with 30,000–45,000 MW consumption. In Chu *et al.* [56], the Taiwan Power Company (Taipower)—through Heat Index (HI)—perform peak load forecasting with values over 33,000 MW. In [51,57,58], the chosen areas for load forecasting are large provinces, which present high electric power consumption. Rejc *et al.* [52] apply a novel short-term active-power-loss forecast method for Slovenia, which has a consumption of 950–1550 MW. Nose-Filho *et al.* [59] analyzed a New Zealand distribution subsystem and performed forecasting using data from several nodes in an electrical network system; consumption data, however, are still high: 150–300 MW. Kebriaei *et al.* [53] present a forecasting method based on fuzzy logic and an *ANN*, and proposes a modified RBF, which uses genetic algorithms to estimate the weights for the network in a Mazandaran area in Iran, with consumption ranging 800–1550 MW.

However, all the publications examined so far [35–59]—regardless of the forecast model and target—have in common that the prediction is calculated for a large geographical area where the electric power load is aggregated and very high.

However, as shown in Figure 1, the features of the aggregated load curve of a large (metropolitan, regional or national) area are much different from the aggregated load curve of a *microgrid*, and therefore, their results cannot be directly extrapolated to *microgrid* environments. While the solutions studied in the literature [35–53,55–59] present sometimes good prediction efficiency figures (normally their MAPEs are around 2%), they deal almost exclusively with big areas, and mainly entire countries, and they are never applied to smaller environments of the size of small cities or *microgrids*. Therefore, they do not give any evidence of how will they behave when applied to highly variable load curves.

Works regarding load curve data processing in *microgrids* have started to appear only recently, such as clustering of load curves in [60], which helps extracting meaningful information by finding groups of similar patterns. Strictly speaking about load forecasting, [61] presents a *STLF* model for a *microgrid* based on *Multiple Classifier Systems* (*MCS*), using data from a similar *microgrid*-sized environment with a similar load curve. *MCSs* are systems combining a set of basic classifiers offering a better performance when operating together than on their own. The base classifiers can include different classification approaches or be trained differently, with different algorithms and data sets, and then combined with a fusion method. This specific work employs four base classifiers (*MLP* or *RBF* are used due to its good generalization ability), dividing the training set into several parts: 24 h, 3 days, 1 week and 1 month before the predicting hours. Dynamic weighting is selected as the fusion method. With a dataset collected from the aggregated load in the city of Hong Kong from September 2008 to August 2010, the MAPE found for this model is 15.66% with a *Generalized Regression Neural Network* (*GRNN-MLP*) and 15.12%, with a *Radial Basis Function Neural Network* (*RBFNN*). These errors are sensibly higher than those reported in works applied to national/regional environments.

4. An Architectural Model for Load Forecasting in *Microgrids*

A *microgrid* is capable of controlling electric power loads that will range between thousands of kW to hundreds of MW. Consequently, while traditional grids supply electric power to a whole country, *microgrids* supply electric power to small cities and villages. Disaggregated data are known to produce load peaks and valleys that are more difficult to forecast, and thus traditional methods are not directly applicable if accurate results are required. This section presents not only a prediction algorithm, but a complete *ANN*-based system for forecasting electric load in *microgrids*. For implementation and testing of this system, real world electric load data from Soria, a small Spanish city with a size that could be considered similar to that of a *microgrid*, has been employed.

4.1. Dataset

The real data used in this study were provided by the Spanish electric power utility company Iberdrola (Bilbao, Spain). The historical record provided spans from 1 January 2008 to 31 December 2010 (for a total of 1096 daily records sliced in 15-minute reports) corresponding to a substation located in Soria, Spain, that supplied electricity to this small city. The data provided included information about day of the month, month, year and hourly electric loads making up the daily load curve. This dataset has been enriched with calendar information (day of the week, day type—workday/public holiday) and daily aggregated load. Loads ranged between 7 and 39 MW, which is a load similar to that of a *microgrid*, rather than to a large area or country. A total of 70% of the data available were employed for *ANN* training, and the remaining 30% were used for the validation/testing phase.

4.2. Top Level Architecture of the Forecasting System

The aim of the system is to operate in real time within a *microgrid* environment, receiving data from data concentrators connected to smart meters and other smart data sources present in the grid. The architecture of the predictor is shown in Figure 3.

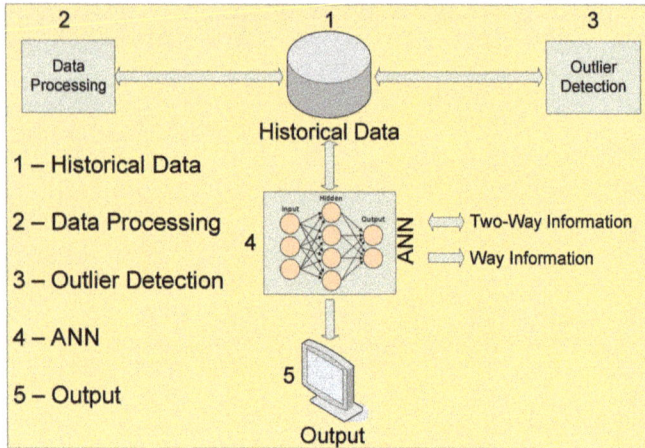

Figure 3. System architecture.

The different components are:

1. *Historical Data*: a database containing all the data handled by the system. This includes raw and filtered load data (processed by modules 2 and 3) in periods of 15-minute and 1 h, and the forecasting reports produced by the *ANN*.

2. *Data Processing*: this module implements three algorithms carrying out the following operations: a) to detect missing data produced by faults in the data retrieval system, completing them via interpolation when possible; and b) to cluster 15-minute samples so as to get hourly and daily loads.

3. *Outlier Detection*: this module tries to identify faulty data (potentially caused by malfunctions in sensors or communications) and remove them from the database. To complete this task, the outlier detector searches for abnormal data (meaning data which is outside the typical values of a given magnitude). Therefore, it is necessary to distinguish between abnormal values that are correct—as in the case of low electric power demand in a public holiday as compared to the demand in a workday—and errors that might be caused by a technical failure, which are the ones that must be identified and removed. For the detection of outliers, the *Principal Component Analysis (PCA)* is employed [62], which is a mathematical procedure that uses an orthogonal transformation to convert a set of observations of possibly correlated variables into a set of values of uncorrelated variables called principal components. Figure 4 shows the results yielded by *PCA* with components (components are the eigenvectors of the correlation matrix and are different from the covariance matrix) 8 and 9. Out of the 1096 daily patterns available in the dataset, a total of 53 patterns were marked as outliers.

4. *ANN*: the *ANN* receives data from 1 and, once forecast is performed, the information obtained is sent to 5 to be distributed among the different elements of the grid and to 1 to be stored for future use.

5. *Output*: this module is called after forecast in 4 is completed. Its main task is to send data to different devices where it is displayed, as an operator's screen, a mobile device, *etc.*

Figure 5 shows the on-line operation scheme of the predictor. Internal processes are distinguished from external processes. Internal processes are those performed by a predictor during operation. External processes are those dependent on external events or on interaction with external devices. In this figure we can see that in order to perform a forecast for day *d*, the information stored in the *Historical Data* are presented to the trained *ANN*, which produces an hourly forecast for day *d*.

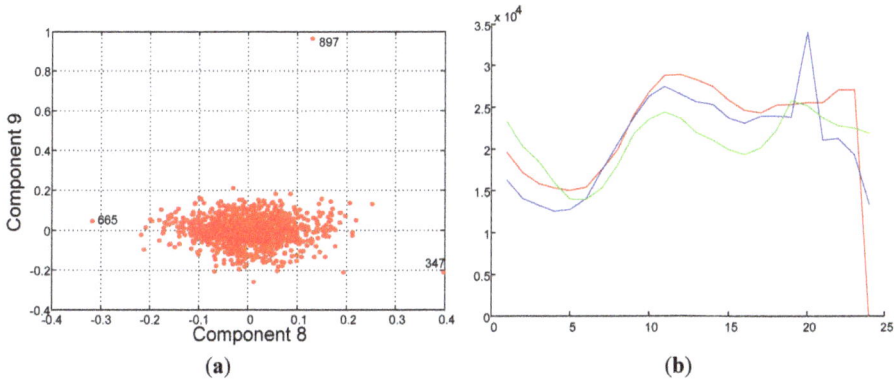

Figure 4. (a) *PCA* Analysis of the dataset showing components 8 and 9; (b) Example of identified outlier daily consumption patterns: Blue and red are faulty data; Green is a correct (but abnormal) pattern.

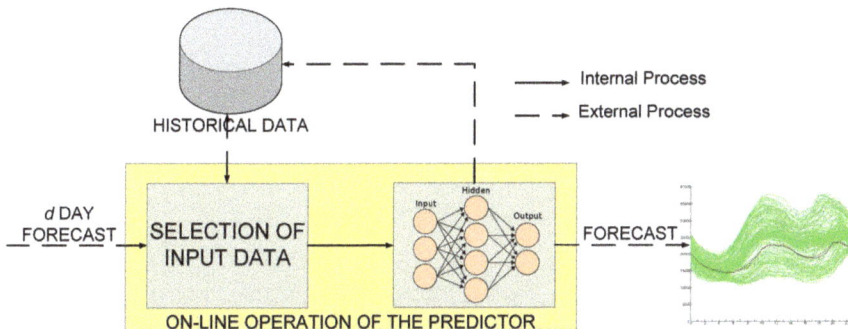

Figure 5. On-line operation of the predictor.

4.3. Artificial Neural Network Design

The *ANN*-based system presented works over the hypothesis that the daily electric load pattern is related to the pattern of the previous day and other calendar data. More specifically:

- Electric consumption highly depends on the hour of the day, and the load curve of the previous day. This previous day load curve actually packs a lot of information about other conditions (season and weather, as shown by Hernández *et al.* [63]) that are not explicitly fed into the system in this work.

- There are many next-day total-load forecasting models, the 24 h-ahead forecast of the aggregated total load for the day. This is a very valuable input data for the *ANN* which packs a lot of information.

- Therefore, load forecasting is performed on the basis of previous-day hourly load curve, aggregated daily load forecast, and calendar variables (day of the week, month, *etc.*)

- Periodic variables are supplied to the network in the form of values of sines and cosines, as it has been demonstrated that this transformation significantly improves the performance of the *ANN*, as shown Drezga *et al.* [64]. Day of the week and month, which are essential for the *ANN* to detect weekly, monthly and seasonal patterns, are entered as sine and cosine, because the cyclical variables are best understood by *ANN*, as shown in [65,66].

- While previous studies on load patterns—as the *Red Eléctrica de España* (*REE*) study [67]—have demonstrated that the type of day—workday or public holiday—has a clear effect on electric

131

load, during the testing phase it was found that the accuracy of the forecast did not improve with the information provided by the type of day. The reason for this could be that the input variables used for the load curve of the previous day and the aggregated load forecast for the forecast day are enough for the network to understand the type of forecast day.

■ Electric load highly varies between workdays and weekends; electric demand in a public holiday is similar to that on Sundays.

■ The seasonality of electric demand is evident, as it significantly varies throughout the year.

The architecture employed in this study follows the next model: to perform a load forecasting for day d, when day $d - 1$ ends and the data for that day are available, the system can perform the load forecast for day d. The architecture implemented is shown in Figure 6, a three-layer *MLP*: an input layer, a hidden layer, and an output layer.

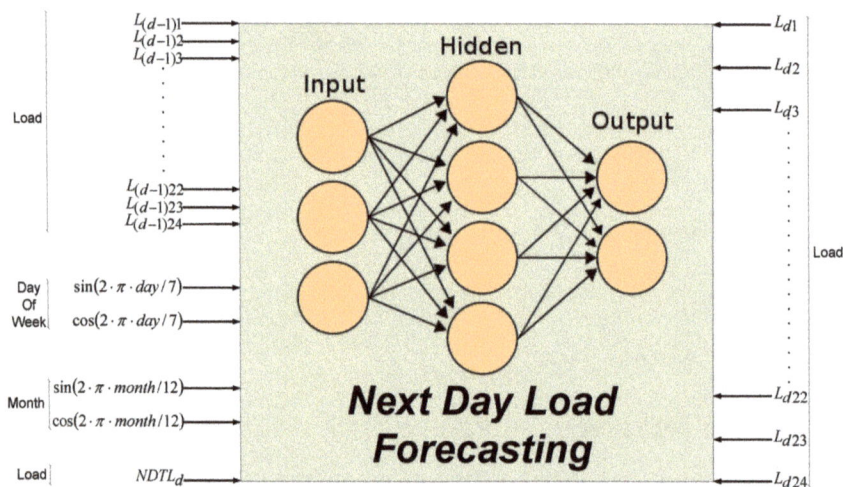

Figure 6. *MLP* architecture. The Figure shows the variables of the input and output layers.

Input:

■ $L_{(d-1)1}, L_{(d-1)2}, L_{(d-1)3}, \ldots, L_{(d-1)24}$: represent the 24 values for the load curve of the previous day.

■ Day of the week $d - 1$: this variable is presented as two variables expressed as sine and cosine by $\sin[(2\cdot\pi\cdot day)/7]_{(d-1)}$ and $\cos[(2\cdot\pi\cdot day)/7]_{(d-1)}$, with *day* from 0 to 6 (*Sunday* = 0, *Monday* = 1, *Tuesday* = 2, *Wednesday* = 3, ..., *Saturday* = 6).

■ Month $d - 1$: this variable is presented as two variables expressed as sine and cosine by $\sin[(2\cdot\pi\cdot day)/12]_{(d-1)}$ and $\cos[(2\cdot\pi\cdot day)/12]_{(d-1)}$, *month* from 1 to 12 (*January* = 1, *February* = 2, ..., *November* = 11, *December* = 12).

■ $NDTL_d$: Next Day's Total Load, which can be easily estimated with an error ranging ±2% using for instance the model proposed by Hsu *et al.* [68].

Output:

■ $L_{d1}, L_{d2}, L_{d3}, \ldots, L_{d24}$ represent the 24 values of the load curve for the forecast day.

Hidden:

■ The neurons of the hidden layer are fully connected with input and output layer neurons.
■ There are 16 neurons in the hidden layer.

Prior to operation, the *ANN* has to be trained. During this training stage, the *ANN* network is confronted with a series of inputs coupled with the real expected output, that is, a set of inputs is associated to the real load curve that the system would have had to forecast. During this training, the internal weights of the *ANN* are adjusted to produce the appropriate outputs.

ANN optimization—both to determine the number of neurons in the hidden layer and to establish the best training algorithm—is usually performed by a heuristic method. In our case, we decided to use an automated script where all parameters were modified (number of neurons in the hidden layer, training function, network performance function during training, *etc.*), calculating the estimation error for several test runs for each combination of parameter values. The best results were obtained with a total of 16 neurons in the hidden layer, the *Bayesian Regulation Backpropagation* training function and the *Sum Squared Error* network performance function.

4.4. Error Calculation

Models and forecast accuracy were validated by *MAPE*, which is widely recommended in the field of research and is expressed as:

$$MAPE = 100 \times \frac{\sum_{i=1}^{n} \left| \frac{L(i) - \hat{L}(i)}{L(i)} \right|}{n} \tag{1}$$

where $L(i)$ represents the measured value for $t = i$, $\hat{L}(i)$ represents the estimated value and n represents the test sample size.

Once the $MAPE_d$ for each of the days of the testing set is obtained, the mean error for all days is estimated by means of:

$$ERROR_{OP} = \frac{\sum_{i=1}^{k} MAPE_d}{k} \tag{2}$$

To examine how the prediction error is reflected on the load curve, error is displayed on a graphic including all forecasted days in the testing set; using this method, the forecast mean error for each of the 24 h is obtained by means of:

$$MAPE\_h_i = 100 \times \frac{\sum_{k=1}^{n} MAPE_{i,k}}{n} \tag{3}$$

with $i = 1, 2, \ldots , 24$; n stands for the sample size in the testing set and $MAPE_{i,k}$ the hourly error i for the day k.

5. Results

This section provides the errors per day, *Probability Density Function (PDF)* curve errors, errors per hour, *PDF* curve errors per hour, and the forecasts of several days with low mean error, when our system is running.

5.1. Results

Once the network is trained, a forecast is performed for the testing set; a forecast load curve is generated for each datum and the daily average error is estimated; average errors are displayed in Figure 7 together with the mean value, mean ± standard deviation and mean ± 2× standard deviation. The mean error of the whole testing phase yielded a value of 2.4037%. The figure uses a specific nomenclature with the format "*A B/C – D E*". *A* represents the day type of the previous day (2: workday, 1: holiday); *B* is the month number; *C* is the day of the month; *D* is the day of the week (Monday, Thursday), and *E* the day type (workday/holiday).

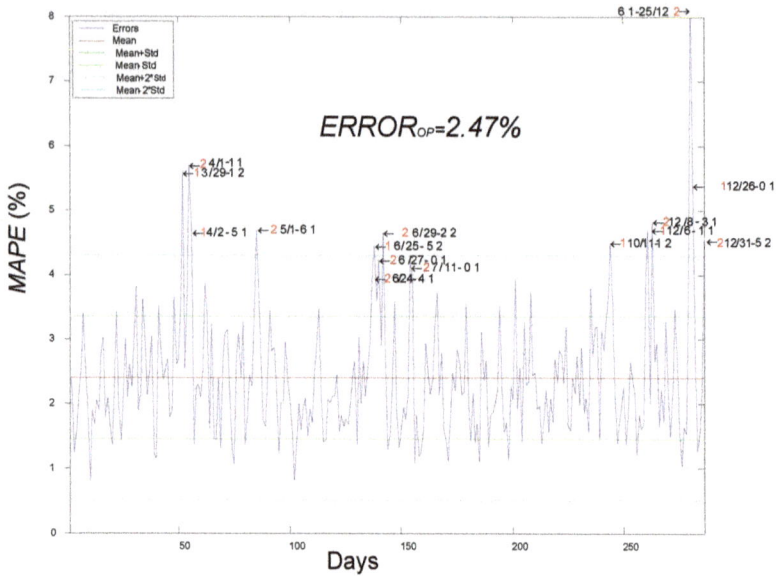

Figure 7. Errors per day (without bad measures). In the *x-axis* are the days. In the *y-axis* are the errors by Equation (1).

In Figure 8 errors are expressed as *PDF*, where the intervals between the mean and mean ± standard deviation, and mean and mean ± 2× standard deviation are displayed. As the figure shows, most errors (72%) correspond to the first interval, as shown in the percentages displayed in Table 1. Over and below that interval, errors have a similar distribution.

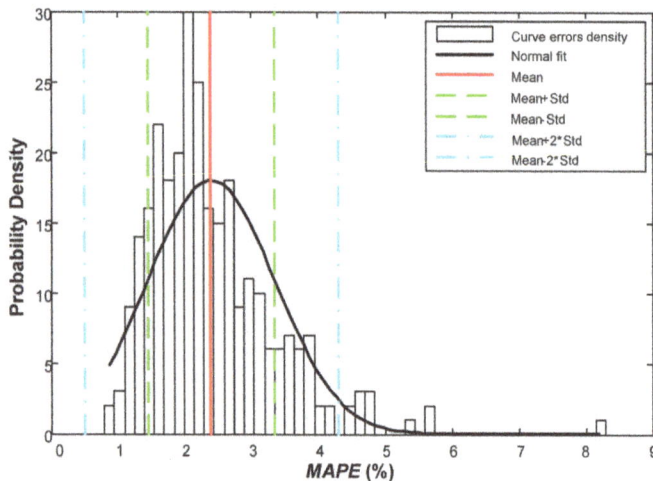

Figure 8. Curve errors data. In the *x-axis* are the errors by Equation (1). In the *y-axis* are the probability densities.

Table 1. Error distribution per day.

Variable	Value	Percentage
Mean	0.024	2.40%
Standard deviation (Std.)	0.0095	0.95%
No. of errors above ×1 Std.	42	14.73
No. of errors between ×1 Std.	206	72.28%
No. of errors below ×1 Std.	37	12.99%
No. of errors above ×2 Std.	12	4.21%
No. of errors between ×2 Std.	273	95.79%
No. of errors below ×2 Std.	0	0.00%

Figure 9 displays errors per hour occurred during the testing phase. Most errors occur in specific parts of the load curve, which normally follows the same topology: from hour 4 to hour 7, the curve starts rising from the first valley; from hour 10 to hour 15, the curve rises until reaching the first peak; the curve starts to drop into the second valley; from hour 18 to hour 21, the curve starts rising again towards the second peak; at the end of the day the curve starts to drop.

Figure 10 displays errors per hour expressed as *PDF* and shows the intervals between the mean and mean ± standard deviation; and mean and mean ± 2× standard deviation. Most errors are concentrated in the first interval—62%—as evidenced by the percentages shown in Table 2. A total of 21% and 17% of errors are above and below the first interval respectively.

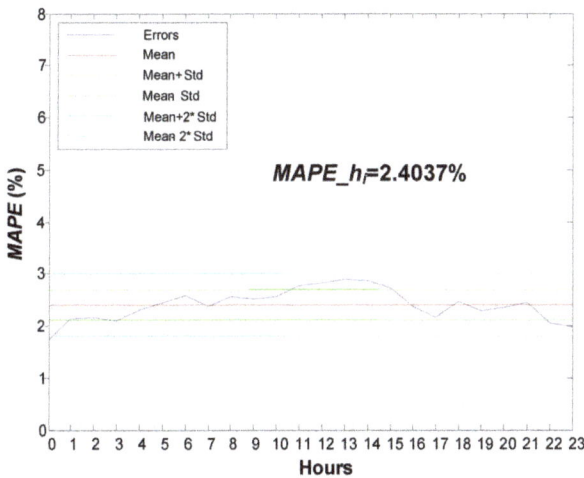

Figure 9. Errors per hour (without bad measures). In the *x-axis* are the 24 h. In the *y-axis* are the errors by Equation (3).

Energies **2013**, *6*, 1385–1408

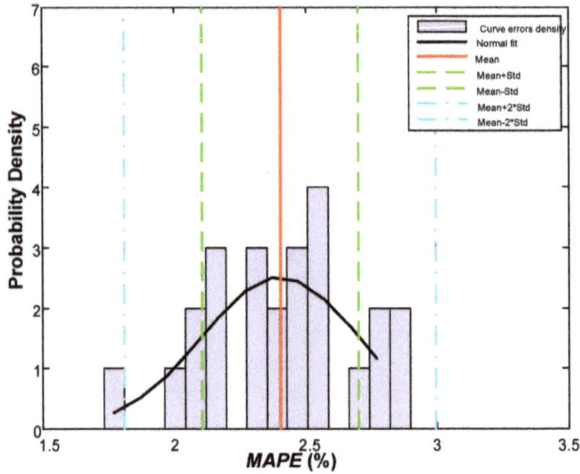

Figure 10. Curve errors per hours. In the *x-axis* are the errors by Equation (3). In the *y-axis* are the probability densities.

Table 2. Error distribution per hour.

Variable	Value	Percentage
Mean	0.024	2.40%
Standard deviation (Std.)	0.0030	0.30%
No. of errors above ×1 Std.	5	20.83%
No. of errors between ×1 Std.	15	62.50%
No. of errors below ×1 Std.	4	16.67%
No. of errors above ×2 Std.	0	0.00%
No. of errors between ×2 Std.	23	95.83%
No. of errors below ×2 Std.	1	4.17%

Figure 11 shows load curve forecasts for three days with a low daily mean error, where (a) represents the forecast for 2/15/2010 with a mean error of 1.20%; (b) represents the forecast for 5/18/2010 with a mean error of 1.10%; and (c) represents 12/21/2010 with a mean error of 1.13%. As we can see, the forecast load curve coincides almost completely with the real load curve.

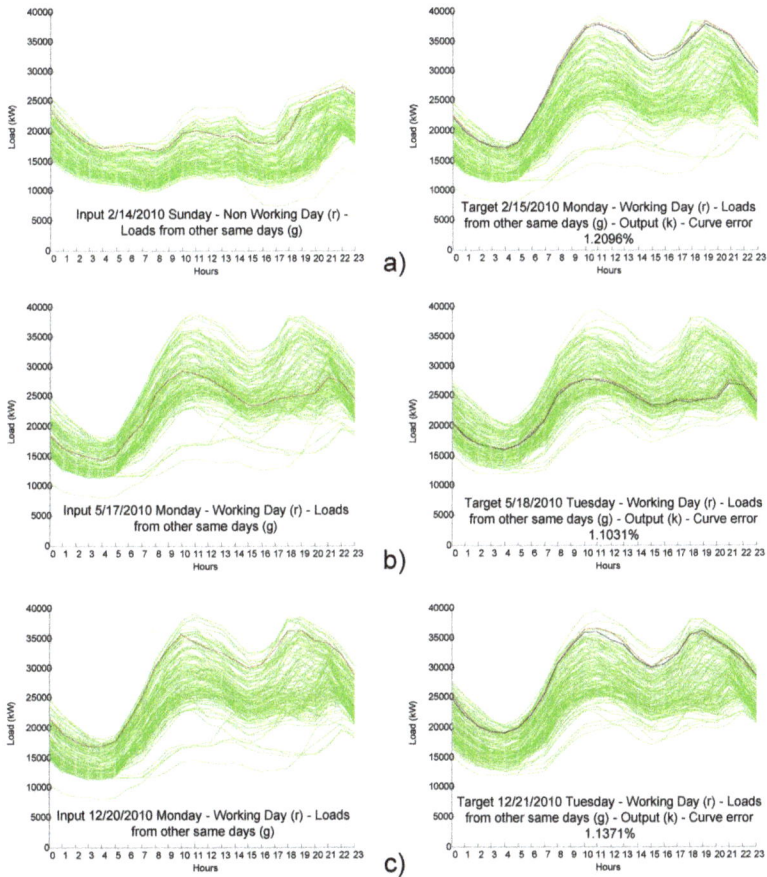

Figure 11. Forecasts with low mean error: (**a**) 2/15/2010; (**b**) 5/18/2010; (**c**) 12/21/2010.

5.2. Computational Cost

As indicated above, MatLab was employed to implement the *ANN* and the rest of scripts developed for additional tasks (error estimation, figures, *etc.*). We used a desktop computer with an Intel Core2 vPro 3.4GHz 2GB RAM processor.

The computation method is as follows: the training set is imported to the Historical Data database, as shown in Figure 3; then *ANN* training is initiated. For this work, the computer used 70% of all data available and took 16 min and 43 s to train the *ANN* model.

When the network is trained as shown in Figure 5, then the data obtained are used to predict the load curve for the forecast day. The computer took 2 min and 49 s to process the testing set (outliers excluded), display the load curves and complete the database. Therefore, approximately, the computer needs 0.59 s to produce a forecast for one day.

6. Result Analysis

6.1. Error Distribution

As shown in Figure 7, and as supported by the results displayed in Figure 8 and Table 1, the daily mean error is within the mean ± standard deviation range, which means that errors ranged between

1.45% and 3.35%, which are fairly good results. Figure 7 also shows that the days with high daily mean errors (above 4%) were special days; further details on this regard are provided below.

By observing the hourly mean error shown in Figure 9, the data displayed in Figure 10 and Table 2, we can see that the most significant errors occur at the turning points of the forecast load curve. This coincidence may suggest that additional information on the form of the curve should be used to improve forecasts and prevent the most serious errors.

6.2. Errors per Day of the Week and Month

Figure 12 represents the evolution of the daily mean error per day of the week. The reason why the highest mean errors occur on Fridays, Saturdays and Sundays is that the training set (load curve) is more scattered; as a result, data uncertainty is higher in weight adjustment after training, and errors increase. In addition, as regards Saturdays and Sundays, their load curves significantly differ—both in demand and form—from those of other days of the week. Fridays are also a special day, as it marks the beginning of the weekend and electric power demand is lower than in the rest of the days.

Figure 13 shows the evolution of daily mean errors by month. October and November include fewer days because of the removed outliers; the mean error per month approximately ranges between 2% and 3%, which evidences the accuracy of forecasts.

6.3. Error Analysis

The purpose of this Section is to present the most significant forecast errors and analyze the reason underlying such errors; finally, this Section summarizes the conclusions drawn from this experience.

The forecast for 4/2/2010, with a mean error of 4.34%. 2 April 2010 is Good Friday (Holy Week) and the previous day is also a public holiday; consequently, a small number of pattern pairs with the same characteristics had been previously fed to the network in the training phase.

The forecast for 5/1/2010, with a mean error of 4.77%. The load in this holiday Saturday is similar to that in the working Saturdays of the same month; as compared with the previous day, the Friday before the holiday Saturday presents half the load; the shape of the curve is irregular, especially at the origin of the curve; however, towards the end, the load is similar to other Fridays of the same month.

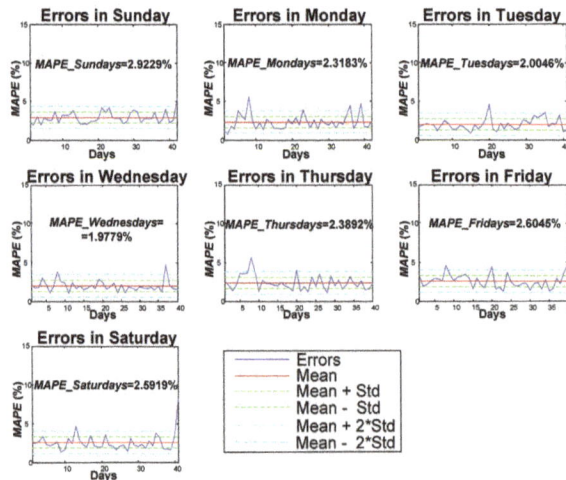

Figure 12. Errors per day. The *y-axis* represents the forecast error by Equation (1). The *x-axis* represents each of the forecast days.

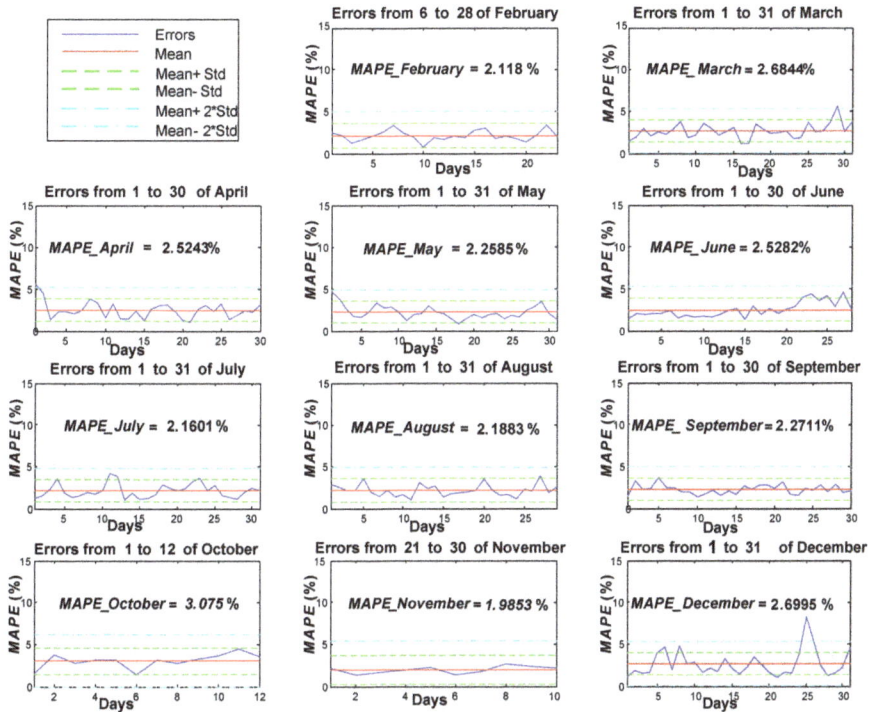

Figure 13. Errors per month. The *y-axis* represents the forecast error by Equation (1). The *x-axis* represents the days of the month.

The forecast for 6/24/2010, with a mean error of 3.92%. That Thursday is a local holiday called Jueves la Saca, and the load is clearly lower than in other holiday Thursdays; in addition, though the Wednesday before Jueves la Saca is not a local holiday, it is included in the holidays and the load is clearly atypical.

The prediction for the 12/8/2010 with a mean error of 4.73%. That day is a holiday Wednesday where the load is lower than in working Wednesdays of the same month; there was only one similar Wednesday in 2008; nevertheless, the *ANN* model predicts a demand rise and the curve starts to rise before the real curve, causing an error. The Tuesday before the holiday Wednesday is lower than that for other Tuesdays of the same month; for this reason, the forecast curve starts low to prematurely drop; the end of the curve is atypical, and is much lower than the real load curve.

The forecast for 12/25/2010, with a mean error of 8.04%. That day is a Christmas Saturday, the load is lower than in other Saturdays of the same month, and the first peak occurs later than usual. The previous day is Christmas Eve, which is a working day; consequently, the load curve is lower than the average load curve for the whole month and than previous years. Although Christmas Eve is not a public holiday, demand is much lower than in normal working days, which leads to significant forecast errors.

The forecast for the 12/31/2010 with a mean error of 4.78%. That day is New Year's Eve, and it was a working Friday; that year, there were only two working Fridays with a similar load curve; that Friday's curve is lower than that of other working Fridays and slightly higher than that of Sundays of the same month. The day before New Year's Eve is a low-profile day as compared to other Thursdays of the same month, and it presents an atypical shape between 11 and 16 h; all these factors together caused the forecast error.

6.4. Association between Errors and Availability of Training Patterns

The association between the mean error obtained during the testing phase, and the number of patterns fed to the *ANN* during the training phase has been analyzed. For such purpose, different numbers of patterns were presented to the same model during the training phase: initially, 150 patterns were fed to the model, then the number of patterns was gradually increased (in steps of 50) until reaching 700 patterns. A forecast was produced for each of the days in the testing set; the optimum architecture for each network was achieved by the following method: firstly, a script was used to test all training and performance functions of a network with three neurons; then, the number of neurons in the hidden layer was increased to four and the network's training and performance functions were tested again; then, the number of neurons was increased to five and so forth, until the network had 9 neurons. The reason for using this method is that the model's architecture is entirely dependent on the number of input patterns, as shown in Table 3.

The data above were entered in MatLab, which yielded a cubic polynomial, as follows:

$$Y = -1e^{-0.08} \times X^3 + 2.2e^{-0.05} \times X^2 - 0.017 \times X + 6.8 \tag{4}$$

where Y is the mean error of the Testing phase, and X is the number of patterns used in the training phase. The association between mean error and the number of patterns is evidenced in Figure 14. Red dots stand for the real error value of the *ANN*, while green dots represent error values according to the polynomial function for a specific number of patterns. Figure 13 evidences that, at some point, the architecture cannot further improve the mean error by increasing the number of training patterns. The *ANN* presented reached its maturity phase, as additional patterns did not appear to improve the mean error. It is worth noting that the mean error for the 730 days in the testing set was 2.40%. The improvement with respect to the results of the test including 700 training patterns is irrelevant. By means of Equation (4), we can estimate that by using 750 and 800 patterns we would obtain a mean error of 2.38% and 2.33% respectively; this improvement is far from being as significant as the improvement achieved between the beginning and middle of the test, as shown in Table 4. To assess how error was improved by increasing the number of neurons, we used Equation (5):

$$\Delta error_{networks\_patterns} = \frac{\left|error_i - error_j\right|}{patterns_i - patterns_j} \tag{5}$$

where $i = 2, \ldots, 12; j = 1, \ldots, 11$; with data in Table 4 $error_i$, $error_j$, $pattern_i$ and $patterns_j$. High values obtained by Equation (5) suggest a significant error improvement against the number of patterns.

Table 3. Correlation between errors and the number of training patterns.

Network number	Patterns	Neurons	Error
1	150	3	4.78%
2	200	4	4.18%
3	250	4	3.94%
4	300	4	3.48%
5	350	6	3.26%
6	400	6	3.04%
7	450	12	2.82%
8	500	12	2.65%
9	550	13	2.62%
10	600	13	2.58%
11	650	15	2.48%
12	700	16	2.41%

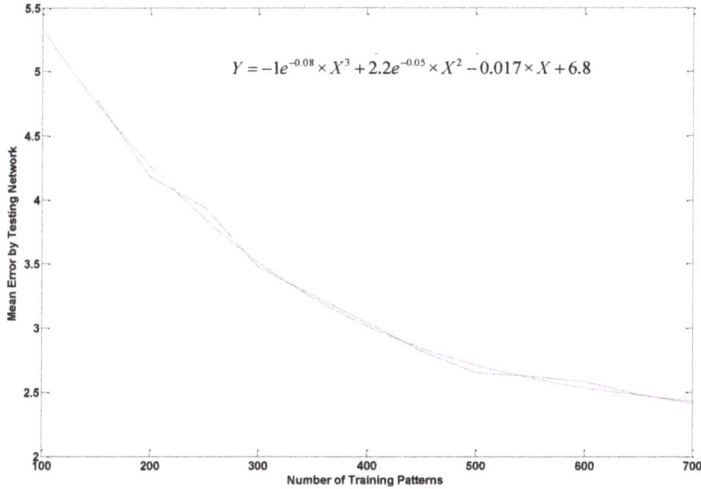

$$Y = -1e^{-0.08} \times X^3 + 2.2e^{-0.05} \times X^2 - 0.017 \times X + 6.8$$

Figure 14. Evolution of mean error in the testing phase with respect to the number of patterns employed in the training phase. Fitting curve is red line and real curve is blue line.

Table 4. Error variation between networks with respect to the number of patterns.

Networks *i–j*	Patterns added	Indicator given by Equation (5)
2–1	50	0.01199
3–2	50	0.00478
4–3	50	0.00428
5–4	50	0.00428
6–5	50	0.00455
7–6	50	0.00440
8–7	50	0.00340
9–8	50	0.00060
10–9	50	0.00080
11–10	50	0.00200
12–11	50	0.00140

6.5. Comparison with Other Solutions

In principle, the results of this work can be only compared directly to other load curve forecasting methods also validated in *microgrid*-sized environments. Like [61], this work also presents a *MLP*-based prediction model. The approach followed in this work employs load curves from day $d − 1$ as an input to predict load curves for day d in the output, which allows a more intimate input-output relationship and a more efficient internal weight adjustment than the models in [61], which use as inputs groups of curves up to three days before the day to forecast. This could explain the better *MAPE* results obtained by the solution employed in this work: 2%–5% against figures around 15%.

When compared to the large-area load forecasting methods studied in Section 3, several differences can be extracted. References [35–39] and [49,50] offer generally short prediction horizons, normally forecasting values in the next hour. While this work employs 29 input variables and 16 neurons in the hidden layer, [44,51–53] use high dimensional input spaces (with a number of input variables ranging from 40 and 50 and neurons in the hidden layer between 24 and 50) and therefore require a bigger training database to reach similar results.

Energies **2013**, *6*, 1385–1408

Finally it is worth mentioning that most of the works in the literature only study daily *MAPEs*. In this paper, however, a detailed hour-by-hour *MAPE* is presented which allows hourly error studies as presented in Section 5.

7. Conclusions and Future Studies

This paper proposes an *ANN*-based model for short-term load forecasting in disaggregated, *microgrid*-sized environments using a simple *MLP*-based architecture. For such purpose, relevant input variables were selected in order to minimize forecast errors. As remarked above, forecasting is more complex in a *microgrid* due to the increased variability of disaggregated load curves. An accurate forecasting in a *microgrid* will depend on the variables employed and the way they are presented to the *ANN*. This study also shows numerically that there is a close relationship between forecast errors and the number of training patterns used, so it is necessary to carefully select the training data to be employed with the system. Finally, this work demonstrates that the concept of load forecasting and the *ANN* tools employed are also applicable to the *microgrid* domain with very good results, showing that small errors around 3% are achievable. This demonstration is backed up by a detailed database containing real information of load curves disaggregated up to city/*microgrid* level running for three entire years.

Acknowledgments: We would like to thank D. Óscar Villanueva, Head of Iberdrola Business Department in Burgos and Soria, Silvia Herrero, PR of Iberdrola in Castilla y León, and Javier Sanjuán and Alvaro González from the University of Zaragoza for their collaboration with this study.

References

1. Booklets European Comission. Your Guide to the Lisbon Treaty 2009. Available online: http://ec.europa.eu/publications/booklets/others/84/en.pdf (accessed on 1 September 2012).
2. Hernández, L.; Baladrón, C.; Aguiar, J.M.; Carro, B.; Sánchez-Esguevillas, A.; Lloret, J.; Chinarro, D.; Gómez, J.J.; Cook, D. A multi-agent system architecture for smart grid management and forecasting of energy demand in virtual power plants. *IEEE Commun. Mag.* **2013**, *51*, 106–113. [CrossRef]
3. *Towards a Low-Carbon Future*; Publications Office of the European Union: Luxembourg, 2010. Available online: http://ec.europa.eu/energy/technology/set_plan/set_plan_en.htm (accessed on 15 May 2012).
4. FUTURED—Spanish Technological Platform for Energy Grids Home Page. Available online: http://www.futured.es/ (accessed on 12 May 2012).
5. European Technology Platform for Electricity Networks of the Future—SmartGrids ETP Home Page. Available online: http://www.smartgrids.eu/ (accessed on 2 May 2012).
6. Kim, H.M.; Kinoshita, T.; Shin, M.C. A multiagent system for autonomous operation of islanded microgrids based on a power market environment. *Energies* **2010**, *3*, 1972–1990. [CrossRef]
7. Kim, H.M.; Lim, Y.; Kinoshita, T. An intelligent multiagent system for autonomous microgrid operation. *Energies* **2012**, *5*, 3347–3362. [CrossRef]
8. Xiao, Z.; Li, T.; Huang, M.; Shi, J.; Yang, J.; Yu, J.; Wu, W. Hierarchical MAS based control strategy for microgrid. *Energies* **2010**, *2*, 1622–1638. [CrossRef]
9. Hong, Y.Y.; Chou, J.H. Nonintrusive energy monitoring for microgrids using hybrid self-organizing feature-mapping networks. *Energies* **2012**, *5*, 2578–2593. [CrossRef]
10. Douglas, A.P.; Breiphol, A.M.; Lee, F.N.; Adapa, R. The impacts of temperature forecast uncertainty on Bayesian load forecasting. *IEEE Trans. Power Syst.* **1998**, *13*, 1507–1513. [CrossRef]
11. Sadownik, R.; Barbosa, E.P. Short-term forecasting of industrial electricity consumption in Brazil. *J. Forecast.* **1999**, *18*, 215–224. [CrossRef]
12. Huang, S.R. Short-term load forecasting using threshold autoregressive models. *IEE Proc. Gener. Transm. Distrib.* **1997**, *144*, 477–481. [CrossRef]
13. Infield, D.G.; Hill, D.C. Optimal smoothing for trend removal in short term electricity demand forecasting. *IEEE Trans. Power Syst.* **1998**, *13*, 1115–1120. [CrossRef]
14. Sargunaraj, S.; Sen Gupta, D.P.S.; Devi, S. Short-term load forecasting for demand side management. *IEE Proc. Gener. Transm. Distrib.* **1997**, *144*, 68–74. [CrossRef]

15. Yang, H.T.; Huang, C.M. A new short-term load forecasting approach using self-organizing fuzzy ARMAX models. *IEEE Trans. Power Syst.* **1998**, *13*, 217–225. [CrossRef]

16. Yang, H.T.; Huang, C.M.; Huang, C.L. Identification of ARMAX model for short term load forecasting: An evolutionary programming approach. *IEEE Trans. Power Syst.* **1996**, *11*, 403–408. [CrossRef]

17. Yu, Z. A temperature match based optimization method for daily load prediction considering DLC effect. *IEEE Trans. Power Syst.* **1996**, *11*, 728–733. [CrossRef]

18. Charytoniuk, W.; Chen, M.S.; Van Olinda, P. Nonparametric regression based short-term load forecasting. *IEEE Trans. Power Syst.* **1998**, *13*, 725–730. [CrossRef]

19. Taylor, J.W.; Majithia, S. Using combined forecast with changing weights for electricity demand profiling. *J. Oper. Res. Soc.* **2000**, *51*, 77–82.

20. Ramanathan, R.; Engle, E.; Granger, C.W.J.; Vahid-Araghi, F.; Brace, C. Short-run forecasts of electricity load and peaks. *Int. J. Forecast.* **1997**, *13*, 161–174. [CrossRef]

21. Soliman, S.A.; Persaud, S.; El-Nagar, K.; El-Hawary, M.E. Application of least absolute value parameter estimation based on linear programming to short-term load forecasting. *Elec. Power Energy Syst.* **1997**, *19*, 209–216. [CrossRef]

22. Hyde, O.; Hodnett, P.F. An adaptable automated procedure for short-term electricity load forecasting. *IEEE Trans. Power Syst.* **1997**, *12*, 84–94. [CrossRef]

23. Ho, K.L.; Hsu, Y.Y.; Chen, C.F.; Lee, T.E.; Liang, C.C.; Lai, T.S.; Chen, K.K. Short term load forecasting of Taiwan power system using a knowledge-based expert system. *IEEE Trans. Power Syst.* **1990**, *5*, 1214–1221. [CrossRef]

24. Rahman, S.; Hazim, O. A generalized knowledge-based short-term load forecasting technique. *IEEE Trans. Power Syst.* **1993**, *8*, 508–514. [CrossRef]

25. Mori, H.; Kobayashi, H. Optimal fuzzy inference for short-term load forecasting. *IEEE Trans. Power Syst.* **1996**, *11*, 390–396. [CrossRef]

26. Bakirtzis, A.G.; Theocharis, J.B.; Kiartzis, S.J.; Satsios, K.J. Short-term load forecasting using fuzzy neural networks. *IEEE Trans. Power Syst.* **1995**, *10*, 1518–1524. [CrossRef]

27. Papadakis, S.E.; Theocharis, J.B.; Kiartzis, S.J.; Bakirtzis, A.G. A novel approach to short-term load forecasting using fuzzy neural nerworks. *IEEE Trans. Power Syst.* **1998**, *13*, 480–492. [CrossRef]

28. Rumelhart, D.; Hinton, G.; Williams, R. Learning internal representations by error propagation. In *Parallel Distributed Processing: Explorations in the Microstructure of Cognition*; Rumelhart, D., McClelland, J.L., Eds.; MIT Press Cambridge: Cambridge, MA, USA, 1996; pp. 318–362.

29. Broomhead, D.S.; Lowe, D. Multivariable functional interpolation and adaptive networks. *Complex Syst.* **1988**, *2*, 321–355.

30. Elman, J.L. Finding structure in time. *Cognit. Sci.* **1990**, *14*, 179–211. [CrossRef]

31. Elman, J.L. Distributed representations, simple recurrent networks, and grammatical structure. *Mach. Learn.* **1991**, *7*, 95–126.

32. Kohonen, T. The Self-organizing Map. *Proc. IEEE* **1990**, *78*, 1464–1480. [CrossRef]

33. Baladrón, C.; Aguiar, J.M.; Calavia, L.; Carro, B.; Sánchez-Esguevillas, A.; Hernández, L. Performance study of the application of artificial neural networks to the completion and prediction of data retrieved by underwater sensors. *Sensors* **2012**, *12*, 1468–1481. [CrossRef] [PubMed]

34. Martínez-Martínez, V.; Baladrón, C.; Gomez-Gil, J.; Ruiz-Ruiz, G.; Navas-Gracia, L.M.; Aguiar, J.M.; Carro, B. Temperature and relative humidity estimation and prediction in the tobacco drying process using artificial neural networks. *Sensors* **2012**, *12*, 14004–14021. [CrossRef] [PubMed]

35. Park, D.C.; El-Sharkawi, M.A.; Marks II, R.J.; Atlas, L.E.; Damborg, M.J. Electric load forecasting using an artificial neural network. *IEEE Trans. Power Syst.* **1991**, *6*, 442–449. [CrossRef]

36. Ho, K.L.; Hsu, Y.Y.; Yang, C.C. Short term load forecasting using a multilayer neural network with and adaptative learning algorithm. *IEEE Trans. Power Syst.* **1992**, *7*, 141–149. [CrossRef]

37. Drezga, I.; Rahman, S. Input variable selection for ANN-based short-term load forecasting. *IEEE Trans. Power Syst.* **1998**, *13*, 1238–1244. [CrossRef]

38. Drezga, I.; Rahman, S. Short-term load forecasting with local ANN predictors. *IEEE Trans. Power Syst.* **1999**, *14*, 844–850. [CrossRef]

39. McMenamin, J.S.; Monforte, F.A. Short-term energy forecasting with neural networks. *Energy J.* **1998**, *19*, 43–61.

40. Lee, K.Y.; Cha, Y.T.; Park, J.H. Short-term load forecasting using an artificial neural network. *IEEE Trans. Power Syst.* **1992**, *7*, 124–132. [CrossRef]
41. Lu, C.N.; Wu, H.T.; Vemuri, S. Neural network based short term load forecasting. *IEEE Trans. Power Syst.* **1993**, *8*, 336–342. [CrossRef]
42. Papalexopoulos, A.D.; Hao, S.; Peng, T.M. An implementation of a neural network based load forecasting models for the EMS. *IEEE Trans. Power Syst.* **1994**, *9*, 1956–1962. [CrossRef]
43. Bakirtzis, A.G.; Petridis, V.; Kiartzis, S.J.; Alexiadis, M.C.; Maissis, A.H. A neural network short term load forecasting model for the Greek power system. *IEEE Trans. Power Syst.* **1996**, *11*, 858–863. [CrossRef]
44. Alfuhaid, A.S.; El-Sayed, M.A.; Mahmoud, M.S. Cascaded artificial neural networks for short-term load forecasting. *IEEE Trans. Power Syst.* **1997**, *12*, 1524–1529. [CrossRef]
45. Lamedica, R.; Prudenzi, A.; Sforma, M.; Caciotta, M.; Cencelli, V.O. A neural network based technique for short-term forecasting of anomalous load periods. *IEEE Trans. Power Syst.* **1996**, *11*, 1749–1756. [CrossRef]
46. Srinivasan, D.; Liew, A.C.; Chang, C.S. Forecasting daily load curves using a hybrid fuzzy-neural approach. *IEE Proc. Gener. Transm. Distrib.* **1994**, *141*, 561–567. [CrossRef]
47. Kim, K.H.; Park, J.K.; Hwang, K.J.; Kim, S.H. Implementation of hybrid short-term load forecasting system using artificial neural networks and fuzzy expert systems. *IEEE Trans. Power Syst.* **1995**, *10*, 1534–1539. [CrossRef]
48. Daneshdoost, M.; Lotfalian, M.; Bumroonggit, G.; Ngoy, J.P. Neural network with fuzzy set-based classification for short-term load forecasting. *IEEE Trans. Power Syst.* **1998**, *13*, 1386–1391. [CrossRef]
49. Senjyu, T.; Mandal, P.; Uezato, K.; Funabashi, T. Next day load curve forecasting using hybrid correction method. *IEEE Trans. Power Syst.* **2005**, *20*, 102–109. [CrossRef]
50. Hsu, Y.Y.; Yang, C.C. Design of artificial neural networks for short-term load forecasting. Part II: Multilayer feedforward networks for peak load and valley load forecasting. *IEE Proc. C Gener. Transm. Distrib.* **1991**, *138*, 414–418. [CrossRef]
51. Rejc, M.; Partos, M. Short-term transmission-loss forecast for the slovenian transmission power system based on a fuzzy-logic decision approach. *IEEE Trans. Power Syst.* **2011**, *26*, 1511–1521. [CrossRef]
52. Wang, Y.; Xia, Q.; Kang, C. Secondary forecasting based on deviation analysis for short-term load forecasting. *IEEE Trans. Power Syst.* **2011**, *26*, 500–507. [CrossRef]
53. Kebriaei, H.; Araabi, B.N.; Rahiminkian, A. Short-term load forecasting with a new nonsymmetric penalty function. *IEEE Trans. Power Syst.* **2011**, *26*, 1817–1825. [CrossRef]
54. Bishop, C.M. Neural networks and their applications. *Review Sci. Instrum.* **1994**, *65*, 1803–1832. [CrossRef]
55. Taylor, J.W.; Buizza, R. Neural network load forecasting with weather ensemble predictions. *IEEE Trans. Power Syst.* **2002**, *17*, 626–632. [CrossRef]
56. Chu, W.C.; Chen, Y.P.; Xu, Z.W.; Lee, W.J. Multiregion short-term load forecasting in consideration of hi and load/weather diversity. *IEEE Trans. Ind. Appl.* **2011**, *47*, 232–237. [CrossRef]
57. Zhang, W.Y.; Hong, W.-C.; Dong, Y.; Tsai, G.; Sung, J.-T.; Fan, G. Application of SVR with chaotic GASA algorithm in cyclic electric load forecasting. *Energy* **2012**, *45*, 850–858. [CrossRef]
58. Hong, W.-C. Electric load forecasting by seasonal recurrent SVR with chaotic artificial bee colony algorithm. *Energy* **2011**, *36*, 5568–5578. [CrossRef]
59. Nose-Filho, K.; Plasencia, A.D.; Minossi, C.R. Short-term multinodal load forecasting using a modified general regression neural network. *IEEE Trans. Power Deliv.* **2011**, *26*, 2862–2869. [CrossRef]
60. Hernández, L.; Baladrón, C.; Aguiar, J.M.; Carro, B.; Sánchez-Esguevillas, A. Classification and clustering of electricity demand patterns in industrial parks. *Energies* **2012**, *5*, 5215–5228. [CrossRef]
61. Chan, P.P.K.; Chen, W.-C.; Ng, W.W.Y.; Yeung, D.S. Multiple Classifier System for Short Term Load Forecast of Microgrid. In Proceedings of the 2011 International Conference on Machine Learning and Cybernetics, Guilin, China, 10–13 July 2011; Volume 3, pp. 1268–1273.
62. Good, R.P.; Kost, D.; Cherry, G.A. Introducing a unified pca algorithm for model size reduction. *IEEE Trans. Semicond. Manuf.* **2010**, *23*, 201–209. [CrossRef]
63. Hernández, L.; Baladrón, C.; Aguiar, J.M.; Calavia, L.; Carro, B.; Sánchez-Esguevillas, A.; Cook, D.J.; Chinarro, D.; Gómez, J. A study of the relationship between weather variables and electric power demand inside a smart grid/smart world framework. *Sensors* **2012**, *12*, 11571–11591. [CrossRef]

64. Drezga, I.; Rahman, S. Phase-Space Short-Term Load Forecasting for Deregulated Electric Power Industry. In Proceedings of International Joint Conference on Neural Networks, Washington, DC, USA, 10–16 July 1999; Volume 5, pp. 3405–3409.

65. Ramezani, M.; Falaghi, H.; Haghifam, M.-R. Short-Term Electric Load Forecasting Using Neural Networks. In The 2005 International Conference on Compute as a Tool (EUROCON 2005), Belgrade, Serbia, 21–24 November 2005; Volume 2, pp. 1525–1528.

66. Razavi, S.; Tolson, B.A. A new formulation for feedforward neural networks. *IEEE Trans. Neural Netw.* **2011**, *22*, 1588–1598. [CrossRef] [PubMed]

67. *Proyecto Atlas INDEL de la Demanda Eléctrica Española [in Spanish]*; Red Eléctrica de España (REE): Madrid, Spain, 1998. Available online: http://www.ree.es/sistema_electrico/pdf/indel/Atlas_INDEL_REE.pdf (accessed on 13 May 2012).

68. Hsu, C.-C.; Chen, C.-Y. Regional load forecasting in Taiwan—Applications of artificial neural networks. *Energy Convers. Manag.* **2003**, *44*, 1941–1949. [CrossRef]

Article

Quantile Forecasting of Wind Power Using Variability Indices

Georgios Anastasiades [1,2] **and Patrick McSharry** [1,2,*]

[1] Mathematical Institute, University of Oxford, 24-29 St Giles', OX1 3LB, Oxford, UK;
georgios.anastasiades@exeter.ox.ac.uk

[2] Smith School of Enterprise and the Environment, University of Oxford, Hayes House, 75 George Street, OX1 2BQ, Oxford, UK

* Author to whom correspondence should be addressed; patrick.mcsharry@smithschool.ox.ac.uk;
Tel.: +44-1865-614-943; Fax: +44-1865-614-960.

Received: 23 November 2012; in revised form: 12 January 2013; Accepted: 22 January 2013;
Published: 5 February 2013

Abstract: Wind power forecasting techniques have received substantial attention recently due to the increasing penetration of wind energy in national power systems. While the initial focus has been on point forecasts, the need to quantify forecast uncertainty and communicate the risk of extreme ramp events has led to an interest in producing probabilistic forecasts. Using four years of wind power data from three wind farms in Denmark, we develop quantile regression models to generate short-term probabilistic forecasts from 15 min up to six hours ahead. More specifically, we investigate the potential of using various variability indices as explanatory variables in order to include the influence of changing weather regimes. These indices are extracted from the same wind power series and optimized specifically for each quantile. The forecasting performance of this approach is compared with that of appropriate benchmark models. Our results demonstrate that variability indices can increase the overall skill of the forecasts and that the level of improvement depends on the specific quantile.

Keywords: wind power forecasting; wind power variability; quantile forecasting; density forecasting; quantile regression; continuous ranked probability score; quantile loss function; check function

---

1. Introduction

Wind power is one of the fastest growing renewable energy sources (Barton and Infield [1]). According to the European Wind Energy Association (EWEA), the wind industry has had an average annual growth of 15.6% over the last 17 years (1995–2011). In 2011, 9616 MW of wind energy capacity was installed in the EU, making a total of 93957 MW, which is sufficient to supply 6.3% of the European Union's electricity. These figures represent 21.4% of new power capacity showing that wind energy continues to be a popular source of energy.

However, due to the large variability of wind speed caused by the unpredictable and dynamic nature of the earth's atmosphere, there are many fluctuations in wind power production. This inherent variability of wind speed is the main cause of the uncertainty observed in wind power generation. Recently, scientists have been directly or indirectly attempting to model this uncertainty and produce improved forecasts of wind power production.

According to Boyle [2], the most important application for wind power forecasting is to reduce the need for balancing the energy and reserve power which are needed to optimize the power plant scheduling. Moreover, wind power forecasts are used for grid operation and grid security evaluation. For maintenance and repair reasons, the grid operator needs to know current and future values of wind power for each grid area or grid connection point. Wind power forecasts are also required for small regions and individual wind farms.

The length of the relevant forecast horizon usually depends on the required application. For example, in order to schedule power generation (grid management), forecast horizons of several hours are usually sufficient, but for maintenance planning forecast horizons of several days or weeks are needed [3].

Since there is no efficient way to store wind energy, the wind power production decreases to zero if wind speed drops below a certain level known as the "cut-in speed". On the other hand, excessively strong winds can cause serious damage to the wind turbines, and hence they are automatically shut down at the "disconnection speed", leading to an abrupt decline of power generation. In addition, the wind power generated is limited by the capacity of each turbine. Therefore, it is important to produce accurate wind power forecasts for enabling the efficient operation of wind turbines and reliable integration of wind power into the national grid.

The literature of wind power forecasting starts with the work of Brown *et al.* [4] where they used autoregressive processes to model and simulate the wind speed, and then estimate the wind power by applying suitable transformations to values of wind speed. Most of the early literature focuses on producing wind power point forecasts, directly, or indirectly in the sense that the focus is on modelling the wind speed and then transforming the forecasts through a power curve [5,6]. The approach of modelling the wind speed series is found to be quite useful because in many situations researchers do not have access to wind power data due to its commercial sensitivity. This approach has as an advantage the fact that the wind speed time series is much smoother than the corresponding wind power time series. An obvious disadvantage is that, since the shape of the power curve may vary with the time of year and different environmental conditions, it is much more difficult to model this type of behaviour.

Recent research has focused on producing probabilistic or density forecasts, because the point forecast methods are not able to quantify the uncertainty related to the prediction. Point forecasts usually inform us about the conditional expectation of wind power production, given information up to the current time and the estimated model parameters. Only a fully probabilistic framework will give us the opportunity to model the uncertainty related to the prediction, and avoid the intrinsic uncertainty involved in a point forecasting calibrated model. Up to now, the number of studies on multi-step quantile/density forecasting is relatively small compared with point forecasting.

Moeanaddin and Tong [7] estimated densities using recursive numerical methods, which are quite computationally intensive. Gneiting *et al.* [8] introduces regime-switching space¨Ctime (RST) models which identify forecast regimes at a wind energy site and fit a conditional predictive model for each regime. The RST models were applied to 2-h-ahead forecasts of hourly average wind speed near the Stateline wind energy center in the U.S. Pacific Northwest. One of the most recent regime-based approaches is the one used by Trombe *et al.* [9], where they propose a general model formulation based on a statistical approach and historical wind power measurements only. The model they propose is an extension of Markov-Switching Autoregressive (MSAR) models with Generalized Autoregressive Conditional Heteroscedastic (GARCH) errors in each regime to cope with the heteroscedasticity.

Pinson [10], by introducing and applying a generalised logistic transformation, managed to produce ten-minute ahead density forecasts at the Horns Rev wind farm in Denmark. Pinson and Kariniotakis [11] described a generic method for the providing of prediction intervals of wind power generation and Sideratos and Hatziargyriou [12] proposed a novel methodology to produce probabilistic wind power forecasts using radial basis function neural networks. Taylor *et al.* [6] used statistical time series models and weather ensemble predictions to produce density forecasts for five wind farms in the United Kingdom. This is a relatively new approach for wind power forecasting that uses ensemble forecasts produced from numerical weather prediction (NWP) methods [6,13]. Moreover, Lau and McSharry [14] produced multi-step density forecasts for the aggregated wind power series in Ireland, using ARIMA-GARCH processes and exponential smoothing models. Jeon and Taylor [15] modelled the inherent uncertainty in wind speed and direction using a bivariate VARMA-GARCH model and then they modelled the stochastic relationship of wind power to wind speed using conditional kernel density (CKD) estimation. This is a rather promising semi-non-parametric model but unfortunately cannot be used as benchmark in this article because we aim to make predictions using only wind power data.

The quantile regression method [16] has been extensively used to produce wind power quantile forecasts, using a variety of explanatory variables among which are wind speed, wind direction, temperature and atmospheric pressure. Recent literature includes papers by Bremnes [17], Nielsen *et al.* [18], and Moller *et al.* [19]. More specifically, Bremnes [17] produced wind power probabilistic forecasts for a wind farm in Norway, using a local quantile regression model. The predictors used for the local quantile regression were outputs from a NWP model (HIRLAM10), and used lead times from 24 to 47 h. Nielsen *et al.* [18] used an existing wind power forecasting system (Zephyr/WPPT) and showed how the analysis of the forecast error can be used to build a model for the quantiles of the forecast error. The explanatory variables used in their quantile regression model include meteorological forecasts of air density, friction velocity, wind speed and direction from a NWP model (DMI-HIRLAM). Moreover, Moller *et al.* [19] presented a time-adaptive quantile regression algorithm (based on the simplex algorithm) which manages to outperform a static quantile regression model on a data set with wind power production. In addition, Pritchard [20], discussed ways of formulating quantile-type models for forecasting variations in wind power within a few hours. Such models can predict quantiles of the conditional distribution of the wind power available at some future time using information presently available.

Davy *et al.* [21], proposed a new variability index that is designed to detect rapid fluctuations of wind speed or power that are sustained for a length of time, and used it as an explanatory variable in the quantile regression model they constructed. Bossavy *et al.* [22] extracted two new indices that are able to recognize and predict ramp events (A ramp event is defined as a large change in the power production of a wind farm or a collection of wind farms over a short period of time.) in the wind power series, and used them to produce quantile estimates with the quantile regression forest method as their basic forecasting system. Finally, Gneiting [23] studied the behaviour of quantiles as optimal predictors and illustrated the relevance of decision theoretic guidance in the transition from a predictive distribution to a point forecast using the Bank of England density forecasts of United Kingdom inflation rates, and probabilistic predictions of wind energy resources in the Pacific Northwest.

This article does not have as a purpose to develop models that can compete with the commercially available models that focus on forecast horizons greater than six hours (and are using NWPs). This is also the main reason we chose a very short forecast horizon (six hours), since it has been shown that statistical time series models may outperform sophisticated meteorological forecasts for short lead times within six hours [24]. In fact, NWPs are not even available (for some regions) for lead times shorter than three hours. So, as mentioned above, our choice of such a short forecast horizon is particularly useful for the assessment of grid security and operation. We would like to investigate the extent to which the use of quantile regression models with endogenous explanatory variables can improve the forecasting performance of probabilistic benchmarks such as persistence and climatology.

In this article we use wind power series from three wind farms in Denmark, to produce very short-term quantile forecasts, from 15 min up to six hours ahead. In order to produce quantile forecasts, we will use a linear quantile regression model, with explanatory variables extracted from the same wind power time series. Modelling the wind power series directly is preferable to a method based on wind speed forecasts because we avoid the uncertainty involved in transforming wind speed forecasts back to wind power forecasts using the power curve. The fact that we use only endogenous explanatory variables is also a very important practical consideration that we have taken on board to ensure the ability to apply our model to all wind farms. Power systems operators will require an approach to forecast a wide range of sites, where a collection of different wind farm owners implies that the only variable that they are guaranteed to have access to is the wind power generation over time.

Four new variability indices will be produced (extracted from the original wind power time series), which serve to capture the volatile nature of the wind power series. These indices, together with some lagged versions of the wind power series, will be used as explanatory variables in the quantile regression model. As for any regression model, we need predictions (point forecasts) for the future values of the explanatory variables in order to produce future quantile estimates. To produce these predictions we will use time series models that are able to model both the mean and the variance of the underlying series.

The motivation behind the chosen model structure is based on understanding the way that the underlying weather variability can affect the conditional predictive density of the wind power generation. We would like to keep the model structure as simple as possible and therefore assume that the probability of observing a value of wind power below a certain level can be written as a function of some local mean plus the local variability involved in observing the specific wind power value. A linear combination of recently observed wind power values seems to be the easiest way to identify a function that can forecast the expected value of a specific quantile, given recent information. It is worth noticing that the model may be linear in parameters but the nonlinearity is attained in the explanatory variable themselves, and especially in the variability indices. In addition, the variability indices can capture the underlying weather variability, and hence help to improve the probabilistic forecasts given a certain weather regime.

The three Danish wind farms were chosen according to their monthly wind power capacity and standard deviation. We choose one high, one low, and one average variability wind farm, in order to understand better the ability of each model to produce probabilistic forecasts under different circumstances.

The indices used will be independently optimized for each of the three wind farms, using a one-fold cross validation technique. In fact, two different optimizations will take place for each wind farm: The first one will aim to minimize the Check Function Score (defined in Section 4.2) produced by a 1-step ahead quantile regression forecast, for each of 19 different quantiles. The second one will aim to minimize the averaged Check Function Score, produced by taking the average over all 24 predicted lead times (equal to six hours), for each quantile. The final forecast results will be compared with those of some widely used benchmark models (persistence distribution and unconditional distribution).

The remainder of the article is presented as follows. In Section 2 we will introduce the wind power data, and the new variability indices will be derived in Section 3. Section 4 will present the methodology behind the various models and explain ways to evaluate the resulting quantile forecasts. In Section 5 we will present the four competing quantile regression models and optimize their quantile forecast performance on the in-sample testing set. In Section 6 the out-of-sample quantile and density forecast performance of the competing quantile regression models will be assessed, and Section 7 will conclude the article.

2. Wind Power Data

We use wind power data recorded at three wind farms in Denmark summarized in Table 1. These wind farms were chosen to have different amounts of wind power variability, located in different geographical regions (The 446 wind farms in Denmark are assigned to 15 different geographical regions, but no further information about the actual locations of the wind farms is disclosed), and have the smallest percentage of missing values among all available wind farms. The percentage of missing values (mostly isolated points) is found to be less than 0.025% for all three wind farms, and missing values were imputed using linear interpolation. For such a small percentage of missing values, the smoothing effect caused by using linear interpolation to impute the missing values is practically negligible.

Table 1. The Danish wind farms used in this study.

Wind farm station name	Wind power variability	Wind farm rated capacity (kW)
DØR	Low	1000
ALB	Medium	25,500
VES	High	2195

Our data sets contain wind power measurements recorded every 15 min for four years, from 1 January 2007 to 31 December 2010. The data of each wind farm is bounded between zero and the maximum capacity of the wind farms. The zero value is attained in the case of excessively strong wind, where the turbines shut down in order to prevent them from damage, or in the case of very weak wind (the cut-in wind speed, usually 3–4 ms^{-1} according to Pinson [10]). In order to facilitate comparisons between the data sets of different capacities, we normalize the wind power data of each wind farm by dividing by the total (rated) capacity, which is constant over the four years period. Hence, the data is now bounded within the interval [0,1].

We dissect the data of each farm into a set of exactly two years (2007 and 2008) for in-sample model training and calibration, and an out-of-sample testing set (the remaining two years) for out-of-sample testing and model evaluation. The in-sample set is dissected again into two sub-sets, a training set and a testing set. For the in-sample training set we use the first 1.5 years and for the in-sample testing set the remaining half year. This way, we can use a *one-fold cross validation technique* to optimize the indices introduced in Section 3, and test the performance of our final chosen model using the out-of-sample testing set.

The time series plots for the year 2010, together with the monthly mean power output and standard deviation, are shown in Figure 1. The monthly mean power output and monthly standard deviation were generated by taking the mean and standard deviation of wind power, respectively, for each month over the entire four year period. As we observe, the three wind farms have different wind power variability. More specifically, the first and last wind farms of Figure 1 have the lowest and highest possible wind power variability for all four years (from all the available wind farms in Denmark), without having any significant changes (Wind power variability may change from year to year by addition of new turbines or removal (maybe for maintenance) of existing ones.) in the capacity from year to year. The second wind farm of Figure 1 was chosen to have an average (medium) wind power variability compared with the other two farms, but again without having any significant changes in the monthly capacity from year to year.

3. Indices of Wind Power Variability

Davy *et al.* [21] proposed a variability index that is designed to detect rapid fluctuations of wind speed or power that are sustained for a length of time. They defined this variability index as the standard deviation of a band-limited signal in a moving window, and they constructed such an index for a wind speed time series. This variability index depends on four parameters: the order of the filter (integer greater than one), the upper and lower frequencies of the extracted signal, and the width of the moving window. We would like to use such an index as an explanatory variable in our quantile regression, but a proper optimization of this is too computationally expensive because of the number of parameters involved.

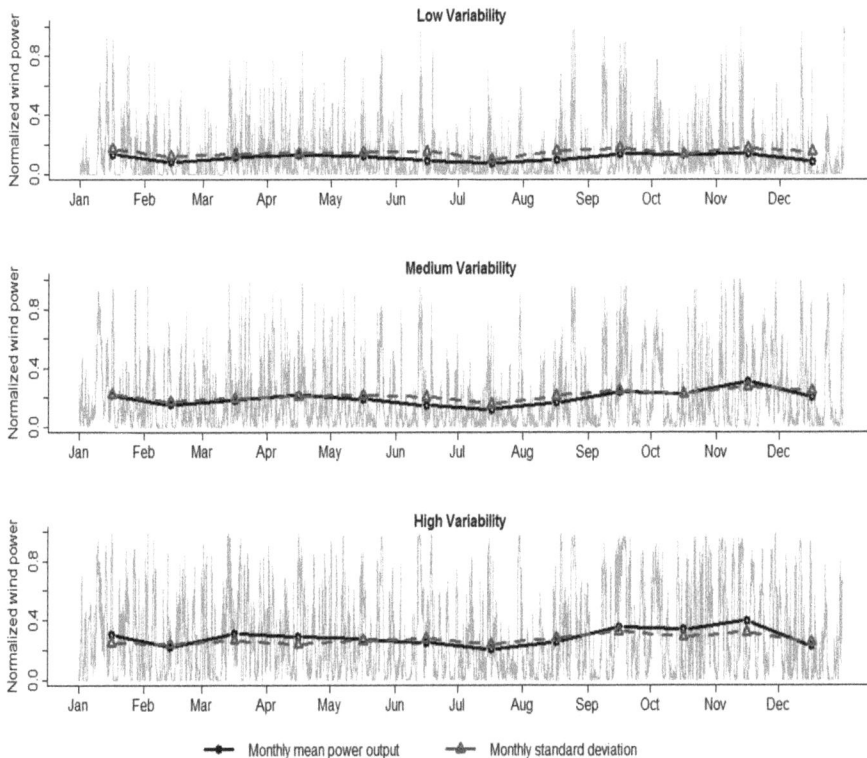

Figure 1. Time series plots of normalized power data for the three chosen Danish wind farms, for the year 2010. Please note that the point on the time axis labelled Jan refers to 00:00 on 1 January and similarly for every month.

Instead, we propose a parsimonious variability index which depends only on two parameters, (m, n) where $m, n \in \mathbb{N}_0 \setminus \{1\}$, and is constructed as follows. Firstly we smooth our original wind power series using an averaging window of size m, in order to obtain the smoothed wind power series,

$$r_t = \begin{cases} \frac{1}{m} \sum\limits_{i=1}^{m} y_{t-i+1} & \text{if } m > 1 \\ y_t & \text{if } m = 0 \end{cases} \tag{1}$$

for $t \geq m$. Note that this series behaves in a fully retrospective way, in the sense that each point of the series depends only on the historical values of the original series. Since the smoothed series is $m - 1$ points smaller than the original series, we set $r_t = r_m$, for $t = 1, 2, ..., m - 1$.

Finally, the new variability index is just the standard deviation of the extracted smoothed wind power series in a moving window of width n. So, if r_t is a given point of the smoothed series, we define the new index as

$$SD_t = \begin{cases} \sqrt{\frac{1}{n-1} \sum\limits_{i=1}^{n} \left(r_{t-i+1} - \frac{1}{n} \sum\limits_{j=1}^{n} r_{t-j+1} \right)^2} & \text{if } n > 1 \\ r_t & \text{if } n = 0 \end{cases} \tag{2}$$

for $t \geq n$. Again, we impute the first $n - 1$ points of the series by setting $SD_t = SD_n$, for $t = 1, 2, ..., n - 1$. This index can be optimized much more easily than the one proposed by Davy *et al.* [21], since it has only two parameters: the smoothing parameter m, and the variability parameter n.

By similar reasoning, we create another three variability indices. We create the smoothed wind power series, r_t, as defined by Equation (??), and then instead of finding the standard deviation we find the sample interquartile range (IQR), the 5% and the 95% sample quantiles of the smoothed series over a moving variability window (different for each series) of width n.

There are many different ways to define the quantiles of a sample. We use the definition recommended by Hyndman and Fan [25] and presented as follows. Let $R_t = \{r_{t-n+1}, ..., r_{t-1}, r_t\}$ for $t \geq n > 1$, denote the order statistics of R_t as $\{r_{(1)}, ..., r_{(n)}\}$ and let $\hat{Q}_{R_t}(p)$ denote the sample p-quantile of R_t with proportion $p \in (0,1)$. We calculate $\hat{Q}_{R_t}(p)$ (for a chosen proportion p) by firstly plotting $r_{(k)}$ against p_k, where $p_k = \frac{k-1/3}{n+1/3}$ and $k = 1, .., n$. This plot is called a quantile plot and p_k a plotting position. Then, we use linear interpolation of $\left(p_k, r_{(k)}\right)$ to get the solution $\left(p, \hat{Q}_{R_t}(p)\right)$ for a chosen $0 < p < 1$. Therefore, the three new indices can be defined as:

$$IQR_t = \begin{cases} \hat{Q}_{R_t}(0.75) - \hat{Q}_{R_t}(0.25) & \text{if } n > 1 \\ r_t & \text{if } n = 0 \end{cases} \tag{3}$$

$$Q05_t = \begin{cases} \hat{Q}_{R_t}(0.05) & \text{if } n > 1 \\ r_t & \text{if } n = 0 \end{cases} \tag{4}$$

$$Q95_t = \begin{cases} \hat{Q}_{R_t}(0.95) & \text{if } n > 1 \\ r_t & \text{if } n = 0 \end{cases} \tag{5}$$

for $t \geq n$. We also impute their values for $t = 1, ..., n - 1$ in a similar way as we did for the SD index. An example of the construction of the three variability wind power indices is shown in Figure 2. A first observation is that the IQR and SD indices behave similarly, but the IQR index has higher peaks than the SD index, and hence gives more emphasis to the high variability regions of the wind power series. Moreover, the Q05 and Q95 indices also behave quite similarly, capturing the two tails of the wind power distribution over a predefined window.

These indices will be properly optimized and will be used, together with some lagged values of the original power series, as explanatory variables in the quantile regression introduced in the next section. It is worth mentioning that the choice of firstly smoothing the wind power series is taken in order to take into consideration the fact that any noise may hide or alter the pattern of the underlying weather regime we wish to capture. By choosing $m = 0$ we do not remove any of the underlying noise, and hence we assume that the weather variability is fully captured by using the original wind power time series.

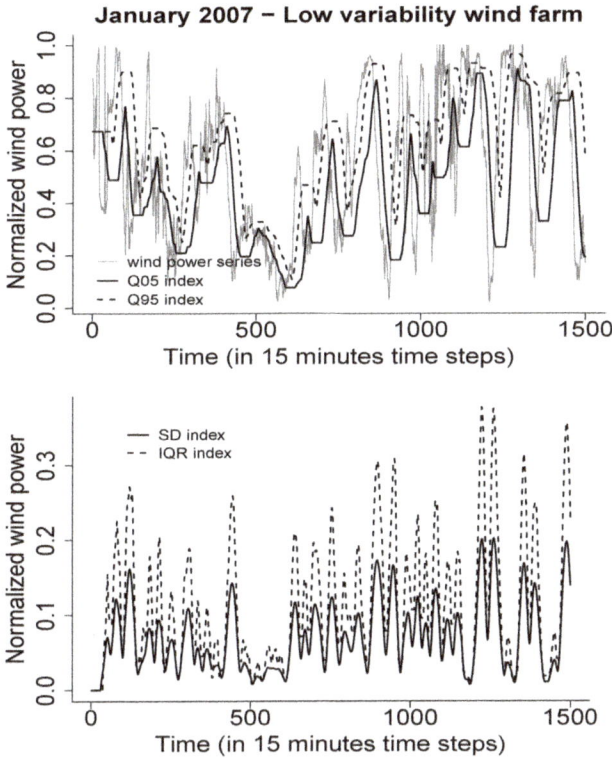

Figure 2. Wind power time series plot of the low variability farm, together with the four variability indices (Q05, Q95 on upper plot, and SD, IQR on lower plot). The parameters are chosen to be the same for all indices to facilitate comparison ($m = 30$ and $n = 30$).

4. Quantile Regression, Forecasting, and Evaluation Methodology

In order for the paper to be self-consistent, we include the theory of linear quantile regression in Section 4.1. In Section 4.2 we introduce the methodology we will use to evaluate the produced quantile and density forecasts.

4.1. Quantile Regression

Given a random variable, y_t, and a strictly increasing continuous CDF, $F_t(y)$, the α_i-quantile, $q_t^{(\alpha_i)}(y)$, with proportion $\alpha_i \in [0, 1]$ is defined as the value for which the probability of obtaining values of y_t below $q_t^{(\alpha_i)}$ is α_i:

$$\mathbb{P}\left(y_t < q_t^{(\alpha_i)}\right) = \alpha_i \, or \, q_t^{(\alpha_i)} = F_t^{-1}(\alpha_i) \tag{6}$$

Note that the notation y_t is used for denoting both the stochastic state of the random variable at time $t = 1, 2, ..., T$, and the measured value at that time for a training set of size T.

Quantile regression, introduced by Koenker and Bassett [16], models $q_t^{(\alpha_i)}$ for $\alpha_i \in [0,1]$, as a linear combination of some given explanatory variables (also called regressors or predictors). So, the α_i-quantile is modelled as:

$$
\begin{aligned}
q_t^{(\alpha_i)} &= \gamma_0^{(\alpha_i)} + \gamma_1^{(\alpha_i)} x_{t,1} + \dots + \gamma_p^{(\alpha_i)} x_{t,p} \\
&= \gamma_0^{(\alpha_i)} + \sum_{j=1}^{p} \gamma_j^{(\alpha_i)} x_{t,j}
\end{aligned}
\tag{7}
$$

where $\gamma_j^{(\alpha_i)}$ are unknown coefficients depending on α_i, and $x_{t,j}$ are the p known explanatory variables. In quantile regression, a regression coefficient estimates the change in a specified quantile of the response variable produced by a one unit change in the corresponding explanatory variable.

We define the *quantile loss function* [16], also known as the *check function*, for a given proportion $\alpha_i \in [0,1]$ as:

$$
\begin{aligned}
\rho_{\alpha_i}(u) &= \left(\alpha_i - 1_{\{u<0\}} \right) u \\
&= \begin{cases} \alpha_i u, & u \geq 0 \\ (\alpha_i - 1)u, & u < 0 \end{cases}
\end{aligned}
\tag{8}
$$

where u is a given value. Then, the sample α_i-quantile can be calculated by minimizing $\sum_{t=1}^{T} \rho_{\alpha_i}(y_t - q)$ with respect to q. Hence, we can estimate the unknown coefficients, $\gamma_j^{(\alpha_i)}$, by replacing q with the right-hand side of Equation (7):

$$
\hat{\gamma}^{(\alpha_i)} = \text{argmin}_{\gamma} \sum_{t=1}^{T} \rho_{\alpha_i}\{y_t - (\gamma_0 + \gamma_1 x_{t,1} + \dots + \gamma_p x_{t,p})\}
\tag{9}
$$

where $\hat{\gamma}^{(\alpha_i)}$ is a vector containing the unknown coefficients. Usually, these estimates are calculated using linear programming techniques as in Koenker and D'Orey [26].

In this article we will use quantile regression to forecast the values of quantiles with nominal proportion $\alpha_i = \{0.05, 0.10, \dots, 0.95\}$, for forecast horizons $k = 1, 2, \dots, 24$, measured in time steps of 15 min. We denote the forecast for the quantile with nominal proportion α_i issued at time t for forecast time $t + k$, by $\hat{q}_{t+k|t}^{(\alpha_i)}(y)$. In order to produce these forecasts, we use Equation (7), and the estimated coefficients, $\hat{\gamma}^{(\alpha_i)}$:

$$
\begin{aligned}
\hat{q}_{t+k|t}^{(\alpha_i)}(y) &= \gamma_0^{(\alpha_i)} + \gamma_1^{(\alpha_i)} \hat{x}_{t+k|t,1} + \dots + \gamma_p^{(\alpha_i)} \hat{x}_{t+k|t,p} \\
&= \gamma_0^{(\alpha_i)} + \sum_{j=1}^{p} \gamma_j^{(\alpha_i)} \hat{x}_{t+k|t,j}
\end{aligned}
\tag{10}
$$

where $\hat{x}_{t+k|t,j}$ for $j = 1, \dots, p$ denote the forecasts of the explanatory variables $x_{t,j}$, issued at time t with lead time $t + k$.

The random variable y_t will represent the normalized wind power time series, (y_t), and the explanatory variables will be represented by time series, $(x_{t,j})$, extracted from the normalized wind power series. In order to produce the forecasts, $\hat{x}_{t+k|t,j}$, we will fit suitable time series models to the variables $(x_{t,j})$, and then predict from these models up to $t + k$ values ahead.

It is worth mentioning that by producing quantile forecasts using quantile regression, we may end up with some quantile forecasts crossing each other. This is a not very common phenomenon for so few quantile forecasts (19 in our case), but monitoring its occurrence is very important. In our analysis, whenever this phenomenon happens (it occurs very rarely because we fit the models to a large amount of data) we just shift the crossing quantile forecasts in order to keep $\hat{F}_{t+k}\left(\hat{q}_{t+k|t}^{(\alpha_i)}\right) = \alpha_i$, for $\alpha_i = \{0.05, 0.10, \dots, 0.95\}$, a strictly increasing function.

4.2. Quantile and Density Forecast Evaluation

The evaluation of the quantile forecasts, for each quantile, $\alpha_i = \{0.05, 0.10, ..., 0.95\}$, will be undertaken using the quantile loss function:

The **quantile loss function**, also known as the **check function** [3,27] is used to define a specific quantile of the distribution and was defined in Section 4.1, Equation (8). Hence, given a testing set of size N, we can estimate a particular quantile, $\hat{q}^{(\alpha_i)}$, with proportion α_i, using

$$\hat{q}^{(\alpha_i)} = \min_q \sum_{t=1}^{N} \rho_{\alpha_i}(y_t - q) \tag{11}$$

and therefore we can evaluate a series of quantile forecasts, $\hat{q}^{(\alpha_i)}_{t+k|t}$, issued at time t with lead time $t + k$ and nominal proportion α_i, using:

$$QL(k, \alpha_i) = \frac{1}{N} \sum_{t=1}^{N} \rho_{\alpha_i}\left(y_{t+k} - \hat{q}^{(\alpha_i)}_{t+k|t}\right) \tag{12}$$

This is the average over the whole testing set of the *check function score*, $\rho_{\alpha_i}\left(y_{t+k} - \hat{q}^{(\alpha_i)}_{t+k|t}\right)$, for the quantile α_i, for a k-step ahead prediction. From now on we will call this function the **Check Function (CF)**, and the its score the **Check Function Score (CFS)**.

Using the different quantile forecasts we can also reconstruct the whole probability / cumulative forecasted distribution. We use the **Continuous Ranked Probability Score (CRPS)** in order to evaluate the density forecasts for each forecast horizon:

The *crps* [28] is computed by taking the integral of the Brier scores for the associated probability forecasts at all real valued thresholds,

$$crps\left(\hat{F}_{t+k|t}(y), y_{t+k}\right) = \int_{-\infty}^{+\infty}\left(\hat{F}_{t+k|t}(y) - 1_{\{y \geq y_{t+k}\}}\right)^2 dy$$
$$= \int_0^1 QS_{\alpha_i}\left(\hat{F}^{-1}_{t+k|t}(\alpha_i), y_{t+k}\right) d\alpha_i$$

where $\hat{F}_{t+k|t}(y)$ corresponds to the CDF forecast, and y_{t+k} to the corresponding verification. $1_{\{y \geq y_{t+k}\}}$ is an indicator function that equals one if $y \geq y_{t+k}$ and zero otherwise. The quantile score, QS_{α_i} [29], is defined by

$$QS_{\alpha_i}(q, y) = 2\left(\alpha_i - 1_{\{y < q\}}\right)(y - q) \tag{15}$$

Hence, the average of these *crps* values over each forecast-verification pair gives the CRPS for each forecast horizon k:

$$CRPS(k) = \frac{1}{N} \sum_{t=1}^{N} crps\left(\hat{F}_{t+k|t}(y), y_{t+k}\right)$$
$$= 2 \int_0^1 QL(k, \alpha_i) d\alpha_i$$

where $QL(k, \alpha_i)$ is CF defined in Equation (12). Representation (17) is useful to produce a rough estimate of the in-sample CRPS for each forecast horizon, using the CFS for each quantile. This is a rather poor approximation of the CRPS, because the number of quantiles used in this article (19 quantiles), is not large enough to produce an accurate approximation of the integral in Equation (17).

In order to find the out-of-sample CRPS for each k, we will use the following alternative representation of the *crps*, introduced by Gneiting and Raftery [29]:

$$crps\left(\hat{F}_{t+k|t}(y), y_{t+k}\right) = \mathbb{E}_F|X - y_{t+k}| - \frac{1}{2}\mathbb{E}_F|X - X'| \tag{18}$$

where X and X' are independent copies of a random variable with CDF $\hat{F}_{t+k|t}$. This representation is particularly useful when $\hat{F}$ is represented by a sample, as in our case. Then, the CPRS for each forecast horizon k is given by Equation (16).

Moreover, it will be necessary to quantify the gain/loss of some forecasting models with respect to a chosen reference model. Following McSharry *et al.* [3], this gain, denoted as an improvement with respect to the considered reference forecast system, is called a *Skill Score* and is defined as:

$$\text{SkillScore}(k) = \frac{\text{SCORE}_{\text{ref}}(k) - \text{SCORE}(k)}{\text{SCORE}_{\text{ref}}(k)} = 1 - \frac{\text{SCORE}(k)}{\text{SCORE}_{\text{ref}}(k)} \tag{19}$$

where k is the lead time of the forecast and SCORE is considered the evaluation criterion score (such as CRPS or CFS). By using the above definition we can also introduce the *Average Skill Score*. This is just the Skill Score with the scores of the competing and reference models averaged over all forecast horizons. It is defined as:

$$\text{AverageSkillScore} = 1 - \frac{\sum_{k=1}^{k_{\max}} \text{SCORE}(k)}{\sum_{k=1}^{k_{\max}} \text{SCORE}_{\text{ref}}(k)} \tag{20}$$

So, when we are talking about Score, the lower the value the better the performance; but, when we are talking about Skill Score (or Average Skill Score), the higher the value the better, since we are comparing the candidate model to the reference model. Please note that the reference model will be different each time, and chosen according to the comparison we wish to make.

In order to formally rank and statistically justify any possible difference in the CRPS and CFS of the competing models with respect to the reference models, we will use the Amisano and Giacomini test [30] of equal forecast performance. This test is based on the statistic

$$t_{N,k} = \sqrt{N} \frac{\text{SCORE}(k) - \text{SCORE}_{\text{ref}}(k)}{\hat{\sigma}_{N,k}} \tag{21}$$

where SCORE again is considered the evaluation criterion score such as the CRPS or CFS, N is the out-of-sample size, and

$$\hat{\sigma}_{N,k}^2 = \frac{1}{N-k+1} \sum_{j=-(k-1)}^{k-1} \sum_{t=1}^{1+N-k-|j|} \delta_{t,k}\delta_{t+|j|,k} \text{ where } \delta_{t,k} = S(t+k|t) - S_{\text{ref}}(t+k|t) \tag{22}$$

The functions S and S_{ref} represent the before averaging scores (such as the *crps* of Equations (13) or (18) and *check function score* defined just after Equation (12)) of the competing and reference models, respectively. Assuming suitable regularity conditions, according to Amisano and Giacomini [30], the statistic $t_{N,k}$ is asymptotically standard normal under the null hypothesis of zero expected score differentials. Small *p*-values of this test provide evidence that the difference in the forecast performance of the two forecasting (given a specific evaluation score) is statistically significant.

5. Optimization of the Variability Indices

In this section we will introduce four different quantile regression models, and using one-fold cross validation try to optimize their probabilistic forecasting performances. Our main goal is to evaluate whether or not the four variability indices (introduced in Section 3) can help to provide trustworthy quantile forecasts of wind power, when used as explanatory variables in the quantile regression model (7). For this purpose, we have to find the optimal set of parameters (m, n) of these indices, which provides the best quantile forecast performance, for each individual quantile. We do that using the following procedure.

For each index, we sample different combinations of parameters from the range $m, n = \{0, 8, 16, ..., 192\}$, in order to produce 625 different realizations of each index, for each wind farm. A preliminary analysis showed that creating a moving window larger than 192 time-points wide (2880 min *i.e.*, 2 days) did not increase the performance of the indices.

Then, for each set of parameters, we fit the following four different quantile regression models on the in-sample training set (of each wind farm), for each of the 19 quantiles $\alpha_i = \{0.05, 0.1, ..., 0.95\}$:

$$
\begin{aligned}
\text{SDmodel} &: q_t = \gamma_{01} + \gamma_{11} y_{t-1} + \gamma_{21} y_{t-2} + \gamma_{31} y_{t-3} + \gamma_{41} SD_t^{(\alpha_i)} \\
\text{IQRmodel} &: q_t = \gamma_{02} + \gamma_{12} y_{t-1} + \gamma_{22} y_{t-2} + \gamma_{32} y_{t-3} + \gamma_{42} IQR_t^{(\alpha_i)} \\
\text{Q05model} &: q_t = \gamma_{03} + \gamma_{13} y_{t-1} + \gamma_{23} y_{t-2} + \gamma_{33} y_{t-3} + \gamma_{43} Q05_t^{(\alpha_i)} \\
\text{Q95model} &: q_t = \gamma_{04} + \gamma_{14} y_{t-1} + \gamma_{24} y_{t-2} + \gamma_{34} y_{t-3} + \gamma_{44} Q95_t^{(\alpha_i)}
\end{aligned}
\tag{23}
$$

where $q_t \equiv q_t^{(\alpha_i)}$ is defined in Equation (6), $\gamma_{hl} \equiv \gamma_{hl}^{(\alpha_i)}$ are the regression coefficients, and y_{t-j} are lagged wind power series. The choice of the number wind power series lags used as explanatory variables was taken by considering the AIC (a prediction based criterion according to Akaike [31]) of different quantile regression models which have different numbers of lags as explanatory variables. We also investigated the improvement obtained by adding to the right hand side of Equation (23) a combination of variability indices. Due to collinearity effects, the SD and IQR indices cannot coexist in the same equation. Any other combinations of the variability indices did not provide reduction to the AIC for more than 14 out of 19 quantile regression equations, at any of the three wind farm sites. Hence, we considered examining the effect that each individual variability index will provide by being included as an explanatory variable to the quantile regression equations, as defined by Equation (6).

Moreover, we also considered adding to the right hand sides of Equation (23) a trigonometric function (also introduced in Equation (24) below) which uses two pairs of harmonics to regress wind power quantile, q_t, on the 15 min time step of the day. The addition of this function, which is used to model the diurnal component of each quantile of the wind power production at each wind farm, was not found to provide reduction to the AIC of 17 out of 19 quantile regression equations, at any of the three wind farm sites. Hence, in order to obtain parsimonious models we excluded these functions from the final models. Nevertheless we must acknowledge the fact that a diurnal effect may be relevant and very important for wind farms in other locations or countries.

The models in Equation (23) are regression models, and hence, in order to predict their responses, $q_t^{(\alpha_i)}$, we need predictions for their explanatory variables. These are just lagged versions of the original wind power series, and the different variability indices. All of these explanatory variables have similar characteristics as they result from the original wind power series. The lagged versions of the wind power series are certainly non-stationary and all 4×625 different realizations of the variability indices (for every wind farm), even though they can be much smoother (for large values of m, n) than the original wind power series, are also non-stationary.

The predictions (point forecasts) of the explanatory variables are produced using ARIMA and ARIMA (in mean)-GARCH (in variance) models. By modelling the mean of the series using an ARIMA model, we allow for its non-stationary nature, and by modelling the variance using a GARCH process we allow for its heteroskedastic nature. Due to the fact that the wind power series (and the resulting variability indices) is bounded and does not follow any known parametric distribution, one may argue that an ARIMA or an ARIMA-GARCH model may not be appropriate. A modified (This version of ARIMA-GARCH model limits the forecasts to be bounded between two specific values (zero and one in our case) ARIMA/ARIMA-GARCH model with limiter (as proposed by Chen *et al.* [32]) is used to deal with the problem of the data being bounded. Moreover, the empirical density of the differenced series is close to a Student's *t*-distribution density. Hence, we fit an ARMA/ARMA-GARCH model to the transformed series, (w_t) (or differenced variability index), assuming those data come from a Student's *t*-distribution whose parameters are estimated for each series. We incorporate this distributional assumption by assuming the resulting residual series (white noise) follows a Student's *t*-distribution.

The next step is to produce point forecasts from 15 min up to 6 h ahead ($k = 1, 2, ..., 24$), from each point of the in-sample testing set, by fitting ARIMA$(1, 1, 1)$ models to each realization of the four variability indices of the above regressions. Our choice of ARIMA$(1, 1, 1)$ model may seem unappealing and arbitrary, but was made mainly for simplicity after exploring the forecast performances of various time series models. Choosing the best ARIMA-GARCH model (according to AIC) for each of the 625 different realizations of each index (for each wind farm) is extremely computationally expensive and hence we have to make some simplifications in order to make our optimization process computationally feasible. An ARIMA(1,1,1) is able to capture the non-stationary nature of the indices, and avoid overfitting at the same time. In order to assess the goodness of the fits, we use the Ljung–Box test, and restrict our selection to the fits that do not reject the null hypothesis of this test (so the corresponding residuals are consistent with white noise).

Modelling the variance of the indices using ARCH/GARCH models (in combination with an ARIMA model for the mean) does not provide a consistent and significant improvement of the RMSE (We used the Root Mean Square Error to evaluate the point forecast performance of various time series models.) of the point forecasts. This is mainly because of the very small forecast horizon we have, and hence it suffices to use a simple ARIMA model with limiter. In order to produce point forecasts of the lagged wind power series, the model solution using AIC (results are also the same using BIC) identified an ARIMA$(0, 1, 2)$ - GARCH$(1, 1)$ model for the low variability farm, an ARIMA$(1, 1, 3)$ - GARCH$(1, 1)$ for the medium variability farm, and an ARIMA$(2, 1, 1)$ - GARCH$(1, 1)$ for the high variability farm. These models have the ability to capture the heteroskedastic effects that the wind power series have, taking into account the non-linear nature of the variations. Also, these forecasts are calculated only once for all different realizations of the quantile regression models, and hence there is no point in this case to sacrifice the (small) accuracy gain for simplicity and computational efficiency. Table 2 shows the selected time series models for each wind farm and the two tests that assess their fit.

Table 2. Best fitted models for the three wind power time series according to the AIC, with Ljung–Box and LM tests *p*-values.

Wind farms time series	Selected model based on AIC	LM test *p*-values for lags 5, 15, 25	LB test *p*-values for lags 5, 15, 25
Low Var.	ARIMA$(0, 1, 2)$-GARCH$(1, 1)$	1.00, 1.00, 1.00	0.87, 0.99, 1.00
Medium Var.	ARIMA$(1, 1, 3)$-GARCH$(1, 1)$	1.00, 1.00, 0.95	0.53, 0.98, 1.00
High Var.	ARIMA$(2, 1, 1)$-GARCH$(1, 1)$	1.00, 1.00, 1.00	1.00, 1.00, 1.00

After producing quantile forecasts for 24 different forecast horizons, we evaluate them (i) using the CFS of only the first step ahead forecasts; and (ii) using the CFS averaged over all forecast horizons. The results justify our inspection of better forecast performance for the models with small (smoothing and variability) moving windows. We repeat the above procedure by restricting the range of our parameters even more for each variability index, and sample every different combination of parameters from the range $m, n = \{0, 1, 2, ...50\}$.

We end up with distinct sets of parameters (for each model and wind farm) that minimize the averaged and 1-step ahead CFS of each different quantile. The CFS minimization results are shown in Tables 3–6. In general, we cannot distinguish any particular parameter pattern, but there are some features that are worth mentioning. For all the models, it is more common to have the smoothing window width (m) smaller than the variability window width (n), especially for quantiles less than or equal to the median. This pattern changes for the upper quantiles (larger than the median) where we do not observe a clear pattern. Also, on average, the parameters for the averaged over 24-steps ahead optimization are smaller than the corresponding ones of the 1-step ahead optimization.

Table 3. 1-step and averaged over 24-steps CFS optimization results for the SD model of Equation (23).

α_i	Low Var.		Med Var.		High Var.		Low Var.		Med Var.		High Var.	
	m	n	m	n	m	n	m	n	m	n	m	n
	1-step optimization of SD model						24-steps optimization of SD model					
0.05	0	9	0	18	2	7	2	5	3	3	2	7
0.10	0	4	0	20	2	7	0	6	3	2	2	4
0.15	0	4	0	21	0	7	0	4	2	2	3	2
0.20	0	4	0	21	0	7	3	2	2	2	2	2
0.25	2	3	0	21	2	4	2	2	0	3	2	2
0.30	2	3	2	4	2	4	0	2	0	2	0	2
0.35	2	3	2	2	0	7	0	2	0	2	0	2
0.40	5	3	2	2	5	3	0	2	0	2	0	3
0.45	5	3	0	4	6	3	0	2	0	2	7	0
0.50	28	2	6	3	9	0	0	0	7	0	7	0
0.55	14	2	0	0	2	2	0	0	0	0	0	3
0.60	0	8	13	2	2	2	0	0	0	12	0	2
0.65	0	8	2	11	2	2	5	3	0	3	0	2
0.70	0	9	0	12	3	2	5	3	0	2	0	2
0.75	0	9	0	12	0	2	5	3	2	2	2	2
0.80	0	15	0	9	0	2	3	2	2	2	2	2
0.85	0	8	0	12	0	2	2	3	3	2	3	2
0.90	0	9	2	9	0	3	0	12	3	2	3	2
0.95	2	8	0	12	0	10	3	7	0	15	0	14

Table 4. 1-step and averaged over 24-steps CFS optimization results for the IQR model of Equation Equation (23).

α_i	Low Var.		Med Var.		High Var.		Low Var.		Med Var.		High Var.	
	m	n	m	n	m	n	m	n	m	n	m	n
	1-step optimization of IQR model						24-steps optimization of IQR model					
0.05	2	4	0	9	0	5	2	6	2	4	2	4
0.10	0	4	0	5	0	4	2	3	3	2	2	3
0.15	0	3	0	5	0	4	2	3	2	2	3	2
0.20	0	4	0	5	0	4	3	2	2	2	2	2
0.25	2	3	0	4	0	4	2	2	0	3	2	2
0.30	2	3	0	4	2	3	2	2	0	2	0	2
0.35	2	3	2	2	5	3	0	2	0	2	0	2
0.40	5	3	2	2	5	3	0	2	0	2	0	3
0.45	5	3	0	3	6	3	0	2	0	2	7	0
0.50	28	2	6	3	9	0	0	0	7	0	7	0
0.55	14	2	0	0	2	2	0	0	0	0	0	3
0.60	11	2	13	2	2	2	0	0	0	7	0	2
0.65	2	4	0	4	2	2	5	2	0	3	0	2
0.70	0	11	0	7	3	2	5	3	0	2	0	2
0.75	0	11	0	7	0	2	5	2	2	2	2	2
0.80	0	11	0	7	0	2	3	2	2	2	2	2
0.85	0	11	0	2	0	2	2	3	3	2	3	2
0.90	0	11	0	12	0	3	0	9	3	2	3	2
0.95	2	7	0	12	4	2	5	4	0	7	0	4

Table 5. 1-step and averaged over 24-steps CFS optimization results for the Q05 model of Equation Equation (23).

	Low Var.		Med Var.		High Var.		Low Var.		Med Var.		High Var.	
	m	*n*	*m*	*n*	*m*	*n*	*m*	*n*	*m*	*n*	*m*	*n*
α_i	1-step optimization of Q05 model						24-steps optimization of Q05 model					
0.05	0	18	48	48	2	10	0	14	0	14	7	4
0.10	0	14	15	4	2	10	0	12	3	6	3	9
0.15	0	11	25	3	2	7	0	8	0	7	0	7
0.20	0	11	35	6	2	7	0	6	0	7	0	7
0.25	0	11	24	3	0	11	0	6	0	6	0	5
0.30	0	17	24	2	2	7	0	6	0	3	0	5
0.35	2	11	24	3	2	7	0	6	0	3	0	5
0.40	2	17	25	2	0	8	0	6	0	3	0	7
0.45	2	16	24	3	2	5	0	6	0	3	5	3
0.50	2	16	13	12	6	2	0	6	0	2	5	3
0.55	21	0	0	0	6	2	2	5	0	0	5	3
0.60	21	0	2	14	9	0	0	0	7	0	0	18
0.65	17	0	2	14	0	7	2	5	7	0	0	16
0.70	16	0	2	11	0	7	2	5	7	0	0	15
0.75	16	0	2	11	0	5	2	5	0	18	0	15
0.80	15	0	2	12	0	5	2	5	0	14	0	15
0.85	16	0	0	5	0	5	14	5	0	14	0	15
0.90	12	0	0	5	0	4	14	7	0	14	0	15
0.95	12	0	0	4	0	4	0	4	0	11	0	3

Table 6. 1-step and averaged over 24-steps CFS optimization results for the Q95 model of Equation (23).

	Low Var.		Med Var.		High Var.		Low Var.		Med Var.		High Var.	
	m	*n*	*m*	*n*	*m*	*n*	*m*	*n*	*m*	*n*	*m*	*n*
α_i	1-step optimization of Q95 model						24-steps optimization of Q95 model					
0.05	0	4	0	6	0	4	0	4	0	4	0	8
0.10	0	6	0	6	0	4	0	6	0	5	0	5
0.15	0	6	0	6	0	5	0	6	0	5	0	5
0.20	0	8	0	6	0	5	0	10	0	10	0	8
0.25	0	9	2	18	0	7	0	10	0	10	0	8
0.30	0	10	2	13	0	7	0	10	5	19	2	8
0.35	0	45	2	13	0	7	17	0	3	22	2	8
0.40	0	45	12	14	8	0	4	6	6	3	3	7
0.45	21	0	16	11	8	0	4	6	6	3	6	2
0.50	21	0	32	3	9	0	4	6	5	3	6	2
0.55	0	18	0	0	9	0	0	0	5	3	0	6
0.60	0	18	0	0	0	10	3	7	5	3	0	6
0.65	0	11	0	16	0	9	2	8	5	3	0	4
0.70	2	11	2	9	0	9	3	7	3	5	0	4
0.75	0	15	2	9	0	11	2	8	3	5	0	4
0.80	0	15	2	9	0	11	0	12	3	5	0	6
0.85	3	7	2	9	0	11	0	12	3	5	0	6
0.90	3	7	2	10	2	11	0	12	3	6	0	10
0.95	2	13	3	10	2	11	2	11	2	9	0	14

6. Out-of-Sample Forecast Performance Results

In this section, after fitting the four optimized models of Equation (23) to the whole two years in-sample learning set (for each farm), we will produce quantile forecasts from 15 min up to six hours ahead from each point of the out-of-sample forecasting set, and assess their forecast performance using the CFS, and the CRPS. In short, the CFS will be used to assess the skill of individual quantile forecasts, and the CRPS to assess the skill of the density forecasts (produced by using all 19 quantile forecasts).

In order to facilitate the comparison of forecast performance across different models, we will introduce two widely used *probabilistic benchmarks*:

- *Persistence distribution*: It is defined as the distribution of the last n observations. The persistence benchmark is independently optimized (by estimating n) for each wind farm, by using the same optimization methods as for the variability indices: 1-step ahead CFS minimization, and averaged over 24-steps CFS minimization. So, when the persistence is optimized using one of the two CFS minimization methods, different values of n are chosen to forecast each quantile.
- *Unconditional distribution*: We construct this benchmark by using all the past observations of the time series. This benchmark assumes that the time ordering of the observations is not relevant when attempting to predict the distribution of the response. It is also referred to as *climatology*.

The third benchmark used in this article is the quantile regression model with only the three lags of wind power series as explanatory variables. This benchmark will help us to identify the gain in forecast performance acquired by using the four variability indices and is defined as the *3-lagged series* benchmark.

Predictive distributions are often taken to be Gaussian even though the wind power series is bounded and non-negative. Moreover, in our record of wind power measurements we have values of exactly 0 and 1 and hence the predictive distributions may require point masses at 0 and 1. A convenient way to embed this property is through the use of cut-off normal predictive distributions as achieved by Sanso and Guenni [33], Allcroft and Glasbey [34], Gneiting *et al.* [35] and Pinson [10]. The fourth benchmark of this article uses a cut-off normal predictive density, $\mathcal{N}^{0,1}\left(\mu_{t+k|t}, \sigma^2_{t+k|t}\right)$, and a fitted diurnal trend component to the three wind power series. The parameters $\mu_{t+k|t}$ and $\sigma_{t+k|t} > 0$ for $k = 1, ..., 24$ are called the location parameter (or predictive centre) and scale parameter (or predictive spread) of the cut-off normal density with point masses at 0 and 1. Please note that a truncated normal predictive distribution (with cut-offs at 0 and 1) has also been considered, with results very similar but worse than those of the cut-off normal predictive distribution benchmark.

The procedure to construct the fourth benchmark used in this article (also described in Gneiting *et al.* [35] and Gneiting *et al.* [8]) is as follows. At each of the three sites we firstly fit a trigonometric function,

$$y_t = a_0 + a_1 \sin\left(\frac{2\pi d(t)}{96}\right) + a_2 \cos\left(\frac{2\pi d(t)}{96}\right) + a_3 \sin\left(\frac{4\pi d(t)}{96}\right) + a_4 \cos\left(\frac{4\pi d(t)}{96}\right) \qquad (24)$$

where y_t represents the normalised wind power for each farm at time t, and $d(t)$ is a repeating function that numbers the time variable (in 15 min steps) from 1 to 96 within each day. We then remove the ordinary least square (OLS) fit from each wind power series and use the resulting residual series, denoted by ϵ^r_t, to determine the predictive centre and predictive mean of the cut-off normal predictive distribution.

More specifically we introduce the following linear autoregressive system

$$\epsilon^r_t = b_0 + b_1 \epsilon^r_{t-1} + b_2 \epsilon^r_{t-2} + b_3 \epsilon^r_{t-3} \qquad (25)$$

and use this to determine the forecasts $\hat{\epsilon}^r_{t+k|t}$ in a straightforward way, for each $k = 1, ..., 24$ (from 15 min up to 6 h ahead). Then, the predictive centre of the cut-off normal distribution is modelled as

$$\mu_{t+k|t} = \hat{y}_{t+k|t} + \hat{\epsilon}^r_{t+k|t} \qquad (26)$$

where $\hat{y}_{t+k|t}$ is the forecast issued at time t with forecast horizon k for the fitted diurnal trend of Equation (24).

Finally, in order to model the predictive spread we introduce, following Gneiting *et al.* [8], the volatility function at time *t*:

$$v_t = \left(\frac{1}{2} \sum_{i=0}^{1} (y_{t-i} - y_{t-i-1})^2 \right)^{\frac{1}{2}} \qquad (27)$$

So this benchmark allows for conditional heteroskedasticity by modelling

$$v_t = c_0 + c_1 v_{t-1} \qquad (28)$$

and setting the predictive spread as the forecast of v_t issued at time t for a forecast time $t + k$:

$$\sigma_{t+k|t} = \hat{v}_{t+k|t}. \qquad (29)$$

These four benchmarks will be used as the reference models mentioned in Section 4.2. In the following tables we will present the evaluation results of the four models, for each evaluation criterion and optimization type. As the relative performances of the methods are similar for each of the three locations, following Taylor *et al.* [6], we will present the averaged results over the three wind farms. Moreover, we will present only the Skill and Average Skill Scores of each evaluation criterion, as we are particularly interested to quantify and statistically test (using the Anisano–Giacomini test [30]) the relative increase in forecast performance of the four competing models with respect to the four benchmarks (reference models).

6.1. Out-of-Sample Model Comparison and Evaluation-Quantile Forecasting

In this subsection we compare the out-of-sample forecast performance of the competing models for each quantile and model optimization method. We have a total of 19 quantile forecasts for each model and for two different optimization methods. Please note that in order to avoid presenting any unnecessary information, we summarise the results on the forthcoming tables by including results of 11 out of 19 quantiles (0.05, 0.10, 0.20,...,0.80, 0.90, 0.95 quantiles). Firstly, we present the results obtained using the *1-step ahead CFS* optimization, followed by the results obtained using the *averaged over 24-steps CFS* optimization. For both optimization methods, the scores will be averaged over the three wind farms because the relative performance of the models is similar across the wind farms.

6.1.1. Quantile Forecasting: 1-Step Ahead CFS Optimization

Since the models in this subsection are optimized using a 1-step ahead CFS optimization method, it makes sense to present results for the first lead time only, for each quantile and for each model. Table 7 shows the Skill CFS (as defined by Equation (19)) of the best performing model among the four competing ones and its percentage gain/loss with respect to the four reference (benchmark) models, for each quantile. Moreover, the asterisks next to the scores indicate the level of statistical significance (obtained using the Amisano–Giacomini test of Section 4.2) of the corresponding gain/loss in performance with respect to the four reference models.

Table 7. The best performing model among the four competing ones, and its performance gain/loss with respect to the four reference (benchmark) models, for each quantile. Reference models: 3-lagged series (column 3), Cut-off normal (column 4), Persistence (column 5) and Climatology (column 6). These results are outcomes from a *1-step ahead CFS* optimization, and we use the CFS only for the first predicted step. The asterisks indicate the statistical significance of the gain/loss according to the Amisano and Giacomini test with the following significance codes for the p-value of the test: ***: $p \leq 0.01$, **: $0.01 < p \leq 0.05$, *: $0.05 < p \leq 0.1$.

		Skill CFS (%)			
α_i	Best model	3-lagged series	Cut-off normal	Persistence benchmark	Climatology benchmark
0.05	Q95	4.01***	88.04***	54.82***	65.63***
0.10	Q95	2.57***	81.70***	53.25***	71.89***
0.20	Q95	1.23**	73.88***	55.61***	78.14***
0.30	SD	0.81	69.53***	57.81***	81.56***
0.40	SD	0.22	67.07***	59.18***	83.70***
0.50	Q95	−0.21	65.93***	59.82***	85.18***
0.60	Q05	−0.15	65.84***	59.53***	86.17***
0.70	Q05	0.59	67.21***	58.62***	86.79***
0.80	SD	2.74***	71.12***	57.57***	87.20***
0.90	Q05	2.38***	78.00***	54.48***	86.46***
0.95	Q05	3.44***	84.64***	54.46***	85.05***

A general observation is that for almost all quantiles (except the 0.50–0.60 quantiles scores which have negative signs), the best forecast performance is achieved by one of the four competing models and not by the four benchmarks. The 0.05–0.10 and 0.90–0.95 quantiles form the two tails of the predictive density, and represent the rare events (such as ramps, cut offs) of a wind power series. As we observe from this table, both tails of the predictive density are quite well captured by the Q05 and Q95 models. Out of the four competing models, the lower tail of the predictive density is better predicted by the Q95 model and the upper by the Q05 model, but assuming the structure of the two variability indices used in these models, we might intuitively expect the opposite to happen.

This phenomenon can be explained by having a look at Figure 3. Figure 3(a) shows the probability density function (PDF) of the medium variability wind farm, together with the function values when the normalized wind power is equal to zero and one. Figure 3(b) shows an example of a wind power curve as presented by McSharry *et al.* [3]. On this plot we mark the "cut-in speed" (w_1), the "nominal speed" (w_2) and the "disconnection speed" (w_3). So, for very low wind speeds (<w_1) the wind power production is almost zero, for wind speeds greater than w_2 but less than w_3 the normalized wind power production is equal to one, and for wind speeds greater than w_3 the turbines shut down in order to prevent damage, and hence the wind power production falls again to zero. By combining these two plots, we can plot a rough estimate of the normalized wind power PDF versus the wind speed.

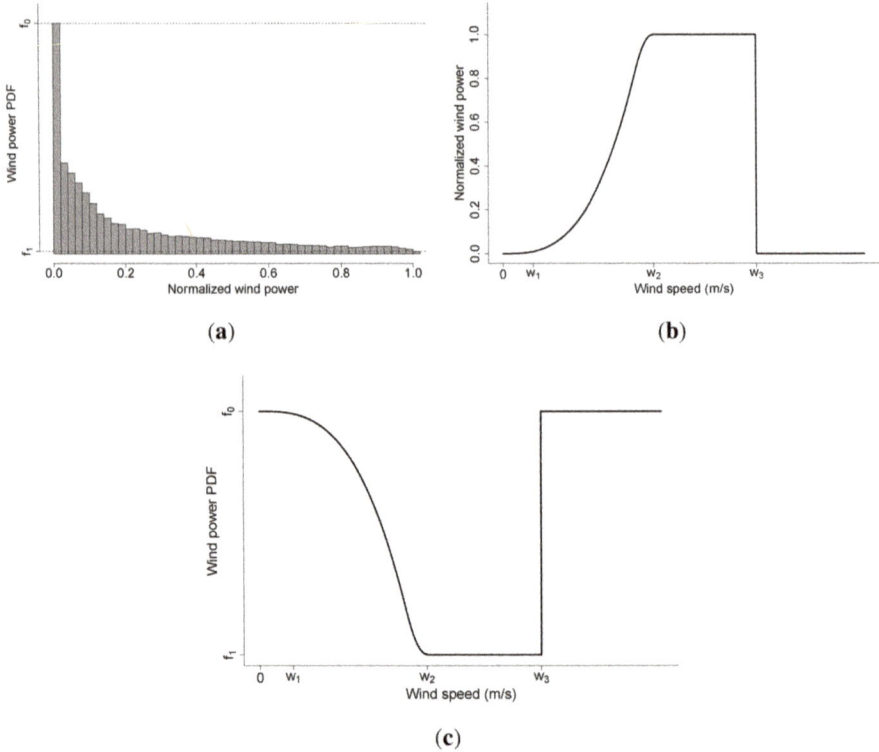

Figure 3. (a) Histogram of normalized wind power data (b) Deterministic power curve (c) Probability density function of normalized wind power data (PDF) *versus* wind speed. The equations used to reproduce (b) and (c) were taken from McSharry *et al.* [3].

We expect the 0.95 quantile of the unconditional density (not to be confused with the predictive density) to be close to the nominal (normalized) wind power value of one. But the produced wind power is driven by the actual wind speed at any given time, and hence falling from the nominal wind power production (one) to zero can happen unexpectedly (Exceeding w_3 can happen unexpectedly, given that we do not have any information about the wind speed at any given time.) if we exceed the disconnection speed w_3 (Figure 3(b)). This results in a sudden jump from the 0.95 quantile to the 0.05 quantile of the unconditional probability density and is represented by the lower tail of the predictive density. Hence, the Q95 index which captures this sort of events can provide some extra information about the lower tail of the predictive distribution that the 3-lagged series and the Q05 index do not describe. Similarly, if the wind speed falls below w_3, we are suddenly jumping from the 0.05 to the 0.95 quantile of the unconditional density and these kind of rare events (jumping from low to high values) are represented by the upper tail of the predictive density. Therefore, the Q05 index can provide some extra information about the upper tail of the conditional predictive distribution.

In addition, Table 7 shows that the strongest benchmark for all quantiles is the 3-lagged series model. The biggest and statistically significant improvement with respect to this benchmark is achieved near the tails of the predictive density, and decays as we move towards the median. This has important practical applications because it is exactly these extreme fluctuations that are of interest to transmission system operators (TSOs). More specifically, we get a performance gain of up to 4.01% (for the 0.05 quantile, achieved by the Q95 model), which is certainly not negligible. Unfortunately, the performance gain by using one of the competing models with respect to this benchmark in order to forecast the quantiles 0.30–0.70 is negligible (statistically insignificant), and is in the range of −0.21% to 0.81%. Furthermore, there is no gain for the quantiles close to the median of the predictive density (0.50–0.60).

Since the 3-lagged series model is our strongest benchmark and the performance gain with respect to it is only worth mentioning near the tails of the predictive density, it makes sense to focus on the performance gain achieved with respect to the last two benchmarks only for quantiles near the tails. Table 7 shows that the increase in forecast performance with respect to the cut-off normal benchmark is at least 78%, which shows that the cut-off normal is not capturing the tails of the predictive distribution as well as our competing models.

Moreover, we get more than 53.25% increase in the forecast performance with respect to the persistence benchmark when we use one of the four competing models. At the tails, where the Q05 and Q95 models are more suitable, we have a gain with respect to the persistence benchmark of up to 54.82%. By using the climatology benchmark as reference model, we observe that the maximum performance gain at the tails goes up to 86.46% (for the 0.90 quantile, achieved by the Q05 model), and in general the Q05 and Q95 models manage to maintain the performance gain (at the tails) above 65.63%.

6.1.2. Quantile Forecasting: Averaged over 24-Steps CFS Optimization

This subsection has similar structure to the preceding one, but now we present the results for the models which minimize the averaged (over six hours) CFS. Because the models are optimized on their forecast behaviour over all 24 forecast horizons, it makes sense to present results with the scores averaged over the 24 horizons, for each of the 11 selected quantiles.

Table 8 is analogous to Table 7, but here we provide the averaged over 24-steps CFS optimization results. It presents the Average Skill CFS (instead of Skill CFS) as defined by Equation (20). Once more, a general observation is that for almost all quantiles (except the first two and the median), the best forecast performance is achieved by our competing models and the strongest benchmark is the 3-lagged series model.

Table 8. The best performing model among the four competing ones, and its performance gain/loss with respect to the four reference (benchmark) models, for each quantile. Reference models: 3-lagged series (column 3), Cut-off normal (column 4), Persistence (column 5) and Climatology (column 6). These results are outcomes from an *averaged over 24-steps CFS* optimization, and the CFS are also averaged over all 24 forecast horizons. The asterisks indicate the statistical significance of the gain/loss according to the Amisano and Giacomini test with the following significance codes for the *p*-value of the test: ***: $p \leq 0.01$, **: $0.01 < p \leq 0.05$, *: $0.05 < p \leq 0.1$.

		Average Skill CFS (%)			
α_i	Best model	3-lagged series	Cut-off normal	Persistence benchmark	Climatology benchmark
0.05	Q95	0.71	43.82***	−102.14***	−103.89***
0.10	IQR	1.99**	31.81***	−29.08***	−29.91***
0.20	IQR	5.95***	22.46***	14.54***	21.00***
0.30	SD	4.57***	15.25***	18.86***	38.16***
0.40	IQR	2.56***	10.22***	20.47***	46.64***
0.50	Q95	−0.11	6.03***	20.13***	50.95***
0.60	Q05	0.07	5.27***	20.05***	53.81***
0.70	IQR	1.67***	7.15***	19.48***	54.67***
0.80	SD	3.71***	11.94***	17.86***	52.16***
0.90	SD	2.66***	19.55***	12.86***	38.12***
0.95	Q05	1.09**	28.78***	8.29***	11.94***

Table 8 also shows that the lower tail of the predictive density is quite poorly captured by our competing models, and the last two benchmarks (persistence, climatology) are performing much better than any other model. On the other hand, for all the other quantiles, the SD and IQR models have quite similar performances and manage to outperform all the benchmarks. Moreover, the performance gain by using one of the four competing models to forecast the quantiles near the median (0.50–0.60) is statistically negligible or does not exist. A final general observation is that, as mentioned for the previous optimization method, the 0.05 quantile is better predicted by the Q95 model and the 0.95 quantile by the Q05 model.

By using one of the SD or IQR models (which perform almost identically) we get a performance gain with respect to the 3-lagged series benchmark of up to 5.95% (for the 0.20 quantile, achieved by the IQR model), which is statistically significant with a *p*-value less than 0.001. In addition all the competing models are outperforming the cut-off normal model by at least 5.27% and attain the maximum increase in forecast performance near the tails of the predictive density (up to 43.82% achieved by the Q95 model for the 0.05 quantile).

Table 8 also shows that we have up to 20.47% (for the 0.40 quantile, achieved by the IQR model) increase of the forecast performance with respect to the persistence benchmark. The SD and IQR models maintain the gain over the persistence benchmark above 8.29% for all quantiles larger than 0.10. By using the climatology benchmark as reference model, we observe that the maximum performance gain goes up to 54.67% (for the 0.70 quantile, achieved by the IQR model), and in general the SD and IQR models can maintain the percentage performance gain with respect to the climatology benchmark above 11.94% for all quantiles larger than 0.10.

6.2. Out-of-Sample Model Comparison and Evaluation-Density Forecasting

In this subsection, we evaluate the out-of-sample density forecast performance of the competing models, for each optimization method. We will use the quantile forecasts obtained from each optimization method to reconstruct the whole predictive density, and assess its skill using the Skill CRPS or the Average Skill CRPS. Firstly, we present the results obtained using the *1-step ahead CFS* optimization, followed by the results obtained using the *averaged over 24-steps CFS* optimization. Moreover, because the relative performance of the models is similar across the wind farms, the scores will be averaged over the three wind farms.

6.2.1. Density Forecasting: 1-Step Ahead CFS Optimization

In this subsection, the models' forecast performance is optimized for only the first predicted step, so it makes sense to focus (initially) on the first lead time and present the out-of-sample Skill CRPS for the first step ahead.

Table 9 presents the out-of-sample Skill CRPS (%) for the *1-step ahead CFS* optimized models, together with significance codes for the Amisano–Giacomini test of equal forecast performance. This table shows that the best benchmark model is the 3-lagged series. That was expected because this benchmark was also the strongest one (for most quantiles) when we were looking at the quantile forecast results for the same optimization method (Section 6.1.1). The SD and IQR models behave almost identically and manage to outperform all the other benchmarks. The SD model performs slightly better than the IQR model, and managed to outperform the 3-lagged series model by 1%, the cut-off normal model by 1.48%, the persistence benchmark by 58.38% and the climatology benchmark by 84.23%.

Table 10 shows the best performing model among the four competing ones, and its performance gain/loss with respect to the four reference (benchmark) models, for a collection of forecast horizons. For simplicity, we present the results for seven of the 24 forecast horizons. The SD model is outperforming the 3-lagged series for the first 16 forecast horizons (except for the second one) where the improvements in forecast performance are also statistically significant for a 90% significance level. For the second forecast horizon we get the maximum forecast performance gain over the 3-lagged series model (equal to 1.96%) achieved by the IQR model. The SD model also manages to outperform the cut-off normal benchmark for all forecast horizons, with all improvements in forecast performance being statistically significant for a 99% significance level.

Table 9. Out-of-sample Skill CRPS (%) (averaged over all wind farms) for the *1-step ahead CFS* optimized models. The scores are just for the first lead time. The asterisks indicate the statistical significance of the gain/loss according to the Amisano and Giacomini test with the following significance codes for the *p*-value of the test: ***: $p \leq 0.01$, **: $0.01 < p \leq 0.05$, *: $0.05 < p \leq 0.1$.

Performance Gain/Loss - Skill CRPS (%)				
Reference model	SD model	IQR model	Q05 model	Q95 model
3-lagged series	1.00**	0.93**	0.29	0.20
Cut-off normal	1.48***	1.41***	0.77*	0.69*
Persistence	58.38***	58.35***	58.08***	58.04***
Climatology	84.23***	84.22***	84.12***	84.11***

Table 10. The best performing model among the four competing ones, and its performance gain/loss with respect to the four reference (benchmark) models, for forecast horizon *k* (measured in 15 min steps). Reference models: 3-lagged series (column 3), Cut-off normal (column 4), Persistence (column 5) and Climatology (column 6). These results are outcomes from a *1-step ahead CFS* optimization. The asterisks indicate the statistical significance of the gain/loss according to the Amisano and Giacomini test with the following significance codes for the *p*-value of the test: ***: $p \leq 0.01$, **: $0.01 < p \leq 0.05$, *: $0.05 < p \leq 0.1$.

Skill CRPS(%)					
k	Best model	3-lagged series	Cut-off normal	Persistence	Climatology
1	SD	1.00**	1.48***	58.38***	84.23***
2	IQR	1.96***	6.79***	44.59***	76.78***
3	SD	1.81***	13.02***	37.00***	71.34***
4	SD	1.71***	14.09***	31.97***	66.73***
8	SD	1.15***	13.95***	20.88***	51.62***
16	SD	0.63*	15.12***	10.94***	27.55***
24	SD	0.40	15.70***	5.87***	8.24***

When the persistence and climatology benchmarks are used as reference models, Table 10 shows that the gain in forecast performance by using the SD model is at least 5.87% and 8.24%, respectively. Moreover, the noted density forecast improvements are statistically significant for all forecast horizons, for a 99% level of significance.

In addition to the above results, we carried out a marginal calibration analysis and investigated how the CRPS evolves conditional to some wind power levels. More specifically, Table 11 presents the marginal Skill CRPS (%) conditional to the normalized wind power being ≤ 0.20 or ≥ 0.80, for a collection of seven forecast horizons. We choose to focus on these specific wind power levels, because these form the two tails of the unconditional wind power density (not to be confused with the predictive density).

Table 11. The best performing model (according to the Marginal Skill CRPS) among the four competing ones, and its performance gain/loss with respect to the four reference (benchmark) models, for forecast horizon k (measured in 15 min steps). Reference models: 3-lagged series (column 3), Cut-off normal (column 4), Persistence (column 5) and Climatology (column 6). These results are outcomes from a *1-step ahead CFS* optimization. The asterisks indicate the statistical significance of the gain/loss according to the Amisano and Giacomini test with the following significance codes for the p-value of the test: ***: $p \leq 0.01$, **: $0.01 < p \leq 0.05$, *: $0.05 < p \leq 0.1$.

k	Best model	3-lagged series	Cut-off normal	Persistence	Climatology
\multicolumn{6}{c}{Marginal Skill CRPS(%) conditional to the normalized wind power being ≤ 0.20}					
1	Q05	1.32***	3.44***	60.21***	84.09***
2	IQR	1.81***	7.44***	44.92***	75.51***
3	IQR	1.83***	11.68***	36.81***	69.40***
4	IQR	1.63***	13.28***	31.48***	64.29***
8	IQR	1.04**	14.71***	20.74***	47.64***
16	IQR	0.56	17.94***	11.58***	20.09***
24	IQR	0.36	19.27***	6.40***	−4.30***
\multicolumn{6}{c}{Marginal Skill CRPS(%) conditional to the normalized wind power being ≥ 0.80}					
1	IQR	4.95***	1.31***	57.74***	93.52***
2	IQR	3.04***	13.51***	48.16***	91.11***
3	SD	3.44***	18.99***	44.76***	89.57***
4	SD	2.40***	22.23***	41.11***	87.88***
8	SD	2.25***	19.97***	32.79***	81.42***
16	SD	1.60***	17.60***	23.59***	68.09***
24	SD	1.75***	15.42***	15.47***	53.93***

Given that the normalized wind power is less than or equal to 0.20, the IQR seems to be the best performing model for all except the first forecast horizon (where the Q05 model is performing better). For small forecast horizons we observe statistically significant improvements over all competing models. These improvements (with the exception of the cut-off normal benchmark) are getting smaller as we move to larger forecast horizons, which is perfectly reasonable because the results are the outcome of a 1-step ahead optimization. Conditioning on power levels which belong to the upper tail of the unconditional wind power density, we observe that the IQR and SD models seem to provide the largest performance gain according to the CRPS. These two models are outperforming all the benchmarks with improvements that are also statistically significant for all forecast horizons, for a 99% level of significance.

6.2.2. Density Forecasting: Averaged over 24-Steps CFS Optimization

Now we would like to assess the out-of-sample density forecast performance of the four competing models, for the *averaged over 24-steps CFS* optimization method. Our assessment criterion will be the out-of-sample Skill CRPS or Average Skill CRPS.

Initially, it makes sense to have a look at the out-of-sample Average Skill CRPS (Equation (20)) with the four benchmarks as reference models (Table 12). The IQR model outperforms the 3-lagged series benchmark by 2.45%, a considerably larger improvement than for the 1-step ahead results given in Table 9. This model also outperforms the cut-off normal benchmark by 16.13%, the persistence benchmark by 12.77% and the climatology benchmark by 39.15%. Moreover, the density forecast performance of the SD model is quite close to that of the IQR model.

Table 12. Out-of-sample Average Skill CRPS (%) (also averaged over all wind farms) for the *averaged over 24-steps CFS* optimized models. The asterisks indicate the statistical significance of the gain/loss according to the Amisano and Giacomini test with the following significance codes for the *p*-value of the test: ***: $p \leq 0.01$, **: $0.01 < p \leq 0.05$, *: $0.05 < p \leq 0.1$.

Performance Gain/Loss - Average Skill CRPS (%)				
Reference model	SD model	IQR model	Q05 model	Q95 model
3-lagged series	2.37***	2.45***	−0.07	0.42
Cut-off normal	16.06***	16.13***	13.96***	14.38***
Persistence	12.70***	12.77***	10.51***	10.96***
Climatology	40.91***	40.96***	39.43***	39.73***

Since our optimization considers all 24 forecast horizons, it will be interesting to investigate how the four competing models perform in producing density forecasts for each forecast horizon, *k*, from 15 min up to 6 h ahead. As for the 1-step ahead optimization case, we present the results for only a collection of seven out of 24 forecast horizons. The best competing model together with the performance gain obtained for each forecast horizon *k* with respect to the four benchmarks can be found in Table 13. Clearly, the best performing benchmark is the 3-lagged series model, and the IQR is the best performing model out of the four competing models.

The competing models' performances are disappointing for the first lead time, where the 3-lagged series benchmark offers a performance gain (of at least 1.57%) with respect to these models. On the other hand, for predictions larger than 30 minutes ahead (second predicted step), Table 13 shows that the IQR model manages to maintain the gain in density forecast performance with respect to the 3-lagged series model above 2.14%, with a recorded maximum of 3.59% (achieved at the fourth predicted step). Moreover, all the scores (except the first two) produce *p*-values which give strong evidence to reject the null hypothesis of equal forecast performance between the competing and reference model. Hence, the observed gain in forecast performance is statistically significant for a 99% significance level. The gain in forecast performance with respect to the cut-off normal model is at least 4.96% (excluding the first lead time) and attains a maximum of 17.18% for the 24th predicted step.

Table 13. The best performing model among the four competing ones, and its performance gain/loss with respect to the four reference (benchmark) models, for forecast horizon *k* (measured in 15 min steps). Reference models: 3-lagged series (column 3), Cut-off normal (column 4), Persistence (column 5) and Climatology (column 6). These results are outcomes from an *averaged over 24-steps CFS* optimization. The asterisks indicate the statistical significance of the gain/loss according to the Amisano and Giacomini test with the following significance codes for the *p*-value of the test: ***: $p \leq 0.01$, **: $0.01 < p \leq 0.05$, *: $0.05 < p \leq 0.1$.

			Skill CRPS(%)		
k	Best model	3-lagged series	Cut-off normal	Persistence	Climatology
1	Q95	−1.57***	−1.07**	59.08***	83.82***
2	IQR	0.03	4.96***	44.84***	76.32***
3	IQR	3.28***	14.33***	38.53***	71.77***
4	IQR	3.59***	15.74***	33.21***	67.37***
8	IQR	3.34***	15.86***	20.02***	52.69***
16	IQR	2.59***	16.79***	6.80***	28.98***
24	IQR	2.14***	17.18***	-0.44	9.85***

If we consider the persistence benchmark as the reference model (column 4 of Table 13), we note that the Skill CRPS of the best model starts at 59.08% (Q95 model) and then decays to meet approximately the performance of the persistence benchmark for the last forecast horizon. When the climatology benchmark is used as a reference model (column 5 of Table 13), we again observe a decay of the skill scores, with the performance gain remaining above 9.85% for all forecast horizons (for the IQR model).

From the results presented we conclude that this optimization method is found to produce models (mainly the IQR model) that can substantially outperform the density forecast performance of the widely used benchmarks (persistence, climatology) and fully parametric models such us the cut-off normal benchmark. Moreover the gain used by including a variability index such as the (IQR) improves considerably the performance (up to 3.59%) of a quantile regression model which uses only autoregressive terms as explanatory variables (3-lagged series benchmark).

Finally, as for the 1-step ahead optimization case, we present some marginal calibration analysis results by investigating how the CRPS evolves conditional to some wind power level. Table 14 presents the marginal Skill CRPS(%) conditional to the normalized wind power being ≤ 0.20 or ≥ 0.80, for a collection of seven forecast horizons.

Given that the normalized wind power is less or equal to 0.20, the IQR model is outperforming all the benchmarks for forecast horizons larger than two steps ahead (except the last forecast horizon of the persistence and climatology benchmarks). The Q05 seems to be the best performing model for the first two steps ahead, but still cannot outperform the 3-lagged series benchmark for the first step ahead. The second part of this table shows that, given normalized power levels greater or equal to 0.80, the SD model is the overall best model among all the others. It manages to outperform all the benchmarks for all forecast horizons, with improvements that are also statistically significant using a 99% level of significance.

Table 14. The best performing model (according to the Marginal Skill CRPS) among the four competing ones, and its performance gain/loss with respect to the four reference (benchmark) models, for forecast horizon k (measured in 15 min steps). Reference models: 3-lagged series (column 3), Cut-off normal (column 4), Persistence (column 5) and Climatology (column 6). These results are outcomes from an *averaged over 24-steps* CFS optimization. The asterisks indicate the statistical significance of the gain/loss according to the Amisano and Giacomini test with the following significance codes for the p-value of the test: ***: $p \leq 0.01$, **: $0.01 < p \leq 0.05$, *: $0.05 < p \leq 0.1$.

k	Best model	3-lagged series	Cut-off normal	Persistence	Climatology
	Marginal Skill CRPS(%) conditional to the normalized wind power being ≤ 0.20				
1	Q05	−0.13	2.02***	59.10***	83.86***
2	Q05	1.20***	6.87***	42.75***	75.36***
3	IQR	3.45***	13.14***	34.89***	69.90***
4	IQR	3.93***	15.32***	29.05***	65.13***
8	IQR	3.81***	17.10***	15.50***	49.10***
16	IQR	3.12***	20.06***	1.77***	22.15***
24	IQR	2.71***	21.18***	−6.61***	−1.84***
	Marginal Skill CRPS(%) conditional to the normalized wind power being ≥ 0.80				
1	SD	6.27***	2.68***	66.58***	93.61***
2	SD	4.92***	15.18***	58.01***	91.28***
3	SD	1.78***	17.60***	52.60***	89.39***
4	SD	1.33***	21.38***	49.18***	87.75***
8	SD	1.95***	19.73***	39.04***	81.36***
16	SD	1.95***	17.89***	27.42***	68.20***
24	SD	2.13***	15.75***	17.99***	54.11***

7. Conclusions

In this paper we showed how to produce wind power quantile and density forecasts, for lead times from 15 minutes up to six hours ahead, using three different univariate wind power series. This was achieved by introducing innovative variability indices, which are able to capture the volatile behaviour of the wind power series.

We used linear (in parameters) quantile regression as our main tool for producing quantile forecasts for 19 different quantiles, with three lagged versions of the wind power series as the main explanatory variables. Four models were proposed, each one having as a fourth explanatory variable one of the four extracted variability indices.

In order for the final results to be consistent, we used data from three wind farms in Denmark, each one chosen to have different wind power variability (low, medium and high). We investigated four years of wind power data, with a 15 min resolution, for each wind farm. The first two years were used for estimating the parameters of the models, and the final two years for out-of-sample forecast evaluation.

All four quantile regression models were optimized using the in-sample training data set, in order to find their specific set of indices' parameters, (m, n), which minimizes (i) the first lead time CFS and (ii) the Average CFS over all forecast horizons, for each individual quantile.

Our main goal was to evaluate how well these models performed compared with the cut-off normal, persistence and unconditional distribution (climatology) probabilistic benchmarks. It is worth mentioning that persistence is a strong yet simple benchmark for very short forecast horizons, and was optimized using the same cost (optimization) functions as the four regression models. The use of a cut-off normal benchmark provided a good comparison between a fully parametric model (as the cut-off normal model) and the non-parametric quantile regression models used in this article.

The fourth and strongest benchmark used was a quantile regression model with three lags of the original series as explanatory variables. The comparison of the competing models with this benchmark provides evidence of how useful our extracted variability indices are for forecasting wind power production. The individual (out-of-sample) quantile forecasts were evaluated using the Skill or Average Skill CFS for direct comparison between the competing models and the benchmarks. The density forecasts of the models were evaluated using the Skill or Average Skill CRPS.

In the following we summarize the quantile and density forecasts results found using the two different types of model optimization:

Quantile forecasting: 1-step ahead CFS optimization

- The best competing models are the Q05 and Q95 models, which outperform our best benchmark (3-lagged series) by a maximum of 3.44% (0.95 quantile) and 4.01% (0.05 quantile), respectively.
- The largest gain in performance with respect to the best benchmark is noticed when forecasting the quantiles which form the tails of the conditional predictive density. In addition, the Q05 model performs better for the upper tail, and the Q95 model for the lower tail.
- The best quantile regression models for each forecast horizon manage to maintain the performance gain with respect to the cut-off normal, persistence and climatology benchmarks above 65.73%, 53.25% and 65.63%, respectively.

Quantile forecasting: Averaged over 24-steps CFS optimization

- The SD and IQR models have the best quantile forecast performance, with similar CFS. They manage to maintain the performance gain with respect to the best benchmark (3-lagged series) above 1.99% for 11 out of 19 quantiles. The maximum Skill CFS is 5.95%, and is achieved by the IQR model for the 0.20 quantile.
- The SD and IQR models maintain the performance gain with respect to the cut-off normal, persistence and climatology benchmarks above 5.25%, 12.86% and 21.00%, respectively, for 15 out of 19 quantiles. The performance gain by using one of the two quantile models over the persistence and climatology benchmarks is much lower (or does not exist) for predicting the tails (0.05, 0.10,0.90, 0.95 quantiles) than for predicting the quantiles close to the median of the conditional density.

Density forecasting: 1-step ahead CFS optimization

- The best competing model is the SD model, which has almost equal density forecast performance with the IQR model. It manages to outperform the best benchmark (3-lagged series) by 1.00% (improvement which is statistically significant for a 95% significance level), for the first lead time. All four competing models manage to outperform the cut-off normal, persistence and climatology benchmarks by at least 0.69%, 58.04% and 84.11%, respectively, for the first lead time.
- Across all 24 forecast horizons, the average gain in forecast performance using the SD or IQR model with respect to the best benchmark is statistically significant (using a 90% significance level) for the first 16 forecast horizons. Moreover, these two models manage to outperform the cut-off normal persistence and climatology benchmarks by at least 1.48%, 5.87% and 8.24%, respectively.

Density forecasting: Averaged over 24-steps CFS optimization

- The IQR model is the best competing model, and manages to outperform the best benchmark (3-lagged series) by, on average (over all forecast horizons), 2.45%. It also outperforms the cut-off normal, persistence and climatology benchmarks by, on average, 16.13%, 12.77% and 40.96%, respectively.
- Across all 24 forecast horizons (excluding the first two lead times), the IQR model manages to maintain a performance gain over the best benchmark by more than 2.14%. Moreover, the noted improvements in density forecast performance are statistically significant for 22 out of 24 forecast horizons, for a 99% significance level.

Acknowledgments: The authors would like to thank Energinet.dk for data provision and support. This work has been partly supported by the European Commission under the SafeWind project (ENK7-CT2008-213740), Her Majesty's Government and an IBM Innovation Award.

References

1. Barton, J.; Infield, D. Energy storage and its use with intermittent renewable energy. *Energy Convers. IEEE Trans.* **2004**, *19*, 441–448. [CrossRef]
2. Boyle, G. *Renewable Electricity and the Grid: The Challenge of Variability*; Earthscan: London, UK, 2007.
3. McSharry, P.; Pinson, P.; Gerard, R. *Methodology for the Evaluation of Probabilistic Forecasts*, SafeWind Report. 2009.
4. Brown, B.G.; Katz, R.W.; Murphy, A.H. Time series models to simulate and forecast wind speed and wind power. *J. Appl. Meteorol.* **1984**, *23*, 1184–1195. [CrossRef]
5. Sanchez, I. Short term prediction of wind energy production. *Int. J. Forecast.* **2006**, *22*, 43–56. [CrossRef]
6. Taylor, J.W.; McSharry, P.E.; Buizza, R. Wind power density forecasting using ensemble predictions and time series models. *IEEE Trans. Energy Convers.* **2009**, *24*, 775–782. [CrossRef]
7. Moeanaddin, R.; Tong, H. Numerical evaluation of distributions in non-linear autoregression. *J. Time Series Anal.* **1990**, *11*, 33–48. [CrossRef]
8. Gneiting, T.; Larson, K.; Westrick, K.; Genton, M.G.; Aldrich, E. Calibrated probabilistic forecasting at the stateline wind energy center. *J. Am. Stat. Assoc.* **2006**, *101*, 968–979. [CrossRef]
9. Trombe, P.J.; Pinson, P.; Madsen, H. A general probabilistic forecasting framework for offshore wind power fluctuations. *Energies* **2012**, *5*, 621–657. [CrossRef]
10. Pinson, P. Very short-term probabilistic forecasting of wind power with generalized logit-normal distributions. *J. R. Stat. Soc. C* **2012**, *61*, 555–576. [CrossRef]
11. Pinson, P.; Kariniotakis, G. Conditional prediction intervals of wind power generation. *Power Syst. IEEE Trans.* **2010**, *25*, 1845–1856. [CrossRef]
12. Sideratos, G.; Hatziargyriou, N. Probabilistic wind power forecasting using radial basis function neural networks. *Power Syst. IEEE Trans.* **2012**, *27*, 1788–1796. [CrossRef]
13. Pinson, P.; Madsen, H. Ensemble-based probabilistic forecasting at Horns Rev. *Wind Energy* **2009**, *12*, 137–155. [CrossRef]

14. Lau, A.; McSharry, P. Approaches for multi-step density forecasts with application to aggregated wind power. *Ann. Appl. Stat.* **2010**, *4*, 1311–1341. [CrossRef]

15. Jeon, J.; Taylor, J.W. Using conditional kernel density estimation for wind power density forecasting. *J. Am. Stat. Assoc.* **2012**, *107*, 66–79. [CrossRef]

16. Koenker, R.; Bassett, G. Regression quantiles. *Econometrica* **1978**, *46*, 33–50. [CrossRef]

17. Bremnes, J. Probabilistic wind power forecasts using local quantile regression. *Wind Energy* **2004**, *7*, 47–54. [CrossRef]

18. Nielsen, H.; Madsen, H.; Nielsen, T. Using quantile regression to extend an existing wind power forecasting system with probabilistic forecasts. *Wind Energy* **2006**, *9*, 95–108. [CrossRef]

19. Moller, J.; Nielsen, H.; Madsen, H. Time-adaptive quantile regression. *Proc. Windpower* **2008**, *52*, 1292–1303. [CrossRef]

20. Pritchard, G. Short-term variations in wind power: Some quantile-type models for probabilistic forecasting. *Wind Energy* **2011**, *14*, 255–269. [CrossRef]

21. Davy, R.; Milton, J.; Russell, C.; Coppin, P. Statistical downscaling of wind variability from meteorological fields. *Bound.-Layer Meteorol.* **2010**, *135*, 161–175. [CrossRef]

22. Bossavy, A.; Girard, R.; Kariniotakis, G. Forecasting ramps of wind power production with numerical weather prediction ensembles. *Wind Energy* **2012**, *16*, 51–63. [CrossRef]

23. Gneiting, T. Quantiles as optimal point forecasts. *Int. J. Forecast.* **2011**, *27*, 197–207. [CrossRef]

24. Milligan, M.; Schwartz, M.; Wan, Y. Statistical Wind Power Forecasting Models: Results of U.S. Wind Farms. In Proceedings of 2004 American Meteorological Society Annual Meeting, Seattle, WA, USA, 11–15 January 2004.

25. Hyndman, R.J.; Fan, Y. Sample quantiles in statistical packages. *Am. Stat.* **1996**, *50*, 361–365.

26. Koenker, R.; D'Orey, V. Computing regression quantiles. *Appl. Stat.* **1987**, *36*, 383–393. [CrossRef]

27. Li, Y.; Zhu, J. L1-Norm Quantile Regression. *J. Comput. Graph. Stat.* **2008**, *17*, 1–23. [CrossRef]

28. Matheson, J.; Winkler, R. Scoring rules for continuous probability distributions. *Manag. Sci.* **1976**, *22*, 1087–1096. [CrossRef]

29. Gneiting, T.; Raftery, A. Strictly proper scoring rules, prediction, and estimation. *J. Am. Stat. Assoc.* **2007**, *102*, 359–378. [CrossRef]

30. Amisano, G.; Giacomini, R. Comparing density forecasts via weighted likelihood ratio tests. *J. Bus. Econ. Stat.* **2007**, *25*, 177–190. [CrossRef]

31. Akaike, H. A new look at the statistical model identification. *IEEE Trans. Autom. Control* **1974**, *19*, 716–723. [CrossRef]

32. Chen, P.; Pedersen, T.; Bak-Jensen, B.; Chen, Z. ARIMA-based time series model of stochastic wind power generation. *Power Syst. IEEE Trans. Power Syst.* **2010**, *25*, 667–676. [CrossRef]

33. Sanso, B.; Guenni, L. A nonstationary multisite model for rainfall. *J. Am. Stat. Assoc.* **2000**, *95*, 1089–1100.

34. Allcroft, D.J.; Glasbey, C.A. A latent gaussian markov random-field model for spatiotemporal rainfall disaggregation. *J. R. Stat. Soc. C (Appl. Stat.)* **2003**, *52*, 487–498. [CrossRef]

35. Gneiting, T.; Larson, K.; Westrick, K.; Genton, M.G.; Aldrich, E. *Calibrated Probabilistic Forecasting at the Stateline Wind Energy Center: The Regime-Switching Space-Time (RST) Method*; Technical Report; Department of Statistics, University of Washington: Seattle, WA, USA, 2004.

energies

MDPI

Article

Analysis of Similarity Measures in Times Series Clustering for the Discovery of Building Energy Patterns

Félix Iglesias * and Wolfgang Kastner

Automation Systems Group, Vienna University of Technology, Treitlstr. 1-3/ 4. Floor, Vienna A-1040, Austria;
k@auto.tuwien.ac.at
* Author to whom correspondence should be addressed; vazquez@auto.tuwien.ac.at;
Tel.: +43-1-58801-18320; Fax: +43-1-58801-18391.

Received: 22 November 2012; in revised form: 31 December 2012; Accepted: 11 January 2013;
Published: 24 January 2013

Abstract: Forecasting and modeling building energy profiles require tools able to discover patterns within large amounts of collected information. Clustering is the main technique used to partition data into groups based on internal and a priori unknown schemes inherent of the data. The adjustment and parameterization of the whole clustering task is complex and submitted to several uncertainties, being the *similarity* metric one of the first decisions to be made in order to establish how the distance between two independent vectors must be measured. The present paper checks the effect of similarity measures in the application of clustering for discovering representatives in cases where correlation is supposed to be an important factor to consider, e.g., time series. This is a necessary step for the optimized design and development of efficient clustering-based models, predictors and controllers of time-dependent processes, e.g., building energy consumption patterns. In addition, *clustered-vector balance* is proposed as a validation technique to compare clustering performances.

Keywords: clustering; time-series analysis; similarity measures; pattern discovery; building energy modeling; cluster validity

---

1. Introduction

The classification and modeling of buildings' energy behavior is a core point to improve several emerging applications and services. For instance, the existence of databases with building energy profiles in connection with BIM models (Building Information Modeling) points out to be a key factor to achieve more sustainable building designs as well as more energy efficient urban development [1] (the more accurate and realistic energy profiles are, the better building energy performance calculations become).

A different but obviously related application area is the electricity market, which claims solutions and proposals that bestow flexibility on it. Expected enhancements must allow to smooth the frequent peaks and imbalances that are detrimental to all links in the energy chain, from suppliers to users [2]. Within this scope, demand or consumption habits can be abstracted by energy models that lead us to customized, more effective and fair relationships between energy providers and customers [3,4]. As further examples, energy use models are also found relevant to enhance the exploitation of renewable energy sources [5], or to achieve smart grid operation enhancement [6].

In the introduced scenarios, buildings—or buildings' energy behaviors—are usually represented as time-based profiles or patterns to cluster. Indeed, the modeling and classification of building energy demand and consumption becomes one of the most representative application fields with regard to the *clustering of time arranged data*. As a general rule, in this scope clustering is commonly used to classify energy consumers [7], predict future energy demand [8,9], or detect distinguished, habitually undesired, behaviors (*i.e.*, outliers) [10].

In addition to the purposes referred before, the identification of building energy patterns is also useful to provide context awareness capabilities to home and building control systems. Actually, the present work is motivated by the search of reliability and accuracy in the design of clustering-based controllers, models and predictors that operate with time-related information, which play a very important role in the fields of home and building automation, e.g., [11,12]. The selection of the *building energy* case is due to the wide scope of its application and the fact that the presented experiments could also be conducted with publicly available data.

Therefore, looking for the improvement of clustering-based applications, this paper develops a novel cluster validation method—*clustered-vector balance*—to be the basis of sensitivity analysis for the adjustment of clustering parameters, metrics and algorithms. The selected parameter under test is the similarity measure used to establish resemblance between two isolated samples, as it is a determining factor to assume in time series clustering. Since cluster validity methods also use similarity measures to check clustering solutions, the undertaken task is submitted to bias and uncertainty. The conducted experiments try to cope with such a problem performing a set of tests where similarity distances for clustering, as well as for evaluation and validation, are repeatedly switched. In addition to cluster-vector balance, classic cluster validity techniques are also utilized, as well as evaluations using non-clustered data. The cross comparisons let us infer some *hypothesis* related to the usage of the selected similarity measures for model discovery in univariate time series clustering.

2. Embedding Clustering in Real Applications

The application of *clustering* or cluster analysis usually covers one or more of the following aims: *data reduction, hypothesis generation, hypothesis testing* or *prediction based on groups* [13]. Indeed, the problem scenario that clustering has to face can take thousands of different shapes, but they usually share a common problem description: given a certain amount of input data vectors, characterized by a set of features or variables, an unsupervised knowledge abstraction of the data is required in order to allow its classification and representation.

It means that we always begin from certain *ignorance* concerning how the available or potential set of data can be internally arranged or structured. On the other hand, clustering adjustment and parametrization demand a deep understanding of the problem nature and domain in order to overcome several significant uncertainties [14]. Otherwise, the blind application of clustering techniques leads to trivial, erroneous or inefficient solutions. Therefore, the more previous knowledge about the nature of data exists, the better the clustering solution will be. As you can see, it entails a certain circularity that emphasizes the complex background of clustering.

There are several works intended to support the design of clustering-based applications (e.g., [15]). In addition, it is rather common that the continuous refinement of such applications leads practitioners to progressively reach some better knowledge of the data nature, being *hypothesis generation* an almost unavoidable companion in the careful design of clustering-based processes.

Some of the difficulties or uncertainties to face in the clustering task involve the selection and adjustment of clustering criteria, clustering algorithms, initial number of clusters, the most suitable features, outlier definition and handling, proximity measures, validation techniques, *etc.* Among them, a basic question remains in the *proximity measure*, *i.e.*, similarity or resemblance between two independent vectors. *Euclidean distance* is *de facto* the most applied similarity metric and usually appropriate for applications that do not present directly or necessarily correlation among distinct features. However, time series clustering deploys vectors where the information are time arranged, thus considering correlation in the similarity measures points out to be suitable, or better, even leading to more accurate solutions.

3. Clustering for Pattern Discovery in Time Series

The task of clustering time series for pattern discovery has the aim to find out a set of model profiles or patterns that represent as faithfully as possible the original data set, in a way that every independent vector of this original data can be considered as one of the models submitted to *acceptable* deviations or drifts, or an *outlier* at the most.

The difference between time series and normal clustering is that, in the time series case, the shape of input vectors entails features that are arranged in time. Hence, in univariate time series an input vector is usually the succession of values that a certain variable takes throughout a specific time scope.

Clustering time series is usually tackled twofold: (a) *feature-based* or *model-based*, *i.e.*, previously summarizing or transforming raw data by means of feature extraction or parametric models, e.g., dynamic regression, ARIMA, neural networks [16]; so the problem is moved to a space where clustering works more easily; (b) *raw-data-based*, where clustering is directly applied over time series vectors without any space-transformation previous to the clustering phase. Several works concerning each kind of time series clustering are referred to in detail in [17].

Beyond the obvious loss of information due to *feature-based* or *model-based* techniques, they can also present additional drawbacks; for instance, the application-dependence of the feature selection, or problems associated to parametric modeling. On the other hand, characteristic drawbacks of *raw-data-based* approaches are: working with high-dimensional spaces (*curse of dimensionality* [18]), and being sensitive to noisy input data.

In any case, we focus on the *raw-data-based* option for two reasons: (1) conclusions and hypothesis can be more easily generalized for other behaviour modeling applications (e.g., individual or community profiles for energy, occupancy, comfort temperature, *etc.*); (2) this is the best option to clearly analyze correlated data in clustering. Indeed, selecting the correct distance measure able to evaluate correlation is the main difficulty in this kind of time series clustering.

4. Similarity Measures

We consider *similarity* as the measure that establishes an absolute value of resemblance between two vectors, in principle isolated from the rest of the vectors and without assessing the location inside the solution space.

Considering continuous features, the most common metric is the Euclidean distance:

$$d_E(\vec{x}, \vec{y}) = \sqrt{(\vec{x} - \vec{y})(\vec{x} - \vec{y})'} \tag{1}$$

Note that Euclidean distance is invariant when dealing with changes in the order that time fields/features are presented; it means that it is in principle blind to capture vector or feature correlation. For time series data comparison, where trends and evolutions are intended to be evaluated, or when the shape formed by the ordered succession of features (*i.e.*, the envelope) is relevant, similarity measures based on Pearson's correlation:

$$d_C(\vec{x}, \vec{y}) = 1 - \frac{(\vec{x} - \overline{\vec{x}})(\vec{y} - \overline{\vec{y}})'}{\sqrt{(\vec{x} - \overline{\vec{x}})(\vec{x} - \overline{\vec{x}})'}\sqrt{(\vec{y} - \overline{\vec{y}})(\vec{y} - \overline{\vec{y}})'}} \tag{2}$$

have also been widely utilized, although it is not free of distortions or problems [19]. Mahalanobis distance,

$$d_M(\vec{x}, \vec{y}) = \sqrt{(\vec{x} - \vec{y})C^{-1}(\vec{x} - \vec{y})'} \tag{3}$$

can be seen as an evolution of the Euclidean distance that takes into account data correlation. It utilizes the covariance matrix of input vectors C for weighting the features. Mahalanobis distance usually performs successfully with large data sets with reduced features, otherwise undesirable redundancies tend to distort the results [20].

An interesting measure specially addressed to time series comparison is the Dynamic Time Warping (DTW) distance [21]. This measure allows a non-linear mapping of two vectors by minimizing the distance between them. It can be used for vectors of different lengths: $\vec{x} = x_1, ..., x_i, ..., x_n$ and $\vec{y} = y_1, ..., y_j, ..., y_m$. The metric establishes an *n*-by-*m cost matrix* C which contains the distances (usually Euclidean) between two points x_i and y_j. A warping path $W = w_1, w_2, ..., w_K$, where $\max(m, n) \leq K < m + n - 1$, is formed by a set of matrix components, respecting the next rules:

- Boundary condition: $w_1 = C(1, 1)$ and $w_K = C(n, m)$;
- Monotonicity condition: given $w_k = C(a, b)$ and $w_{k-1} = C(a', b')$, $a \geq a'$ and $b \geq b'$;
- Step size condition: given $w_k = C(a, b)$ and $w_{k-1} = C(a', b')$, $a - a' \leq 1$ and $b - b' \leq 1$.

There are many paths that accomplish the introduced conditions; among them, the one that minimizes the warping cost is considered the DTW distance:

$$d_W(\vec{x}, \vec{y}) = \min \left(\sqrt{\sum_{k=1}^{K} w_k} \right) \tag{4}$$

The main drawback of the measure remains in the effort dedicated to the calculation of the path of minimal cost, in addition to the fact that, actually, it cannot be considered as a metric, *i.e.*, it does not accomplish the triangular inequality.

In the current work, we focus on these four general-purpose popular distances, in spite of the fact that there exist many additional similarity measures. A survey of distance metrics for time series clustering can be found in [17]. Other noteworthy options are the *cosine measure* [22], which is good for patterns with different or variable size or length; or Jaccard and Tanimoto similarity measures, that can also be intuitively understood as a combination of Euclidean distances and correlations assessed by means of the inner product [23].

5. Cluster Validation

We can say that the validation of the results obtained by a clustering algorithm tries to give us a measure about the level of success and correctness reached by the algorithm. Here, we differentiate two ways of checking clustering solutions:

- On one hand, we have *cluster validity* or *clustering validation methods*, which try to evaluate results according to mathematical analysis and direct observation of solutions based on the inherent characteristics owned by the input data set. In a way of speaking, it consists of *idealistic* analysis methods as they focus on the definition given to *a cluster* irrespective of the reason that lead us to deploy clustering (*i.e.*, the final application);
- On the other hand, sometimes clustering solutions can be benchmarked and checked directly by the application or an environment that simulates the application (entitled *clustering evaluation*). It is a *practical* (or engineering) approach, which mainly covers application-based tests. Here, generalizations are riskier; note that we carry corruption and deformations introduced by the application, the boundary conditions and the specific data used for testing.

In both cases, the value of such quantitative measures is always relative, it means that they "are only tools at the disposal of the experts in order to evaluate the resulting clustering" [13].

With regard to clustering validation, three different kinds of criteria are usually considered: *external criteria*, evaluations of how the solution matches a pre-defined structure based on a previous intuition concerning the data nature (e.g., the adjusted Rand index [24]); *internal criteria*, which evaluate the solution only considering the quantities and relationships of the vectors of the data set (e.g., proximity matrix); and *relative criteria*, carried out comparing clustering solutions where one or more parameters have been modified (e.g., cluster *silhouettes* [25]).

In [26], some of these validation methods are introduced, concluding that they usually work better when dealing with compact clusters. This reasoning yields an interesting point that remarks the uncertainty also related to cluster validity; *i.e.*, as it happens with clustering that usually imposes a structure on the input data, cluster validation methods also impose a rigid definition of *what a good cluster is* and develop their assessments according to this particular definition.

Uncertainties, commitments and discussions also appear concerning the foundations of the cluster validity measures, as they must fix some essential concepts. To refer some examples: how clusters must be represented, how to calculate the distance between two clusters, how to calculate the distance between a point and a cluster, or even which kind of metric must be used for the distance measurement. Beyond these aspects, there exists lot of work that compares clustering solutions by distinct techniques. To give some instances: in [27] clustering methods are benchmarked utilizing *Log Likelihood* and *classification accuracy* criteria. In [28], popular algorithms are analyzed from three different viewpoints: *clustering criteria*, or the definition of similarity; *cluster representation* and *algorithm framework*, which stands for the time complexity, the required parameters and the techniques of preprocessing. In [29] the criteria are mentioned "*stability* (Does the clustering change only modestly as the system undergoes modest updating?), *authoritativeness* (Does the clustering reasonably approximate the structure an authority provides?) and *extremity of cluster distribution* (Does the clustering avoid huge clusters and many very small clusters?)".

6. Clustered-Vector Balance

In this paper, we start on the definition of clustering provided by [30], *i.e.*, a group or cluster can be defined as a dense region of objects or elements surrounded by a region of low density. From here, a consequent step is to consider that any output group can be represented by a model individual (existent or nonexistent), which usually will correspond to the gravity center of the respective cluster, named centroid, discovered pattern, representative or model.

Our intended applications mainly use clustering for *pattern or representative discovery*, so we find suitable validity methods that focus on representativeness or give an important role to the representatives [31]. Therefore, we have developed a validity measure called *clustered-vector balance* (or simply *vector balance*) based on the *clustering balance* measurement introduced in [32]. The clustering balance measurement finds the ideal clustering solution when "intra-cluster similarity is maximized and inter-cluster similarity is minimized". In order to extend the comparison to partitioning clustering with other parameters under test in addition to the number of clusters, we introduce substantial modifications to the original equations.

In the clustered-vector balance validation technique, every solution is expressed by a *representative clustered-vector*, which takes Λ_v and Γ_v (intra-cluster and inter-cluster average distance per vector) as component values (Figure 1). The expressions for Λ_v and Γ_v rest as follows:

$$\Lambda_v = \frac{1}{n} \sum_{j=1}^{k} \sum_{i=1}^{n_j} e(p_i^{(j)}, p_0^{(j)}) \tag{5}$$

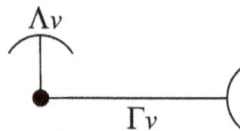

Figure 1. Symbol for a representative clustered-vector. The short segment with the concave arc stands for the average intra-cluster distance, the long segment with the convex arc for the average inter-cluster distance.

$$\Gamma_v = \frac{1}{k(k-1)} \sum_{j=1}^{k} \frac{n_j}{n} \sum_{l \neq j}^{k} e(p_0^{(j)}, p_0^{(l)}) \tag{6}$$

where n is the total number of input vectors, n_j stands for the vectors embraced in cluster j and k is the number of clusters. $p_i^{(j)}$ refers to the input vector i that belongs to cluster j, whereas $p_0^{(j)}$ is the centroid or representative of cluster j. $e(\vec{x}, \vec{y})$ stands for the error function or distance between the vectors $\vec{x}$ and $\vec{y}$. Note that the subindex v denotes the postscript "per vector".

The main differences with respect to [32] remain in the definition of Γ, which now is not related to the distance to an hypothetical global centroid, but to the distances among centroids, individually weighted according to each cluster population. In addition, Λ and Γ are now expressed in connection with a single, representative vector for the whole solution, and this makes both magnitudes comparable. Therefore, Λ_v is the average distance between a clustered vector and its centroid, whereas Γ_v is the average distance between a clustered vector to other clusters (more specifically, to other centroids).

Directly relating Λ_v and Γ_v can lead to doubtful, meaningless absolute indexes. In [32], authors introduce an α weighting factor to achieve a commitment between Λ and Γ. The parameter seems to be arbitrarily defined just to relate to both indexes, being adjusted to $1/2$ by default without providing an appropriate discussion. In our case, we can obviously expect that the best solutions will tend to show lower Λ_v and higher Γ_v, but the relationships among both values, their possible increments and the performance evaluation are not linear and have a high scenario-dependence. Since we lack a priori additional knowledge, the final clustered-vector balance index is proposed to be obtained by relating Λ_v and Γ_v using a previous Z-score transformation (*i.e.*, $z = \frac{x-\mu}{\sigma}$). Means and standard deviations of both Λ_v and Γ_v are obtained considering the total set of solutions to compare. Finally, the best solution maximizes:

$$\mathcal{E}_v(\mathcal{X}) = \Gamma_{v_z} - \Lambda_{v_z} \tag{7}$$

We no longer require α. However, we can consciously add it again if we have a previous biased opinion with respect to *what a good clustering solution is* according to the final application, *i.e.*, whether we want to favor solutions where clusters are compact or we prefer that they are as different/far as possible. Hence it would remain:

$$\mathcal{E}_v(\mathcal{X}) = \alpha \Gamma_{v_z} - (1-\alpha) \Lambda_{v_z} \tag{8}$$

7. Experiments

The conducted experiments have two main objectives:

- To check *clustered-vector balance* as a clustering validity algorithm by means of comparisons with other relative clustering validity criteria;
- To obtain a precedent for the selection of the most appropriate similarity metric for our application case—building energy consumption pattern discovery—which is a significant use case of time series clustering.

To do that, real cases are clustered using different similarity distances. Later on, each clustering solution is *validated* by means of different validation techniques (the similarity measure of the validation algorithm is switched as well). In addition, test vectors (selected at random and not processed by the clustering tool) are utilized to *evaluate* the representativeness of the main patterns of the cluster or centroids, measuring the average distance between the test vectors included in a cluster and the representative of the respective cluster [Equation (9)]. The evaluation also uses all of the diverse similarity distances under test.

$$V = \frac{1}{m} \sum_{j=1}^{k} \sum_{i=1}^{m_j} e(q_i^{(j)}, p_0^{(j)}) \tag{9}$$

m is the total number of vectors put aside for evaluation, m_j stands for the vectors embraced in cluster j. $q_i^{(j)}$ refers to the evaluation vector i that belongs to cluster j. The membership of the evaluation vectors is established according to the proximity to the found patterns p_0. e represents the distance used for evaluation.

In the trivial situation that all similarity measures affect the clustering solution in the same way, or in the hypothetical case that each distance is the most successful at finding a clustering solution with specific characteristics, we should expect that clustering carried out using a specific distance obtains the best results when the same distance has been used for validations or evaluations. Otherwise, we will have arguments to establish better and worse similarity measures for our specific application case.

7.1. Database

For the experiments, information concerning energy consumption of five university buildings has been collected. The buildings are located in Barcelona, Spain, and data cover hourly consumption from 29 August 2011 to 1 January 2012. Data is publicly available in (http://www.upc.edu/sirena). The selected buildings belong to the "Campus Nord", they are: "Edifici A1", "Edifici A4" and "Edifici A5" (university classrooms and laboratories), "Biblioteca" (a library) and "Rectorat" (an office building for administration and rectorship). In Table 1 and Figures 4–6, B1, B2, B3, B4 and B5 identify the presented buildings in the introduced order. The usable spaces of the buildings have the following dimensions: B1, 3966.59 m^2; B2, 3794.95 m^2; B3, 3886.12 m^2; B4, 6644.4 m^2; B5, 5927.21 m^2.

Each building presents 124 days of information, 100 days taken for training and for the cluster validity analysis, and 24 days employed in the evaluation. Input vectors are time series with 24-fields of hourly information concerning the energy consumption in kWh (Figure 2).

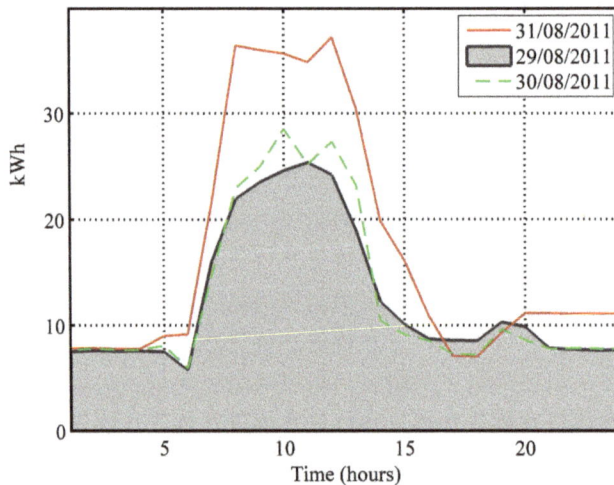

Figure 2. Example of three consecutive consumption days ("Rectorat").

Analysis prior to the clustering processes confirms notable data correlations in all the buildings. Table 1 displays, for each building, statistical data concerning correlation. Taking a daily profile at random, the values of the table show the number of other daily profiles of the same database with which the selected profile will present a Pearson's correlation index higher than 0.8 on average.

Table 1. Given a building, evaluation of the number of daily profiles ($\bar{x} \pm \sigma_x$) that keep $c \geq 0.8$ (Pearson's correlation index) with a daily profile selected at random.

B1	B2	B3	B4	B5
28.0 ± 22.3	31.2 ± 23.0	20.6 ± 15.9	81.6 ± 36.1	47.0 ± 18.1

7.2. Tests and Parameters

The similarity measures under test have been explained in Section 4, they are: (a) Euclidean distance, (b) Mahalanobis distance, (c) distance based on Pearson's correlation and (d) DTW distance. In the first step, the training data is processed by a Fuzzy clustering module that uses the FCM algorithm to compute clusters. As referred to above, the FCM algorithm uses the four distance measures to state vector proximity. In each case, the initial number of clusters has been fixed according to *clustering balance* and Mountain Visualization [33], as well as maintaining the final scenario purposes (*i.e.*, allowing a maximum of 8 energy consumption models).

Since all features correspond to the same phenomenon (electricity consumption), normalization is not carried out feature by feature, but based on the mean μ and standard deviation σ of the whole dataset (*i.e.*, a simple uniform scaling). Failing to ensure that all features move within similar ranges has been addressed as a problem for similarity measures like Euclidean distance, as "features with large values will have a larger influence than those with small values" [34]. In any case, for univariate time series we are confident that the multi-dimensional input space is not distorted and the relationship among features keep the same shape and proportionality.

The clustering solutions are *validated* using: (a) clustering balance with $\alpha = 1/2$ [32], (b) clustered-vector balance (Section 6), (c) Dunn's index [35], (d) Davies–Bouldin index [36], and *evaluated* by means of (e) Equation (9), which checks how representative discovered patterns are by means of data separated for testing.

Therefore, the test process results in: 5 builds. $\times$ 4 clust.(metrics) $\times$ 4 indices $\times$ 4 index(metrics) $=$ 320 validations/evaluations. With all the obtained outcomes the next comparisons are carried out: (a) best clustering solution (best validation), (b) best evaluation, (c) soundness of validation algorithms, and (d) best independent clusters.

The last point refers to the capability of finding good clusters (*i.e.*, dense, regular high similarity) irrespective of the global solution. The best clusters obtain minimum values in the next fitness function:

$$f_j = (1 - m_j) \times \Lambda_j \tag{10}$$

where m_j stands for the membership or amount of population embraced by cluster j (0: none; 1: all input samples) and Λ_j for the intra-similarity of cluster j. Clusters must overcome a membership threshold to be taken into account ($m_j \geq 0.08$, *i.e.*, at least 8% of total population). This limit is a trade-off value established according to the application purposes, which requires a minimum level of representativeness for the discovered patterns.

8. Results

The high number of generated indices leads us to condense results in a meaningful way in some figures and tables. We discuss the obtained findings in separated points.

8.1. Characteristics of the Scenario Under Test

Experiments face quite a demanding scenario where to identify clear clusters is not an easy task and the selection of the similarity measure affects the shape of obtained models. It is obvious when the solution patterns are compared, *e.g.*, Figure 3 shows the representative pattern corresponding to a specific discovered cluster according to every one of the clustering solutions in the case of building "Edifici A1". Note that patterns are similar in shape, but different enough to have a relevant influence in subsequent applications. For instance, a control system that uses the predicted patterns to adjust the supply of energy sources in advance would perform differently in each case, resulting in distinct levels of costs and resource optimization. Moreover, the patterns displayed in the figure represent a different percentage of the input population (Euclidean: 17%, Mahalanobis: 20%, Correlation: 13%, DTW: 24%).

In addition, the demanding nature of the problem is also noticeable in the disagreement detected by the validation techniques (see next point).

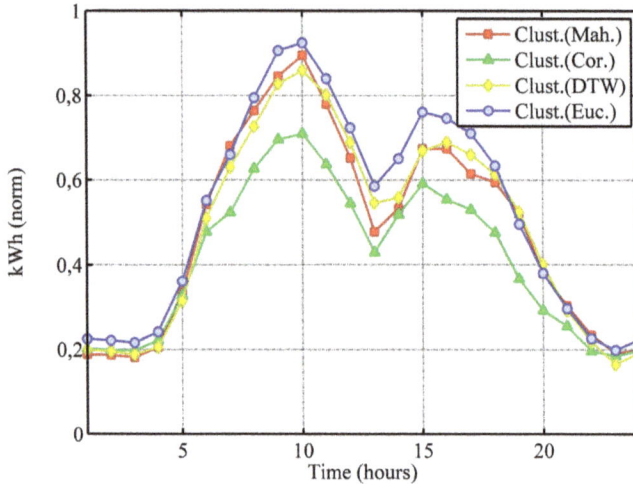

Figure 3. Representative pattern of a specific cluster for building "Edifici A1" according to every clustering solution: using Euclidean (blue circles), Mahalanobis (red squares), based on Pearson's Correlation (green triangles) and DTW (yellow diamonds) similarity metrics.

8.2. Best Validation and Best Evaluation

To establish which similarity measure involves the best clustering performances, we must check all the tests together but separate *validation* from *evaluation* due to the different nature with which they approach the assessment task (see above).

For the evaluation, we use four different validity methods. In order to gain an overall, joined perspective of the obtained results and indices, we ensure that they (validity methods) assign points to the similarity distance that they consider the best for every conducted test for every test, each validity method gives 1/4 points). For example, in building "Rectorat" (B5), in the test where validity methods deploy DTW distance for validation last test in Figure 4), Dunn's index and clustered-vector balance index find that the Euclidean metric is the best, whereas Davies–Bouldin' index bets for Mahalanobis metric, and finally, clustering balance supports the solution based on the DTW distance. Hence, in this example the Euclidean metric gains 1/2 = 1/4 + 1/4 points, Mahalanobis metric 1/4, DTW distance also 1/4 and 0 for Correlation. This way of summarizing results leads to Figure 4. In the figure, tests are ordered from the building "Edifici A1" (B1) to the building "Rectorat" (B5), and starting with Euclidean metric for validation (left area), and finishing with DTW measure for validation (right area). In every test, the clustering solutions using the four different similarity measures are compared and points are given as described above.

What Figure 4 displays is that validity methods are prone to consider clustering solutions based on Euclidean metric as the best, irrespective of the measure used for validation. Moreover, note that the coincidence between the distance for clustering and the distance for validation has no significant influence in the assessments.

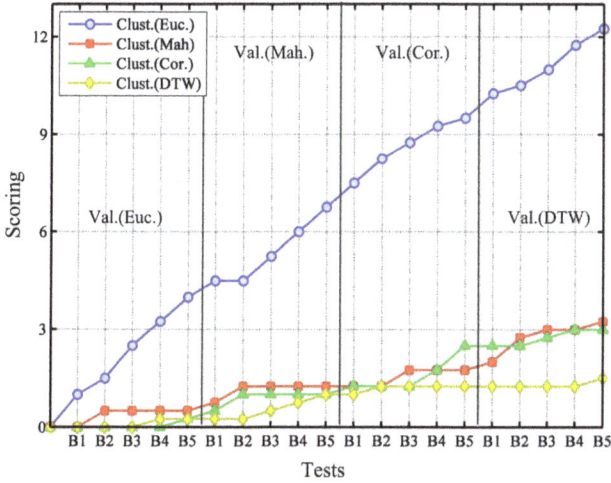

Figure 4. Joined assessment carried out by clustering validity methods.

The case of validation is analogously checked, but here only Equation (9) is used for the assessments. Results are shown in Figure 5. Using data put aside for testing, evaluation reveals that DTW and Euclidean distances compete for the best scoring as measure of similarity for clustering, whereas Mahalanobis and Correlation metrics always perform worse. Curiously enough, DTW distance obtains the worst records in the validation analysis; this issue is dealt with later when validity methods are compared.

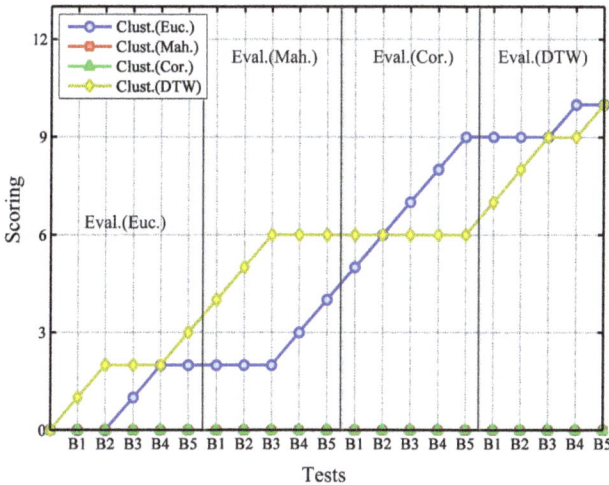

Figure 5. Assessment carried out using data saved for testing.

In short, as far as distances for clustering are compared, *validation* analysis set Euclidean as the best metric for time series clustering, whereas *evaluation* tests favor both DTW and Euclidean similarity distances.

8.3. Validation Algorithms

To review validation algorithms is not an easy task, note that the purpose here is to audit the performance of algorithms that are usually used for checking. In any case, we can reach some conclusions comparing their results to one another as well as looking at the evaluation outcomes. Table 2 displays the trends that validity techniques show when comparing clustering solutions that use different similarity measures. Considering all the tests together, the Mode represents the most typical position taken by the clustering solution that uses the marked distance, standing "1st" for the best evaluation and "4th" for the worst. The Mean contributes to the assessment and gives an impression about how stable the typical scoring is. Therefore, the next points can be reasoned from Table 2:

Table 2. Validity techniques evaluations, statistical Mode and Mean.

	Dunn	D-Boul.	Clust.b.	Vect.b.
	Mode – Mean	Mode – Mean	Mode – Mean	Mode – Mean
Clustering (Euclidean)	1st – 1.4	1st – 1.6	1st – 1.8	1st – 1.3
Clustering (Mahalanobis)	2nd – 2.3	3rd – 2.1	4th – 2.9	4th – 3.2
Clustering (Correlation)	3rd – 2.5	3rd – 2.5	4th – 3.0	4th – 3.2
Clustering (DTW)	4th – 3.9	4th – 3.9	3rd – 2.4	2nd – 2.4

- All methods are in agreement over the measure that achieves the best clustering in general terms, *i.e.*, Euclidean metric;
- Later on, two groups appear:

  - Group 1: Dunn's and David–Bouldin's indices usually scorn solutions based on DTW distance and put it in the worst place, finding that Mahalanobis and Correlation metrics are more suitable for the intended clustering;
  - Group 2: Otherwise, clustering balance and cluster-vector balance give credibility to the DTW distance, placing it before Mahalanobis and Correlation.

The validation tests favors the assessments given by Group 2, so we have arguments to believe that clustering balance and clustered-vector balance are techniques more appropriate to evaluate time series clustering solutions, at least for the current application case. If we look again at Table 2 and compare these two techniques with each other, vector balance seems to be more stable judging distance measures, whereas result comparisons of clustering balance are more variable and case-dependent. In short, there are three factors that opt for clustered-vector balance instead of clustering balance: clustered-vector balance (1) shows higher coincidence with the rest of validity methods, (2) is more stable in the assessments and (3) matches the validation test outcomes better.

Now it is possible to clarify why DTW distance gained such a low score in validation tests, in part due to the rejection of Dunn's and David–Bouldin's indices, but also because of the fact that, although usually showing a very little difference in the evaluations, clustered-vector balance rarely places DTW-based clustering before Euclidean-based (note that in Figures 4 and 5 only the 1st solution obtains points; the 2nd, 3rd and 4th solutions gain no points).

8.4. Best Independent Clusters

Grouping all the clusters generated by the diverse clustering solutions together, the three best clusters according to Equation (10) are highlighted. Again a competition among similarity measures is carried out, and results are displayed in Figure 6. Here, results show no evidence to state that a specific distance measure obtains better compact clusters as a general rule. Again, the type of distance for validation does not significantly affect this measure (except for perhaps the case of Euclidean clustering); instead, the specific case (building) exerts a decisive influence for the selection of the measure to discover compact clusters (low internal dissimilarity). In any case, although results are not discriminative, it is at least worth considering the advantage of DTW and Correlation distances, and the fact that Euclidean metrics receives the lowest results in this aspect.

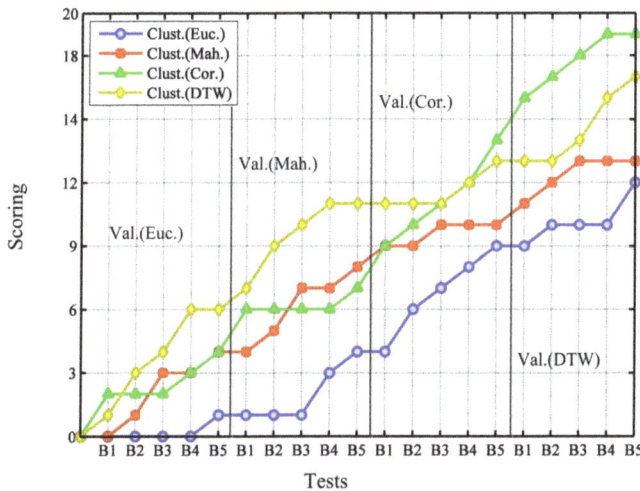

Figure 6. Comparison of the capability to discover *the best individual clusters*.

9. Discussion

In short, the developed experiments place Euclidean distance as the best similarity metric to obtain good general solutions in *raw-data-based* time series clustering. In other words, using Euclidean distance as a similarity metric, the best trade-off, balance solutions are obtained, as it is the most appropriate option to deal with the input space as a whole. Therefore, we hypothesize that Euclidean distance actually considers data correlation in an indirect and fair enough way, suitable for the general clustering solution.

The weights that Mahalanobis provides in the measures in order to favor the appraisal of correlations also introduces a questionable distortion in the input space that causes loss of information or structure and can be even seen as an unnecessary redundancy. On the other hand, distances based on Pearson's correlation, intended to indicate the strength of linear relationships, have trouble correctly interpreting the distribution and relationship of vectors that present low similarity, in addition to being more sensible facing outliers, whether they are vectors or feature values. In the end, Mahalanobis and normal correlation seem to perform well the detection of certain nuclei, but have more problems dealing with intermediate vectors, *i.e.*, the background clouds of vectors with low, variable density. In short, we can consider that these two metrics are biased to find a specific sort of relationship, losing capabilities to manage the space as a whole.

DTW distance deserves a special mention as it has been the most successful in the evaluation test and in finding the best clusters. We can expect that in related prediction applications it performs as good as the Euclidean distance and sometimes even better. If both similarity measures are compared based on the conducted test, the reasons for the different performances can be inferred (Figure 7 shows an example of discovered patterns and embraced input clusters for both clustering solutions, using DTW and Euclidean similarity measures).

On one hand, using DTW distance for clustering also entails a deformation of the input space in order to better capture the representative nuclei. It ensures that the clusters' gravity centers move toward areas where high-correlated samples (or parts of the samples) are better represented, sacrificing capabilities to represent or embrace samples that do not show such high-correlation or coincidence. But, compared with Mahalanobis or Correlation distances, the induced deformation is more respectful with the overall shape or structure that forms the input samples all together. Figure 7 is a good example to check Euclidean clustering, not only compared with DTW, but with all the considered measures that somehow estimate correlation (where DTW has proved to be the most suitable). At first sight, to visually compare between the two clustering solutions is not easy, both seem to capture the essential patterns with minimum variation. DTW distance favors samples that show parts that really match one another, being more lax if the rest of the curve does not fit such coincidence. This can be seen in Figure 7. Note that, as a general rule, the DTW solution shows more dark zones (curves are closer) as a result of the *obsession* to find correlated parts. The two equivalent patterns labeled "3a" and "3b" are a good example to assess this phenomenon. Here, in the DTW case, the effort made to fix the high-correlated first part of the profile is significantly spoiled by the less-coincidental last part of the profile. Otherwise, the group found by the Euclidean solution may display a better trade-off, balanced solution.

In short, and according to the test results, the DTW distance usually better defines the important clusters, losing representativeness in the less correlated ones; otherwise, Euclidean metric could result in main cluster representatives that are not so good, but better in order to define the lower-density ones and to summarize the input space as a whole.

Finally, even though a priori computer resources are not a limiting factor in the introduced application case, the time required by the clustering process in every one of the tested configurations is worthy of consideration. Only by changing the similarity measure, the required time by the clustering task shows a different order of magnitude: Euclidean similarity takes hundredths of seconds (0.0Xs); Mahalanobis metric, tenths (0.Xs); Correlation needs seconds (Xs); and DTW distance, tens of seconds (X0s). These values must not be taken as absolute measures, but only to compare clustering performances with one another. Please, note that the processing time depends on the machine used for computation.

DTW Clustering

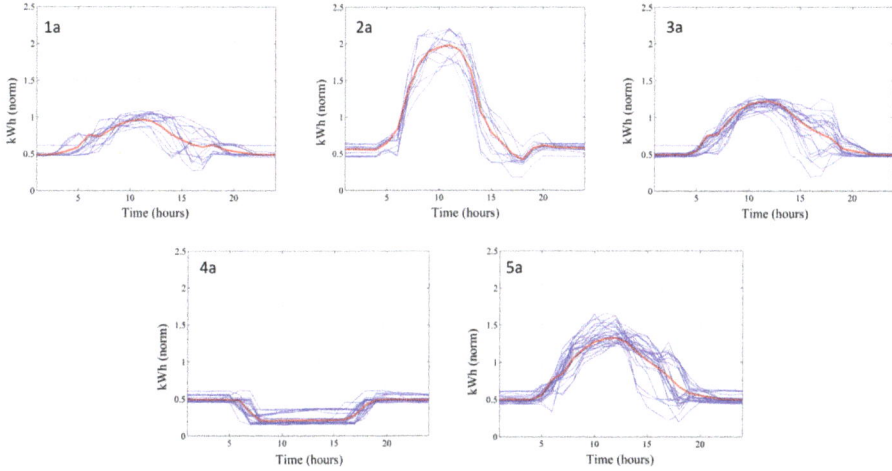

Euclidean Clustering

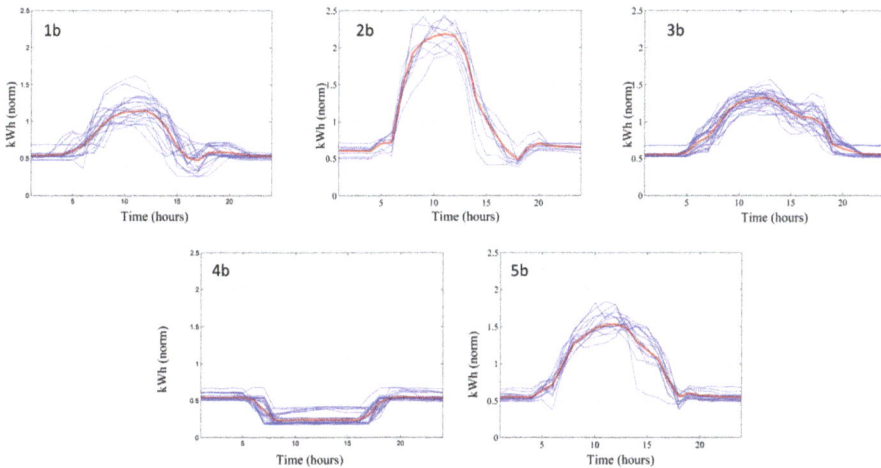

Figure 7. Patterns discovered using DTW and Euclidean measures in "Rectorat" building, and embraced samples.

10. Conclusions

The present paper has introduced and successfully tested *clustered-vector balance*, a validation measure for comparing clustering solutions based on *clustering balance* foundations. This technique is not only useful to improve the adjustment and selection of parameters, algorithms and tools for clustering, but also useful to provide information about the reliability of models obtained from clustering, improving context awareness of predictors and controllers.

On the other hand, popular similarity distances—Euclidean, Mahalanobis, Pearson's correlation distance and DTW—are compared in a time series clustering scenario related to building energy consumption. Although data show strong correlations among vectors and also among features, Euclidean distance is the measure that obtains the best, balanced *general solutions*. However, DTW distance can be considered as an improved alternative in applications that make the most of a better representation of the high-similar nuclei (or parts of the samples), and where losing capabilities to capture and average the not-so-similar samples is not a critic factor. In short, unlike classic considerations, we hypothesize that only a strong correlation in time series clustering does not justify the use of similarity distances that consider data correlation rather than the Euclidean metric.

Seeking for the implementation of accurate controllers and predictors, part of the current ongoing work consists of checking metrics after outlier removal. Here, an outlier is seen as an element that pertains to a group of non-grouped samples or background vectors. Therefore, annoying elements are temporarily removed in order to get a clearer classification. Later on, outliers are relocated in the solution space, identified as background noise or just definitively removed. The definition of outlier itself is a confusing issue. Dealing with outliers entails additional uncertainties and trade-off decisions and also requires improvements in the validation techniques to evaluate the distinct possible performances.

References

1. Crosbie, T.; Dawood, N.; Dean, J. Energy profiling in the life-cycle assessment of buildings. *Manag. Environ. Qual. Int. J.* **2010**, *21*, 20–31. [CrossRef]
2. Kirschen, D. Demand-side view of electricity markets. *IEEE Trans. Power Syst.* **2003**, *18*, 520–527. [CrossRef]
3. Zedan, F.; Al-Shehri, A.; Zakhary, S.; Al-Anazi, M.; Al-Mozan, A.; Al-Zaid, Z. A nonzero sum approach to interactive electricity consumption. *IEEE Trans. Power Deliv.* **2010**, *25*, 66–71. [CrossRef]
4. Räsänen, T.; Ruuskanen, J.; Kolehmainen, M. Reducing energy consumption by using self-organizing maps to create more personalized electricity use information. *Appl. Energy* **2008**, *85*, 830–840. [CrossRef]
5. Yao, R.; Steemers, K. A method of formulating energy load profile for domestic buildings in the UK. *Energy Build.* **2005**, *37*, 663–671. [CrossRef]
6. Jia, W.; Kang, C.; Chen, Q. Analysis on demand-side interactive response capability for power system dispatch in a smart grid framework. *Electr. Power Syst. Res.* **2012**, *90*, 11–17. [CrossRef]
7. Chicco, G.; Napoli, R.; Piglione, F. Comparisons among clustering techniques for electricity customer classification. *IEEE Trans. Power Syst.* **2006**, *21*, 933–940. [CrossRef]
8. Duan, P.; Xie, K.; Guo, T.; Huang, X. Short-term load forecasting for electric power systems using the PSO-SVR and FCM clustering techniques. *Energies* **2011**, *4*, 173–184. [CrossRef]
9. Amjady, N. Short-term hourly load forecasting using time-series modeling with peak load estimation capability. *IEEE Trans. Power Syst.* **2001**, *16*, 798–805. [CrossRef]
10. Li, X.; Bowers, C.; Schnier, T. Classification of energy consumption in buildings with outlier detection. *IEEE Trans. Ind. Electron.* **2010**, *57*, 3639–3644. [CrossRef]
11. Iglesias, F.; Kastner, W. Usage Profiles for Sustainable Buildings. In Procedings of the 15th IEEE Conference on Emerging Techonologies and Factory Automation (ETFA), Bilbao, Spain, 13–16 September 2010; pp. 1–8.
12. Iglesias, F.; Kastner, W.; Reinisch, C. Impact of User Habits in Smart Home Control. In Proceedings of 2011 IEEE 16th Conference on Emerging Technologies Factory Automation (ETFA), Toulouse, France, 5–9 September 2011; pp. 1–8.
13. Theodoridis, S.; Koutroumbas, K. *Pattern Recognition*, 4th ed.; Academic Press: Salt Lake City, UT, USA, 2008.
14. Jain, A.K.; Dubes, R.C. *Algorithms for Clustering Data*; Prentice-Hall, Inc.: Upper Saddle River, NJ, USA, 1988.
15. Jain, A.K.; Murty, M.N.; Flynn, P.J. Data clustering: A review. *ACM Comput. Surv.* **1999**, *31*, 264–323. [CrossRef]
16. Hong, Y.Y.; Wu, C.P. Day-ahead electricity price forecasting using a hybrid principal component analysis network. *Energies* **2012**, *5*, 4711–4725. [CrossRef]
17. Liao, T.W. Clustering of time series data—A survey. *Pattern Recognit.* **2005**, *38*, 1857–1874. [CrossRef]
18. Zervas, G.; Ruger, S. The Curse of Dimensionality and Document Clustering. In Proceedings of 1999 IEE Colloquium on Microengineering in Optics and Optoelectronics (Ref. No. 1999/187), London, UK, 16 November 1999; pp. 19:1–19:3.

Energies **2013**, *6*, 579–597

19. Rodgers, J.L.; Nicewander, W.A. Thirteen ways to look at the correlation coefficient. *Am. Stat.* **1988**, *42*, 59–66. [CrossRef]

20. Maesschalck, R.D.; Jouan-Rimbaud, D.; Massart, D. The mahalanobis distance. *Chemom. Intell. Lab. Syst.* **2000**, *50*, 1–18. [CrossRef]

21. Keogh, E. Exact Indexing of Dynamic Time Warping. In Proceedings of the 28th International Conference on Very Large Data Bases, Hong Kong, China, 20–23 August 2002; pp. 406–417.

22. Huang, A. Similarity Measures for Text Document Clustering. In Proceedings of the 6th New Zealand Computer Science Research Student Conference, Christchurch, New Zealand, 14–18 April 2008; pp. 49–56.

23. Lipkus, A. A proof of the triangle inequality for the Tanimoto distance. *J. Math. Chem.* **1999**, *26*, 263–265. [CrossRef]

24. Hubert, L.; Arabie, P. Comparing partitions. *J. Classif.* **1985**, *2*, 193–218. [CrossRef]

25. Rousseeuw, P.J. Silhouettes: A graphical aid to the interpretation and validation of cluster analysis. *J. Comput. Appl. Math.* **1987**, *20*, 53–65. [CrossRef]

26. Halkidi, M.; Batistakis, Y.; Vazirgiannis, M. Cluster validity methods: Part I. *SIGMOD Rec.* **2002**, *31*, 40–45. [CrossRef]

27. Zheng, X.; Cai, Z.; Li, Q. An Experimental Comparison of Three Kinds of Clustering Algorithms. In Proceedings of 2005 International Conference on Neural Networks and Brain (ICNN&B '05), Beijing, China, 13–15 October 2005; Volume 2, pp. 767–771.

28. Qian, W.; Zhou, A. Analyzing popular clustering algorithms from different viewpoints. *J. Softw.* **2002**, *13*, 1382–1394.

29. Wu, J.; Hassan, A.E.; Holt, R.C. Comparison of Clustering Algorithms in the Context of Software Evolution. In Proceedings of the 21st IEEE International Conference on Software Maintenance 2005 (ICSM'05), Budapest, Hungry, 26–29 September 2005; pp. 525–535.

30. Pamudurthy, S.; Chandrakala, S.; Chandra Sekhar, C. Local Density Estimation based Clustering. In Proceedings of 2007 International Joint Conference on Neural Networks (IJCNN 2007), Orlando, FL, USA, 12–17 June 2007; pp. 1249–1254.

31. Iglesias, F.; Kastner, W. Clustering Methods for Occupancy Prediction in Smart Home Control. In Proceedings of 2011 IEEE International Symposium on Industrial Electronics (ISIE 2011), Gdansk, Poland, 27–30 June 2011; pp. 1321–1328.

32. Jung, Y.; Park, H.; Du, D.Z.; Drake, B. A decision criterion for the optimal number of clusters in hierarchical clustering. *J. Glob. Optim.* **2003**, *25*, 91–111. [CrossRef]

33. Rasmussen, M.; Newman, M.; Karypis, G. *gCLUTO Documentation*, version 1.0.; Department of Computer Science, University of Minnesota: Minneapolis, MN, USA, 2003.

34. De Souto, M.; de Araujo, D.; Costa, I.; Soares, R.; Ludermir, T.; Schliep, A. Comparative study on normalization procedures for cluster analysis of gene expression datasets. In Proceedings of 2008 IEEE International Joint Conference on Neural Networks (IJCNN 2008), Hong Kong, China, 1–8 June 2008; pp. 2792–2798.

35. Dunn, J.C. A fuzzy relative of the ISODATA process and its use in detecting compact well-separated clusters. *J. Cybern.* **1973**, *3*, 32–57. [CrossRef]

36. Davies, D.L.; Bouldin, D.W. A cluster separation measure. *IEEE Trans. Pattern Anal. Mach. Intell.* **1979**, *1*, 224–227. [CrossRef] [PubMed]

Article

Assessing Tolerance-Based Robust Short-Term Load Forecasting in Buildings

Cruz E. Borges [1],*, Yoseba K. Penya [1], Iván Fernández [1], Juan Prieto [2] and Oscar Bretos [2]

[1] DeustoTech-Deusto Technology Foundation, Energy Unit, University of Deusto, Avenida de las Universidades 24, Bilbao 48007, Basque Country, Spain; yoseba.penya@deusto.es (Y.K.P.); ivan.fernandez@deusto.es (I.F.)

[2] Indra, Smart Energy Department, Optimisation and Prevision Area, Parque empresarial Arroyo de la Vega, edificio Violeta 2, Avenida de Bruselas 35, Alcobendas, Madrid 28108, Spain; jprietov@indra.es (J.P.); obretos@indra.es (O.B.)

* Author to whom correspondence should be addressed; cruz.borges@deusto.es; Tel.: +34-94-413-9003 (ext. 2052); Fax: +34-94-413-9166.

Received: 18 January 2013; in revised form: 12 March 2013; Accepted: 18 March 2013; Published: 17 April 2013

Abstract: Short-term load forecasting (STLF) in buildings differs from its broader counterpart in that the load to be predicted does not seem to be stationary, seasonal and regular but, on the contrary, it may be subject to sudden changes and variations on its consumption behaviour. Classical STLF methods do not react fast enough to these perturbations (*i.e.*, they are not robust) and the literature on building STLF has not yet explored this area. Hereby, we evaluate a well-known post-processing method (Learning Window Reinitialization) applied to two broadly-used STLF algorithms (Autoregressive Model and Support Vector Machines) in buildings to check their adaptability and robustness. We have tested the proposed method with real-world data and our results state that this methodology is especially suited for buildings with non-regular consumption profiles, as classical STLF methods are enough to model regular-profiled ones.

Keywords: short term load forecasting; artificial intelligence; statistical methods

1. Introduction

Load forecasting is an essential part of the scheduling, management and operation of a power system. Since electrical energy cannot be stored, it is important to deliver an accurate prediction in order to avoid dispatch problems due to unexpected loads. Moreover, energy market stakeholders also require trustworthy information in order to be more competitive when purchasing electricity. In addition, the use of the data recorded in smart-meters and software tools may help prevent demand peaks in a reliable and efficient fashion.

Short-term load forecasting (STLF) is the prediction of energy demand in a time-span ranging from minutes to several days, being crucial for several smart grid applications. As discussed above, it is important for the economic and secure operation of power grids, but several factors should be considered as well. For instance, the publication of the energy consumption and its conversion to equivalent CO_2 emissions decreases the load that affects future predictions due to its influence on social consciousness. Moreover, it also helps the energy retailer in order to negotiate a better price. In the end, an accurate forecast results in higher savings while helping to maintain the security of the grid.

There exists a large bibliography on STLF (see [1–3] for a comprehensive survey). Most of the methods used can be divided into two main groups depending on the strategy followed: Statistical Methods, which estimate the present value of a given variable depending on the values in the past (*i.e.*, consumption records [2,4,5]), and Artificial Intelligence methods, which have been applied successfully to a wide variety of real-world applications, demonstrating their ability to *learn* the relationships between input and output variables. Moreover, the later have been proven to be ideal when dealing with risk and uncertainty (the main aspects behind prediction). This approach, however, involves the need of an expert whose knowledge can be incorporated into the system capable of making accurate forecasts. The most popular algorithms, according to their efficiency, are Support Vector Machines (SVM) [6,7] and Neural Networks (NN) [8,9]. Note that in this paper we will only use SVM since we have not been able to replicate the results from the literature involving NNs. Moreover, our tests have proven that NNs are slower and do not obtain better results than other techniques (see for example [10]).

Despite the accuracy they provide on forecasting, Artificial Intelligence methods suffer from a great number of disadvantages such as difficult parametrisation, non-obvious selection of variables and the requirement of more historical data to learn than any of the Statistical Methods [2]. Furthermore, they rely on a tedious trial-and-error process to tune them up properly.

This paper focuses on building STLF, a particular case dealing with issuing day-ahead energy consumption predictions in non-residential buildings such as schools, universities, public buildings or companies' facilities. The ideal to reach is the so-called *zero energy building*, (*i.e.*, any construction presenting annually zero net energy consumption and carbon emissions), and, with this objective in mind, there are a number of technologies that must be integrated:

- Weatherproofing, insulation and automatic HVAC (heating, ventilation, and air conditioning).
- Energy re-utilisation (as in co- or tri-generation).
- Use of renewable energy sources and energy storage systems.
- Demand response controllers attached to the HVAC and other loads.

Clearly, STLF is crucial in the last two points. Indeed, maximising the efficiency of the demand response controller requires an accurate forecast for both energy consumption and energy generation (as well as a reliable storage controller).

This branch presents different features. For instance, in normal country-wide STLF, the non-linearity of the load becomes smoothed: expected consumption that does not take place is compensated by non-expected consumption that does (*i.e.*, the consumption curve tends to be seasonal and regular). In contrast, the load profile of a building is more chaotic coinciding with the times it is used. Hence, there is no consumption at night (or it is negligible) and there exists a notable difference between idle and active times. Furthermore, some of these buildings are not yet fully-automated: either the HVAC is manually controlled or it is switched on and off remotely, issues that affect greatly on energy consumption. Another critical aspect is that there is usually scarce (if any) historical data on hourly load and the load profile is sure to vary and evolve over time (just think of the gadgets an office used to have ten years ago compared with nowadays fully-equipped on-line ones). This makes it very difficult to extract the trend component and/or use forecasting methods that need long learning windows (such as ARIMA and their derivatives).

While operating in real environment, it is vital that the forecasting method adapts to changing conditions. If the system cannot react to these changes, the predictions obtained from the system will fail. In order to achieve an optimal prediction, we need a method that evolves over time and is not subject to fixed laws so as to conform to recent data. This is the reason that explains why some models present very good records in a certain situation but fail in others. Meta-learning models address this issue: they belong to a well-established procedure for improving forecasting accuracy [11] and have already been applied in other disciplines (see [12] for a broad survey).

Analysing the buildings' electricity consumption series, we can observe the presence of atypical values that digress from the typical model. They significantly degrade the accuracy of conventional day-ahead estimation even if they appear in a reduced number. These atypical values present two natures:

New Pattern: Some days show a different consumption pattern due to reasons of diverse nature (e.g., long weekends, sport events, election polls, strikes, *etc.*). In the end, the load of these days consists of a new day-type on its own (see Section 3 for more details) and it should be removed from the learning set or classified beforehand.

Scaled Pattern: On the other hand, some days follow profiles similar to those from existing day-types but scaled down or up. Again, these changes can be due to diverse causes (e.g., sudden weather changes, long weekends, special events, works, *etc.*). The load in these days does not consist of a new day-type but can be just an extreme statistical fluctuation or simply represent an underline change in the building (such as new equipment) that may be extended along the time. In the first case, there should be no action, but in the second one, the learning window should be restarted.

Thus, the data series should be treated in a *robust* way, namely, the forecasting method should recognise these atypical values and treat them accordingly in order to avoid burst errors that worsen the overall forecast.

Against this background, we present here a comparison of several robust methods based on threshold values over the errors. We aim at assessing them in order to evaluate whether they can correctly distinguish between the two types of anomalies explained above and fulfil the premises required by a robust algorithm.

The remainder of the paper is organised as follows. Section 2 discusses the related work. Section 3 presents the different algorithms used in the prediction, the features of the used datasets, an explanation of the method and its validation. Section 4 describes the datasets used. Section 5 details the tests and comments the obtained results. Finally, Section 6 summarises our contribution and draws the avenues of future works.

2. Related Work

As previously mentioned, there exists a remarkable work on STLF but, comparatively, not so much related to buildings and the use of meta-learning to provide an appropriate solution in this scope. We have previously researched on this field [13]. Nevertheless, it was not focused on the robustness of the prediction but on the evaluation of model combination and other model-selection methods.

STLF in buildings provides a whole new overview on the paradigm, giving way to several important works such as using a SVM to predict the load of a building complex [14,15], where a NN is tuned up by Automatic Relevance Determination in order to optimise the selected input. In addition, [8] used the temperature data in a feedback NN obtaining a remarkable Mean Absolute Percentage Error (MAPE) of 1.945% (Section 4.3 details the mathematical definition of this type of error measure), but this result was obtained measuring only a single week in a whole year, which is not statistically representative. Finally, all artificial intelligence methods waste most of their efforts in modelling non-linear behaviour of the work calendar [16,17].

In the case of meta-models for normal STLF, research has taken two main directions. The first uses a meta-heuristic to calculate the best set of parameters of a SVM or a NN [18–20], but these works suffer from the same flaw in the single models. The second area has explored the optimal way of combining the output of the single models, usually by assigning weights (see [21] for different approaches to this end). For instance, a very simple but effective approach consists in defining equal weights, which has shown to be surprisingly effective [22,23]. More sophisticated approaches include linear combination [24], dynamic optimal weight combination [25], a genetic algorithm as best model selector [26] or rule-based best model selection [27].

Assessing the robustness of STLF has received very little attention [9,28] and all the efforts focus on the detection of outliers (new patterns of atypical values in our nomenclature) in the time series of national loads.

3. Adaptive STLF in Buildings

3.1. Overall Methodology

In order to ensure the required standard of security and quality in a power system, we need a very reliable and robust forecast. This paper focuses on assessing an adaptive forecasting that tries to mimic the consumption behaviour. Since all methods need a learning period of time, new consumption habits lead to prediction inaccuracies that increase both economic and technical costs.

In this work, we have used the following methodology; we repeat the next steps for every day in the dataset except for the first day of every day-type. Figure 1 shows a sequence diagram of the methodology. Please note that the meter is virtually connected to a real-time processing network as described in [29], to which we will refer from now on as the *Platform*.

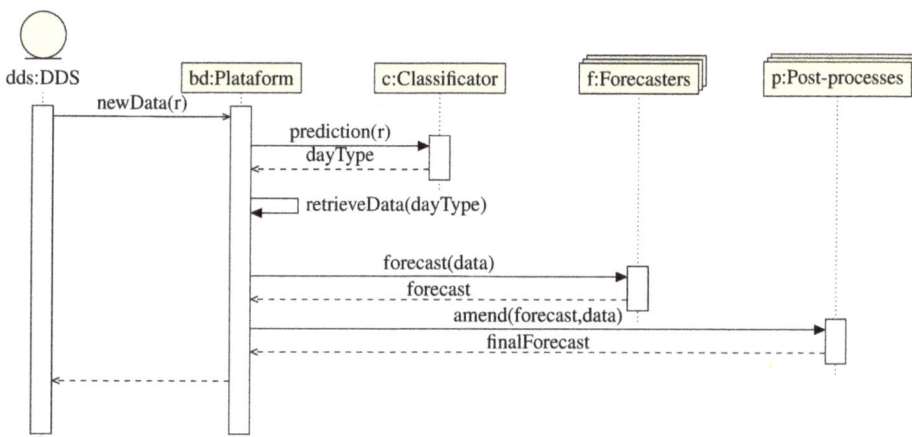

Figure 1. Sequence diagram of the proposed methodology.

Data Distribution Service (DDS): A meter sends a new measure to the Platform through the DDS bus. This measure is stored to then be used to issue a forecast.

Classificator: In this step, the Platform sends the new data to the Classificator and queries it for the prediction of the day-type for the next day. In previous works [30], we have compared several clustering techniques with the use of the local *work calendar*. Our results conclude that the best option is to use the work calendar if it is available. Hence, buildings may present different number of day-types. Specifically, in our tests, there are buildings presenting:

- **Two day-types**: (Weekday and Weekends) such as *c59* or *ashrae* (see Section 4 for more information on the datasets). This building is characterised by the lack of consumption on Saturdays. An example of this behaviour can be seen in Figure 2.
- **Three day-types**: (Weekday, Saturdays and Sundays) such as the four *donosti* datasets, *bilbao$_1$*, *bilbao$_3$* or *bilbao$_4$*. They present a very similar consumption to commercial or service buildings. An example of this behaviour can be seen in Figures 3 and 4.
- **Four day-types**: (Weekday, Saturdays, Sundays and Bank Holidays) such as *bilbao$_2$*. This building shows a special behaviour in Bank Holidays. In Figure 5 can be seen an example of this behaviour.

Forecasters: In the next step, the Platform sends the data of the previous days of the same day-type to the Forecasters. They will adjust the model parameters and then issue a forecast. Note that we have a different model for every day-type, and therefore we must re-train the model for every day.

In this work, we have used AR and SVM models, since according to our experiments [10,30] they produce the best results using this methodology. Moreover, in these works we have optimised the free parameters by means of a *grid search* following the advice given in [31] and we used the results of these test here.

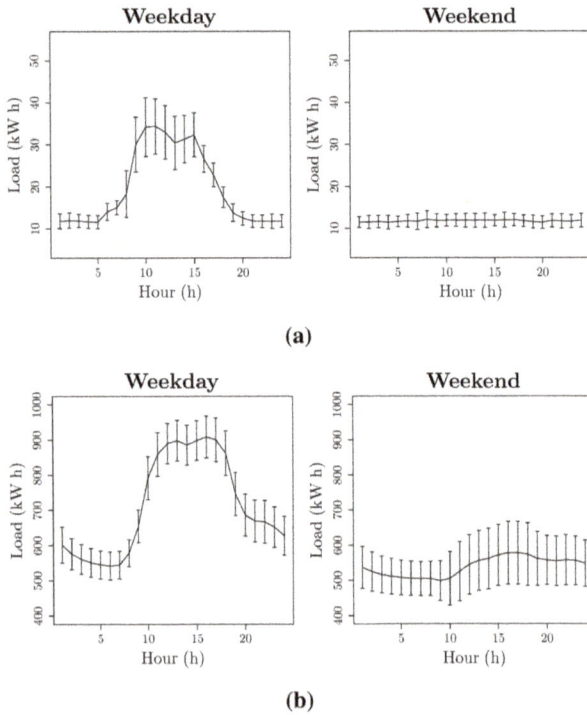

Figure 2. Typical load for the dataset with two day-types. Average daily load (kW h). Error bars denotes ±σ. (**a**) Dataset *c*59; (**b**) Dataset *ashrae*.

Post-processes: Finally, in this stage the Platform sends the forecasts issued by the forecasters to the post-processes in order to improve the results. Examples post-process are:

- **Bias Correction**: some models produce forecast that are systematically biased. We can measure that bias and compensate it. In [10] we have assessed the performance of this post-process.
- **Model Selection**: some models issue a more reliable forecast at certain hours or day-types than others. In [13], we presented a comparison of different strategies to select the best model in every moment.
- **Model Combination**: another option is to group all the predictions issued by the forecasters in order to build a more robust forecast. We have addressed this strategy in [13,32].

Please note that in this work we have not used these three post-processes as we have assessed their performance in previous works. In Section 3.2 we will present two more examples of post-processes that we will analyse in this paper.

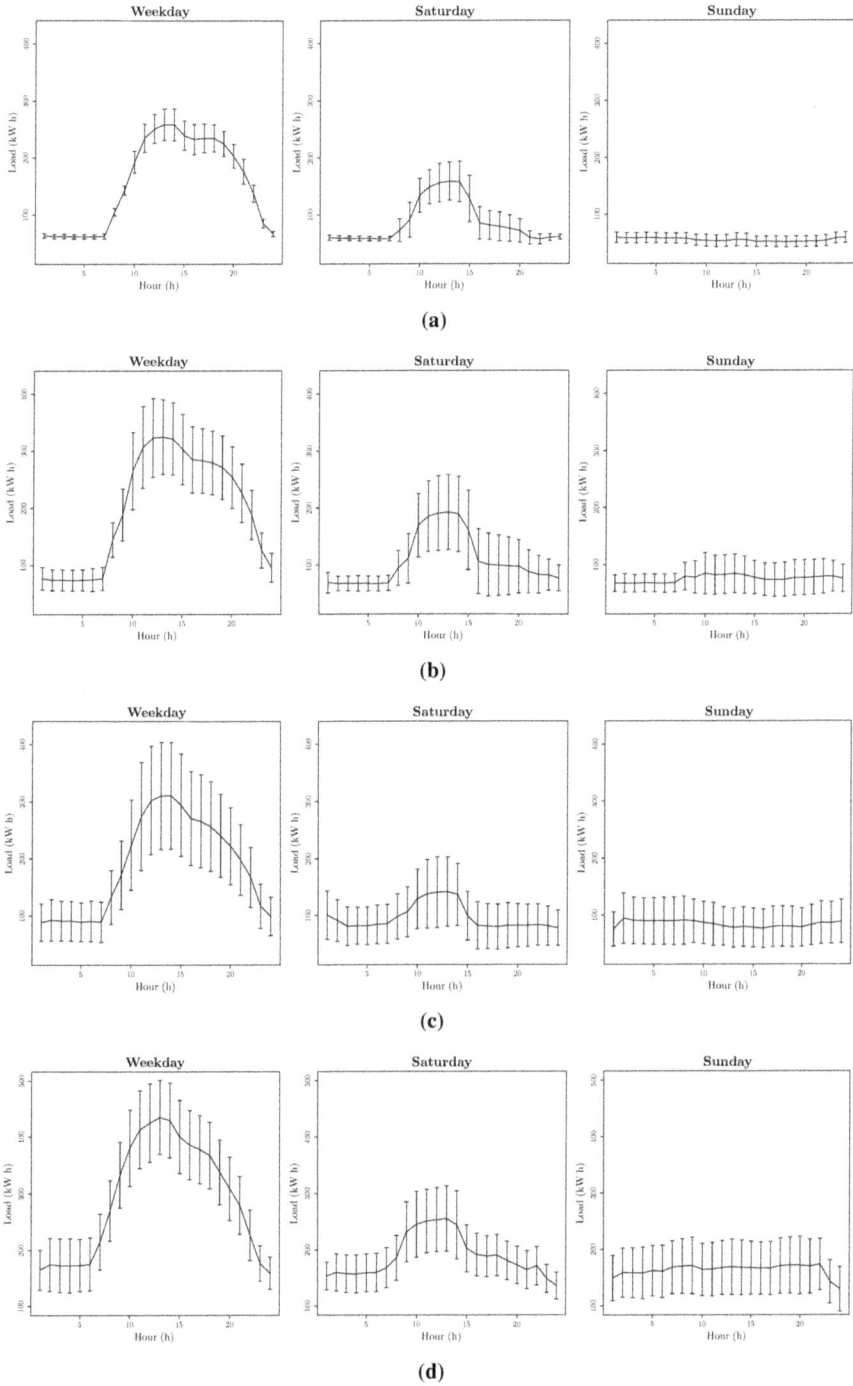

Figure 3. Typical load for the dataset with three day-types. Average daily load (kW h). Error bars denotes $\pm\sigma$. (**a**) Dataset *donosti₁*; (**b**) Dataset *donosti₂*; (**c**) Dataset *donosti₃*; (**d**) Dataset *donosti₄*.

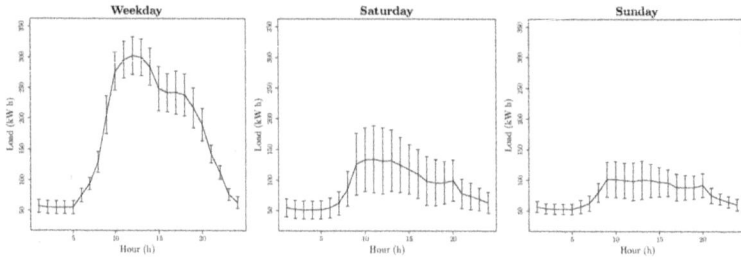

(a)

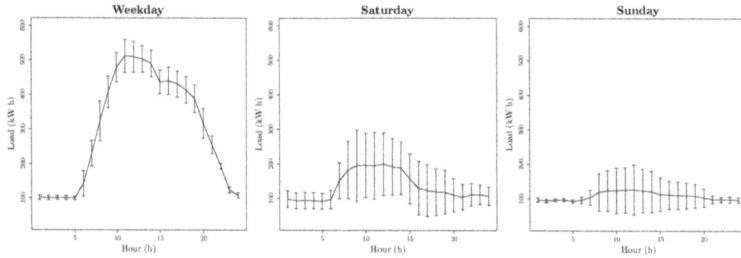

(b)

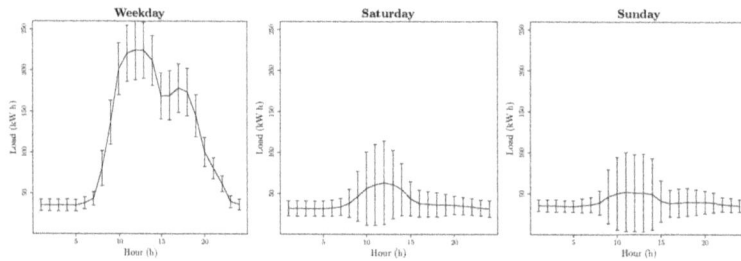

(c)

Figure 4. Typical load for the dataset with three day-types. Average daily load (kW h). Error bars denotes $\pm\sigma$. (**a**) Dataset $bilbao_1$; (**b**) Dataset $bilbao_3$; (**c**) Dataset $bilbao_4$.

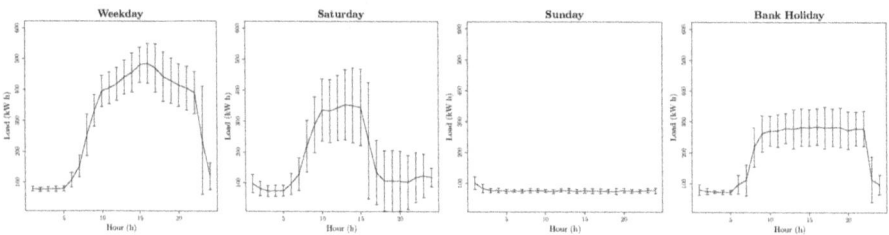

Figure 5. Typical load for the dataset with four day-types. Average daily load (kW h). Error bars denotes $\pm\sigma$. Dataset $bilbao_2$.

3.2. Proposed Post-Process Methods

Despite all the post-process methods, we have detected that some days in the datasets are very different from what their normal profile should be according to their day-type. This abnormal behaviour causes *burst errors* to appear in the forecast until the algorithms manage to adapt to these changes. Thus, we need a faster method in order to avoid large errors in such days.

In this paper, we introduce two post-process methods that check whether the prediction error is bigger than a given threshold value k. Please note that we can use whatever error we define provided that the threshold value is in the same units as the error. As in this paper we are using MAPE error (see Section 4.3 for details), k should be a percentage. Moreover, this value can be fixed a priori or have an adaptive value like:

$$k := \sqrt{\frac{c}{24q} \sum_{d \in Q} \sum_{h=1}^{24} (e_{d,h} - \hat{e})^2} \tag{1}$$

where Q denotes the set of days in the learning window at that moment, q is the number of days in the learning window, $e_{d,h}$ denotes the residuals of the model in day d and hour h and $\hat{e}$ denotes the mean error of the model in the entire learning window. Note that this value is just the variance of the residuals multiplied by the fixed constant c. Please note that this value should be tailored specifically to every dataset. In this work we have made a *grid search* for this value and found that the best option is $c := 3$.

In case the error in one day is above the threshold value k, one of the two post-processes acts.

Learning Window Re-Initialization: In this case, we reboot the learning windows in order to avoid anomalies of the *scaled pattern* type (*i.e.*, we completely delete the data in the dataset and replace it with the newest value). Since the learning windows are very short (just three days in this test, see [10,13] for a broad comparison in this matter) and the models used are sufficiently robust to tailor this degenerated training set, we are able to issue a new forecast in this situation. Moreover, the forecast produced is the obvious one, just the values observed the previous day (*i.e.*, acts like a Random Walk Model), so this adjustment quickly adapts to changes in the dataset.

Skipping Anomalies: In this case we avoid introducing the new pattern in the training data in order to avoid anomalies of the *new pattern* type.

Note that we have not used both methods simultaneously. Moreover, we only present here the results of the Learning Windows Re-Initialization method without using the adaptive threshold value, as both the Skipping Anomalies and the Adaptive Threshold Methods produce much worse results (in any combination).

3.3. Forecasting Models

All models used can be classified as *regression models*. Namely, they follow the equation:

$$LOAD(h) := f(h) + \varepsilon \tag{2}$$

where $h \in [0, 23)$ denotes time, $LOAD(h)$ denotes the load at time h, f is the model used and ε is a random variable.

3.3.1. Time Series Model

The first model is an *Autoregressive Model* commonly used for modelling univariate time series. For every day-type d we have:

$$s^d(h) = \sum_{i=1}^{q} \varphi_i^d r^d(h, i) \tag{3}$$

where $\varphi^d := \left(\varphi_1^d, \ldots, \varphi_q^d \right)$ are the model parameters and $r^d(h, i)$ denote the real load measured at time h of the i-th previous day of day-type d. Namely, in the adjusting step, we have retrieved the q last values of the *same* day type (e.g., with $q = 3$, from a Tuesday, the previous Monday, Friday, Thursday) and not the q last chronological values (e.g., from a Tuesday, the previous Monday, Sunday, and Saturday) and then we have made a *convex combination* with the model coefficients φ^d. In order to give a higher priority to the latest data against the oldest values, the model coefficients are drawn by a *polynomial* or an *exponential* method. The polynomial method produces the following parameters

$$\varphi_i = \frac{(q-i)^l}{\sum_{i=0}^{q}(q-i)^l} \tag{4}$$

whereas the exponential method produces

$$\varphi_i = \frac{l^{(q-i)}}{\sum_{i=0}^{q} l^{(q-i)}} \tag{5}$$

where q is the length of learning window and l can take integer values. In this work, we has chosen the exponential method with parameter $l := 2$. This value has been taken based on empirical experience. Please note that, with this nomenclature, we have fixed the number of learning days to $q := 3$ as previously stated.

3.3.2. Support Vector Machines Model

A SVM constructs a hyperplane, or a set of hyperplanes, in a high or infinite dimensional space, which can be used for classification or regression. SVMs have been previously used for load forecasting in buildings [14]. Here, we have used a ν-SVR. Note that, as we have explained before in Equation (2), the function we regress is the load curve of all day taking only the time as input. As in the previous case, we take a model for every day-type and train every model with the last q days of the same type. The rest of the free parameters are: radial basis function as kernel, threshold $\nu := 0.9$, soft margin parameter $C := 10$ and kernel parameter $\gamma := 1$. The explanation of these parameters is out of the scope of this paper; we encourage the reader to see [33] for an in-depth explanation.

4. Datasets

This study comprises several datasets in order to provide the most representative result possible: ten datasets from seven different buildings' consumption data records and five from Transmission System Operators (TSO) records from different countries. Tables 1 and 2 summarise the main characteristics of these datasets. These tables also contain an estimation of the expected value of the error for every dataset under the hypothesis that ε follows a Gaussian Random Variable (column *Expected MAPE*), which according to our experiments is a fair assumption. Please note that in [32] we present a detailed description of the estimation process. Moreover, some TSOs publish their own STLF so we can calculate their error in the same way as with our predictions. Column *MAPE from Operator* of Table 2 contains this value.

Table 1. Summary of the buildings' features. *N/A* denotes unknown values.

Building	HVAC	Number of day-types	Expected MAPE (%)
donosti$_1$	NO	3	6.01
donosti$_2$	NO	3	8.22
donosti$_3$	YES	3	13.60
donosti$_4$	YES	3	8.74
bilbao$_1$	NO	3	5.89
bilbao$_2$	YES	4	7.22
bilbao$_3$	Partially	3	4.82
bilbao$_4$	NO	3	5.95
ashrae	N/A	2	4.98
c59	N/A	2	3.74

Table 2. Summary of the region's features. *N/A* denotes no available values. Note that TSOs use forecasting models especially tailored to their respective consumption profiles.

Region	Expected MAPE (%)	MAPE from Operator (%)
REE	5.87	1.08
EUNITE	9.20	*N/A*
AP	7.45	5.02
NYC	6.95	1.64
NORTH	2.04	3.57

4.1. Buildings

4.1.1. University of Deusto

We have recorded the energy consumption of several buildings of the University of Deusto in both of its campuses: Donostia-San Sebastián and Bilbao (Basque Country). We have downloaded these data directly from the meter, placed by the Spanish law (54-1997) directly at the transformer, using the IEC 60870-5-102 standard protocol [34].

These buildings present different patterns as each one has its own special features. For example, we have measurements from the Donostia-San Sebastián building complex since March 2009 but it presents three different periods. From March to September, there was only one building with a regular and homogeneous consumption since (among other things) its heating system is not regulated according to the weather: from autumn to spring, it is manually turned on every day at approximately the same time and it works until the building closes at night; therefore, meteorological conditions do not show notable influence on the electricity consumption (season, on the contrary, does) or it is somehow dissolved in the data. These data forms the dataset *donosti*$_1$.

On July 2009, the construction of two more buildings started but this event did not have an impact on the load profile until September 2009 due to the summer holidays. As the behaviour of these three building together are essentially different (a lot more noisy), we have split the dataset and created the *donosti*$_2$ dataset.

Finally, we created the third and fourth dataset (*donosti*$_3$ and *donosti*$_4$) with the rest of the data. Both datasets present the same characteristics but there is a big gap in the records since the utility changed the meter from a GSM based one to an IP based one. The new buildings have an HVAC system for cooling so weather changes might influence the consumption, explaining in this way the high spikes in their loads; however our previous experiments do not show any relationship [30]. The *donosti*$_3$ dataset has a length of 12 months (September 2010–September 2011) while the *donosti*$_4$ dataset has a length of 8 months (April 2012–January 2013).

All builds show quite a regular profile in working days with consumption from 7:00 a.m. to 10:00 p.m. (opening hours go from 8:00 a.m. to 9:00 p.m.). On Saturdays, it shows a peak at noon and on Sundays it is almost flat.

On the other hand, we have also measured the electrical consumptions from four different buildings in the Bilbao Campus of the University of Deusto from September 2012 to January 2013. Three of them, *bilbao*$_1$, *bilbao*$_3$ and *bilbao*$_4$, gather the records from standard university buildings (*i.e.*, classrooms and offices), while *bilbao*$_2$ contains the data of the campus main library. It presents a more hectic activity until late at night as well as during Saturdays and this fact is reflected in the load profile. The rest of the buildings present a similar profile as the ones in the Donostia Campus.

4.1.2. Ashrae competition

The Ashrae competition ([35] dataset *ashrae* in Figure 2b) comes from an unknown building; the data present a quite similar profile as the dataset *donosti*$_1$ and has a length of only 6 months (September 1989 to February 1990) with consumption from 9 a.m. to 9 p.m.

4.1.3. Casaccia Research Centre

Dataset *c59* contains the consumptions records from the C59 Building in the Casaccia Research Centre, Rome, Italy, during the months of September to November 2009 [36]. As with the ASHRAE competition data, no information has been provided about the building but the profile is also quite similar to the *donosti*$_1$ dataset, except that there is no consumption on Saturdays.

4.2. Regions

We have also downloaded public data from several TSO in order to contrast whether this post-process methodology works well when forecasting large regions. We have taken the information from:

REE: the Spanish TSO [37], from January 2007 until October 2011.

AP: the Pennsylvania, Jersey, and Maryland Interconnection (PJM) [38], more accurately from the Allegheny Power (AP) zone, from November 2008 until December 2010.

NYC and *NORTH*: the New York Independent System Operator (NYISO) [39], more accurately from the NYC and NORTH substations. The former contains data from February 2005 until October 2011, the latter from June 2001 to October 2011.

EUNITE: from the Eastern Slovakian TSO (Eunite competition dataset [40]), from January 1997 to December 1998.

As it can be seen in Figure 6, these datasets are essentially different from those of the buildings; they present *homoscedasticity* and only have two slightly-different patterns: one for weekdays and another for holidays. Table 2 presents a summary of the regions' characteristics.

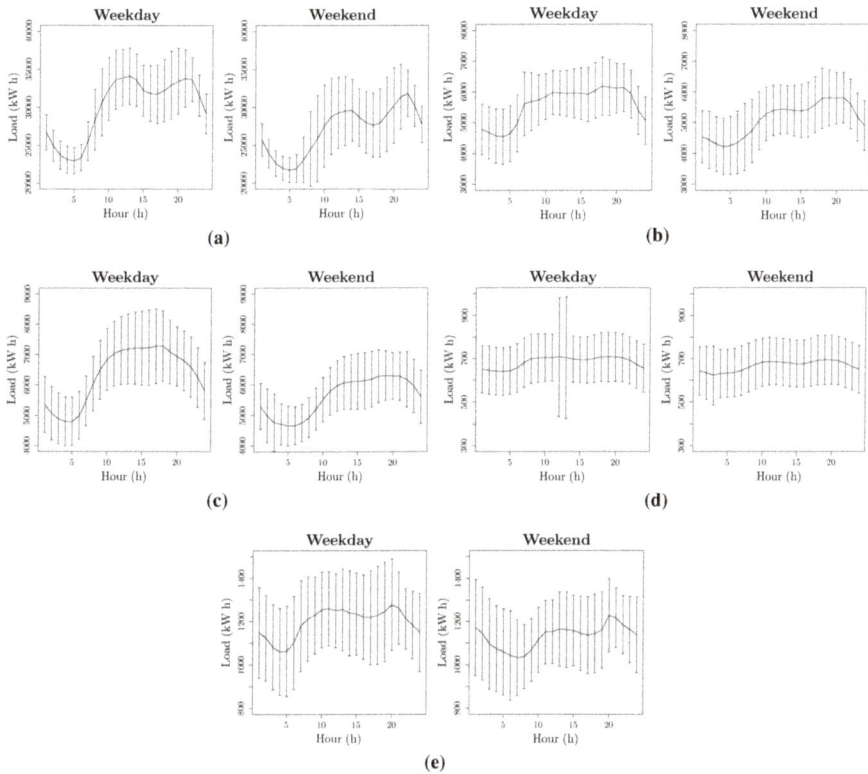

Figure 6. Average daily load (kW h). Error bars denotes $\pm\sigma$. (a) Dataset *REE*; (b) Dataset *AP*; (c) Dataset *NYC*; (d) Dataset *NORTH*; (e) Dataset *EUNITE*.

4.3. Test-Bed and Validation Measurements

In this study, we have tailored the Leave-One-Out Cross-Validation (LOOCV) procedure [41] that tries to mimic the performance in real conditions. In this spirit, for each day in the dataset we have issued a prediction using the last q days' values. Normally, this method cannot be used for the first q days of every day-type but, as we have explained in Section 3, we can already forecast with only one training day.

We use MAPE as error measurement to evaluate performance of the models since it is unit free, which allows comparisons between forecasting errors from different measurement units. Moreover, it is the error measurement most widely used in forecasting despite their problems (see [42] and references therein for an extensive discussion and several solutions to these problems). It is calculated as follows:

$$\text{MAPE} := \frac{1}{days} \sum_{i=1}^{days} \left(\frac{1}{24} \sum_{h=1}^{24} \frac{|r(h,i) - p(h,i)|}{r(h,i)} \right) \times 100 \tag{6}$$

where $p(h,i)$ is the predicted value of the load for the hour h of the day i, $r(h,i)$ refers to the real value and *days* represents the amounts of days in that particular datasets.

Regarding the validation of the post-process, we follow two steps to check the performance of the method. First, we validate the dataset computing the MAPE committed in the whole dataset. The robust method is compared with the normal one in order to assess which one works better over time. Second, we take a closer view by measuring the errors only in the days where the post-process has worked.

5. Experimental Results

The experiments have been carried out on a Core i7 2600 CPU with 16 GB RAM and a Gentoo Linux up-to-date. The AR model has been implemented by means of a home-made java class, whereas the SVM model uses the libSVM library [43].

Please note that this format has been chosen to assess the suitability of the post-process methods defined in Section 3.2 with the two proposed forecasting algorithms, not to compare the AR and SVM models (for such a comparison we refer the readers to [10,13,30,32]).

Figure 7 shows the percentage of days detected as anomalous by the AR and SVM model. These days represent the number of days whose MAPE error is larger than the threshold value k. As expected, the higher the margin error, the lower the amount of anomalous days detected. Further, we can observe a big difference between the number of anomalous days in the two types of datasets: TSO-based ones show much less anomalies than building-based ones. This behaviour stems from their smoother load profile. Note that we cannot expect the method to improve the forecast when it is used often. Finally, as the AR model produces more accurate forecast than the SVM model, it has less anomalies.

Table 3 displays the MAPE results for the day-ahead forecast. The best result for each dataset is shown in bold; grey values correspond to experiments that obtained the same result as the normal method because the dataset did not contain any anomalous days with that parameter configuration. As expected by our previous results, AR outperforms the SVN method globally. In the building datasets, the post-process (slightly) improves the results in almost all of the cases (80% of the cases when using the AR method and 60% when using the SVM). In contrast, only in two of the TSO datasets does the post-process method improve, mainly because these datasets are very regular and large. Still, the difference between the two methods is very small.

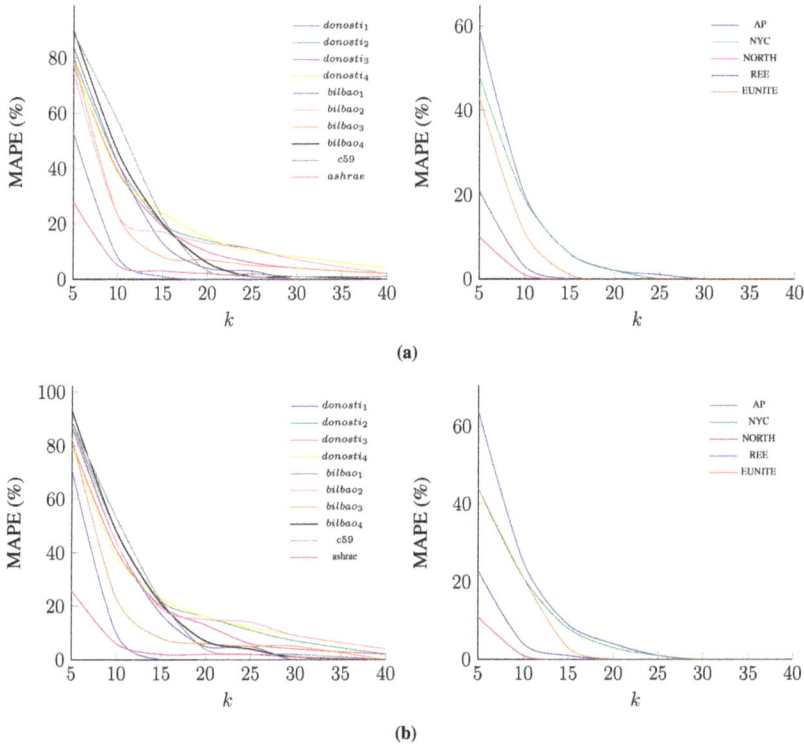

Figure 7. (a) Percentage of anomalous days using the AR model; (b) Percentage of anomalous days using the SVM model.

Note that these tables present the MAPE results for the whole datasets; therefore, the results are dominated by the number of normal days, and the differences are hence not very significant. In addition, we can observe that the MAPE values start increasing below some threshold value k because the post-process considered a normal statistical fluctuation as an error when it was not. Finally, for large values of the k threshold, the error obviously converge to the errors without the post-process as there would not be any anomalous day.

Table 4 displays the MAPE results solely taking into account the records registered when the post-process is working, (*i.e.*, the next two days of the same day-type after an anomalous day is detected). The *Post-process* column lists the MAPE results of the anomalous days when using the post-process method whereas the *Normal* column presents the MAPE result when using the normal method (*i.e.*, without applying any post-process). We only show the results where there has been any improvement.

Table 3. MAPE result for the whole dataset (%). Columns represent the type and the value of the threshold k. The last column has the results without the post-process method. Bold results denote the best results; grey, the absence of anomalous day on that threshold. (**a**) MAPE results of the AR model forecasting (%); (**b**) MAPE result of the SVM model forecasting (%).

Datasets	5	10	15	20	25	30	40	Norm
$donosti_1$	6.42	6.08	6.05	**6.03**	6.05	6.1	0	6.1
$donosti_2$	13.1	12.79	12.52	12.61	12.51	**12.34**	12.35	12.48
$donosti_3$	13.1	12.65	12.19	12.1	12.16	11.97	**11.95**	12.33
$donosti_4$	13.31	13.02	12.85	12.91	12.89	13.02	12.88	**12.86**
$bilbao_1$	10.59	10.23	9.86	**9.77**	9.85	9.95	9.95	9.95
$bilbao_2$	11.68	11.53	**11.48**	11.57	11.51	11.7	11.96	12.04
$bilbao_3$	11.91	11.41	11.18	11.14	11.47	11.07	11.07	**11.07**
$bilbao_4$	11.5	11.18	10.79	**10.49**	10.49	10.49	0	10.64
$c59$	12.11	11.97	11.69	11.14	**11.1**	11.1	11.1	11.24
$ashrae$	4.8	**4.63**	4.63	4.65	4.65	4.65	0	4.66
AP	7.23	7.26	7.25	7.23	7.21	7.2	7.19	**7.19**
NYC	**6.21**	6.34	6.38	6.39	6.39	6.39	0	6.37
$NORTH$	2.82	2.8	2.8	2.8	2.8	2.8	2.8	**2.77**
REE	3.71	3.63	3.63	3.62	3.62	0	0	**3.62**
$EUNITE$	**5.54**	5.86	5.94	5.93	5.93	5.93	0	5.93

(**a**)

Datasets	5	10	15	20	25	30	40	Norm
$donosti_1$	7.14	6.68	6.64	6.64	6.72	6.82	0	**6.38**
$donosti_2$	13.68	13.56	13.34	13.44	13.31	13.05	13.09	**12.99**
$donosti_3$	13.19	12.37	12.41	12.65	13.26	**12.36**	12.36	12.74
$donosti_4$	13.33	13.29	**13.23**	13.48	13.8	14.06	13.59	13.72
$bilbao_1$	10.97	10.82	10.73	**10.37**	10.8	10.85	10.85	10.62
$bilbao_2$	13.84	13.42	13.61	13.51	13.78	14.67	14.73	**13.29**
$bilbao_3$	12	11.47	11.22	10.91	11.66	10.88	10.78	**10.45**
$bilbao_4$	11.87	11.52	11.06	10.89	**10.84**	11.24	0	11.24
$c59$	11.91	11.84	11.88	11.3	**11.26**	11.26	0	11.74
$ashrae$	5	**4.62**	4.64	4.73	4.73	4.73	0	4.8
AP	7.49	7.95	7.88	7.91	7.87	7.81	7.8	**7.78**
NYC	**6.14**	6.22	6.3	6.32	6.32	6.33	6.33	6.31
$NORTH$	2.92	2.91	2.91	2.91	2.92	2.92	2.92	**2.88**
REE	3.88	3.84	3.83	3.83	3.84	0	0	**3.79**
$EUNITE$	**5.73**	6.69	6.75	6.74	6.74	6.74	0	6.75

(**b**)

Table 4. MAPE results taking into account only the days when the post-process is working. Column *Post-process* shows the results using the post-process, while Column *Normal* shows the results without it.

Datasets	AR			SVM		
	k	Post-process	Normal	k	Post-process	Normal
$donosti_1$	20	5.74	9.04	–	–	–
$donosti_2$	30	24.53	25.91	–	–	–
$donosti_3$	40	12.81	22.86	30	18.87	27.68
$donosti_4$	–	–	–	15	15.69	16.59
$bilbao_1$	20	5.54	7.92	20	8.06	12.7
$bilbao_2$	15	10.96	12.52	–	–	–
$bilbao_4$	20	9.7	11.29	25	10.52	17.83
$c59$	25	8.01	11	25	7.97	18.04
$ashrae$	10	8.83	9.08	10	7.93	9.51
NYC	5	6.68	7.07	5	7.1	7.59
$EUNITE$	5	6.18	6.71	5	6.28	7.76

As we can see, regarding *anomalous days* alone, the differences become larger. In this case, the post-processing method performs significantly better, especially in datasets whose loads present large variations as *donosti₃*, where we reduce the error by 50% in those days.

6. Conclusions

In this work, we have assessed a new robust post-processing method for STLF in buildings based on checking whether the prediction error is bigger than a given threshold value. The methodology consists of the use of the work schedule as a helping tool, plus the reinitialisation of the learning set when the MAPE error is bigger than a given threshold value.

We have shown that this methodology prevents burst-errors and manages to adapt quickly to changes in the load curve. Moreover, we have empirically proven that this method improves the building consumption forecast. These results become clearer when examining the error of anomalous days. Unfortunately, this method does not scale very well and our experiments show that the forecasting performance on TSO load curves does not improve (but also does not worsen). Thus, we may conclude that this methodology only pays off in building STLF and other units with non-regular profiles.

Given these results, further works will include trying to improve the results by the combination of STLF algorithms for building, like the ones described in [13], with the robust method explained herewith and test its adaptability to very short-term load forecasting.

Acknowledgments: This work was partially supported by:

- ENERGOS CEN2009-1048 project (funded by Spanish CENIT R&D Programme);
- ITEA2 NEMO&CODED IDI-20110864 and IMPONET TSI-020400-2010-0103 projects (funded by Spanish Industry, Tourism and Commerce Ministry).

References

1. Feinberg, E.; Genethliou, D. Load Forecasting. In *Applied Mathematics for Restructured Electric Power Systems: Optimization, Control, and Computational Intelligence*; Springer: Berlin, Germany, 2005; pp. 269–285.
2. Hinojosa, V.; Hoese, A. Short-term load forecasting using fuzzy inductive reasoning and evolutionary algorithms. *IEEE Trans. Power Syst.* **2010**, *25*, 565–574. [CrossRef]
3. Kyriakides, E.; Polycarpou, M. Short term electric load forecasting: A tutorial. *Stud. Comput. Intell.* **2009**, *35*, 391–418.
4. Alfares, H.; Nazeeruddin, M. Electric load forecasting: Literature survey and classification of methods. *Int. J. Syst. Sci.* **2002**, *33*, 23–34. [CrossRef]
5. Yang, H.; Huang, C. A new short-term load forecasting approach using self-organizing fuzzy ARMAX models. *IEEE Trans. Power Syst.* **1998**, *13*, 217–225. [CrossRef]
6. Jain, A.; Satish, B. Clustering Based Short Term Load Forecasting Using Support Vector Machines. In Proceedings of the IEEE PowerTech Conference 2009, Bucharest, Romania, 28 June–2 July 2009; pp. 1–8.
7. Lin, S.; Lee, Z.; Chen, S.; Tseng, T. Parameter determination of support vector machine and feature selection using simulated annealing approach. *Appl. Soft Comput.* **2008**, *8*, 1505–1512. [CrossRef]
8. González, P.; Zamarreño, J. Prediction of hourly energy consumption in buildings based on a feedback artificial neural network. *Energy Build.* **2005**, *37*, 595–601. [CrossRef]
9. Chakhchoukh, Y.; Panciatici, P.; Mili, L. New Robust Method Applied to Short-Term Load Forecasting. In Proceedings of the IEEE PowerTech Conference 2009, Bucharest, Romania, 28 June–2 July 2009; pp. 1–6.
10. Fernández, I.; Borges, C.; Penya, Y. Efficient Building Load Forecasting. In Proceedings of the 16th IEEE International Conference on Emerging Technologies and Factory Automation (ETFA), Toulouse, France, 5–9 September 2011; pp. 1–8.
11. Hibon, M.; Evgeniou, T. To combine or not to combine: Selecting among forecasts and their combinations. *Int. J. Forecast.* **2004**, *21*, 15–24. [CrossRef]
12. Clemen, R. Combining forecasts: A review and annotated bibliography. *Int. J. Forecast.* **1989**, *5*, 559–583. [CrossRef]

13. Borges, C.; Penya, Y.; Fernández, I. Optimal Combined Short-Term Building Load Forecasting. In Proceedings of the 2011 IEEE PES Innovative Smart Grid Technologies Asia, Perth, Australia, 13–16 November 2011; pp. 1–7.
14. Dong, B.; Cao, C.; Lee, S. Applying support vector machines to predict building energy consumption in tropical region. *Energy Build.* **2005**, *37*, 545–553. [CrossRef]
15. MacKay, D. Bayesian non-linear modelling for the prediction competition. *ASHRAE Trans.* **1994**, *100*, 1053–1062.
16. Darbellay, G.; Slama, M. Forecasting the short-term demand for electricity—Do neural networks stand a better chance? *Int. J. Forecast.* **2000**, *16*, 71–83. [CrossRef]
17. Soares, L.; Souza, L. Forecasting electricity demand using generalized long memory. *Int. J. Forecast.* **2006**, *22*, 17–28. [CrossRef]
18. Wu, C.H.; Tzeng, G.H.; Lin, R.H. A novel hybrid genetic algorithm for kernel function and parameter optimization in support vector regression. *Expert Syst. Appl.* **2009**, *36*, 4725–4735. [CrossRef]
19. Ismail, Z.; Jamaluddin, F. Time series regression model for forecasting Malaysian electricity load demand. *Asian J. Math. Stat.* **2008**, *1*, 139–149. [CrossRef]
20. Ferreira, V.; Pinto-Alves-da Silva, A. Automatic Kernel Based Models for Short Term Load Forecasting. In Proceedings of the 15th International Conference on Intelligent System Applications to Power Systems (ISAP), Curitiba, Brazil, 8–12 November 2009; pp. 1–6.
21. De Menezes, L.M.; Bunn, D.W.; Taylor, J.W. Review of guidelines for the use of combined forecasts. *Eur. J. Oper. Res.* **2000**, *120*, 10–204. [CrossRef]
22. Scott-Armstrong, J. Combining forecasts: The end of the beginning or the beginning of the end. *Int. J. Forecast.* **1989**, *5*, 585–588. [CrossRef]
23. Prudencio, R.; Ludermir, T. Using Machine Learning Techniques to Combine Forecasting Methods. In Proceedings of the 17th Australian Joint Conference on Advances in Artificial Intelligence, Cairns, Australia, 4–6 December 2004; pp. 1122–1127.
24. Song, K.B.; Baek, Y.S.; Hong, D.H. Short-term load forecasting for the holidays using fuzzy linear regression method. *IEEE Trans. Power Syst.* **2005**, *20*, 96–101. [CrossRef]
25. Jin, Y.X.; Su, J. Similarity Clustering and Combination Load Forecasting Techniques Considering the Meteorological Factors. In Proceedings of the 6th World Scientific and Engineering Academy and Society (WSEAS) International Conference on Instrumentation, Measurement, Circuits and Systems, Hangzhou, China, 15–17 April 2007; pp. 115–119.
26. Cortez, P.; Rocha, M.; Neves, J. A Meta-Genetic Algorithm for Time Series Forecasting. In Proceedings of Workshop on Artificial Intelligence Techniques for Financial Time Series Analysis (AIFTSA-01), 10th Portuguese Conference on Artificial Intelligence (EPIA'01), Porto, Portugal, 17–20 December 2001; pp. 21–31.
27. Wang, X.; Smith-Miles, K.; Hyndman, R. Rule induction for forecasting method selection: Meta-learning the characteristics of univariate time series. *Neurocomputing* **2009**, *72*, 2581–2594. [CrossRef]
28. Chakhchoukh, Y.; Panciatici, P.; Bondon, P.; Mili, L. Robust Short-Term Load Forecasting Using Projection Statistics. In Proceedings of 3rd IEEE International Workshop on the Computational Advances in Multi-Sensor Adaptive Processing (CAMSAP), Aruba Island, Dutch Antilles, 13–16 December 2009; pp. 45–48.
29. Penya, Y.K.; Nieves, J.C.; Espinoza, A.; Borges, C.E.; Peña, A.; Ortega, M. Distributed semantic architecture for smart grids. *Energies* **2012**, *5*, 4824–4843. [CrossRef]
30. Penya, Y.; Borges, C.; Agote, D.; Fernandez, I. Short-Term Load Forecasting in Air-Conditioned Non-Residential Buildings. In Proceedings of the 20th IEEE International Symposium on Industrial Electronics (ISIE), Gdansk, Poland, 27–30 June 2011; pp. 1359–1364.
31. Hsu, C.W.; Chang, C.C.; Lin, C.J. A Practical Guide to Support Vector Classification. 2010. Available online: http://www.csie.ntu.edu.tw/~cjlin/papers/guide/guide.pdf (accessed on 19 December 2012).
32. Borges, C.; Penya, Y.; Fernández, I. Evaluating combined load forecasting in large power systems and smart grids. *IEEE Trans. Ind. Inform.* **2012**. [CrossRef]
33. Cristianini, N.; Shawe-Taylor, J. *An Introduction to Support Vector Machines: And Other Kernel-Based Learning Methods*, 1st ed.; Cambridge University Press: Cambridge, UK, 2000.

34. International Electrotechnical Commission. *IEC60870: Telecontrol Equipment and Systems-Part 5: Transmission Protocols-Section 102: Companion Standard for the Transmission of Integrated Totals in Electric Power Systems*; IEC: Geneva, Switzerland, 1996.

35. Kreider, J.; Haberl, J. Predicting hourly building energy usage: The great energy predictor shootout–Overview and discussion of results. *ASHRAE Trans.* **1994**, *100*, 1104–1118.

36. Felice, M.D. Load, Weather and Occupancy Data for C59 ENEA Building. Available online: http://www.matteodefelice.name/research/resources/ (accessed on 21 December 2012).

37. Cancelo, J.; Espasa, A.; Grafe, R. Forecasting the electricity load from one day to one week ahead for the Spanish system operator. *Int. J. Forecast.* **2008**, *24*, 588–602. [CrossRef]

38. PJM Interconnection System Operator. Day-Ahead Scheduling Manual. Aviable online: http://www.pjm.com/sitecore%20modules/web/~/media/training/core-curriculum/ip-gen-201/gen-201-scheduling-process.ashx (accessed on 29 March 2013).

39. New York Independent System Operator. Day-Ahead Scheduling Manual. Aviable online: http://www.nyiso.com/public/webdocs/markets_operations/documents/Manuals_and_Guides/Manuals/Operations/dayahd_schd_mnl.pdf (accessed on 29 March 2013).

40. Eunite Competition Comite. Electricity Load Forecast using Inteligent Adaptative Technology. 2001. Available online: http://neuron.tuke.sk/competition/instructions.php (accessed on 19 April 2011).

41. Arlot, S.; Celisse, A. A survey of cross-validation procedures for model selection. *Stat. Surv.* **2010**, *4*, 40–79. [CrossRef]

42. Hyndman, R.J.; Koehler, A.B. Another look at measures of forecast accuracy. *Int. J. Forecast.* **2006**, *22*, 679–688. [CrossRef]

43. Chang, C.C.; Lin, C.J. LIBSVM: A library for support vector machines. *ACM Trans. Intell. Syst. Technol.* **2011**, *2*, 1–27. Available online: http://www.csie.ntu.edu.tw/~cjlin/libsvm (accessed on 29 March 2013). [CrossRef]

energies

MDPI

Article

Short-Term Power Forecasting Model for Photovoltaic Plants Based on Historical Similarity

Claudio Monteiro [1], Tiago Santos [1], L. Alfredo Fernandez-Jimenez [2], Ignacio J. Ramirez-Rosado [3,* and M. Sonia Terreros-Olarte [2]

[1] Faculty of Engineering, University of Porto, Dr. Roberto Frias, Porto s/n 4200-465, Portugal; cdm@fe.up.pt (C.M.); ee04189@fe.up.pt (T.S.)
[2] Electrical Engineering Department, University of La Rioja, Luis de Ulloa 20, Logroño 26004, Spain; luisalfredo.fernandez@unirioja.es (L.A.F.-J.); maria-sonia.terreros@alum.unirioja.es (M.S.T.-O.)
[3] Electrical Engineering Department, University of Zaragoza, Maria de Luna 3, Zaragoza 50018, Spain
* Author to whom correspondence should be addressed; ignacio.ramirez@unizar.es; Tel.: +34-976-761-929; Fax: +34-976-762-226.

Received: 28 December 2012; in revised form: 15 May 2013; Accepted: 15 May 2013; Published: 22 May 2013

Abstract: This paper proposes a new model for short-term forecasting of electric energy production in a photovoltaic (PV) plant. The model is called HIstorical SImilar MIning (HISIMI) model; its final structure is optimized by using a genetic algorithm, based on data mining techniques applied to historical cases composed by past forecasted values of weather variables, obtained from numerical tools for weather prediction, and by past production of electric power in a PV plant. The HISIMI model is able to supply spot values of power forecasts, and also the uncertainty, or probabilities, associated with those spot values, providing new useful information to users with respect to traditional forecasting models for PV plants. Such probabilities enable analysis and evaluation of risk associated with those spot forecasts, for example, in offers of energy sale for electricity markets. The results of spot forecasting of an illustrative example obtained with the HISIMI model for a real-life grid-connected PV plant, which shows high intra-hour variability of its actual power output, with forecasting horizons covering the following day, have improved those obtained with other two power spot forecasting models, which are a persistence model and an artificial neural network model.

Keywords: power forecasting; solar energy; data mining; genetic algorithm

1. Introduction

The expansion of power plants based on renewable energy sources has experienced an important boost in recent years. Increases in prices of traditional energy sources, the threat of climate change, and policies implemented by national governments have propelled expansions of this kind of power plants. Renewable energies with greater integration in electric power systems are wind energy and solar photovoltaic energy, and they are increasing their integration with time: in 2050, wind energy is expected to provide 12% of the global electricity consumption [1], while PV energy is expected to provide 11% [2]. The integration of the electric energy generated by these power plants into electric power networks is not exempt from problems, which are mainly due to the variability and volatility of renewable resources. Accurate forecasts of power production at wind farms or at PV plants have direct implications on the economic operation of power systems [3,4] and on the economic results of the plants whose generated energy is sold in electricity markets [5]. These economic reasons have driven the development of short-term power forecasting models for wind farms or for relatively large grid-connected PV plants.

The development of numerical weather predictions (NWP) tools has helped in the advance of new power forecasting models for electric plants based on renewable resources, providing new input variables. These NWP tools have the objective, from a set of initial conditions, to supply information regarding the state of the atmosphere for a given time horizon. Models underlying NWP tools can be classified into global models and regional/mesoscale models. Global models simulate the behavior of the atmosphere to a global (worldwide) scale, and regional/mesoscale models simulate the behavior of the atmosphere for more limited areas such as continents, countries or regions. The use of weather forecasted variables, mainly radiation and temperature, can help to improve short-term power forecasting models for PV plants.

PV systems are the most direct way to convert solar radiation into electric power. Traditionally, small PV systems have been used to produce electricity for low power applications in isolated areas (isolated from electric power networks). Installation cost reductions, subsidies, and attractive feed-in tariffs, have propelled constructions of relatively large PV plants connected to electric grids. PV plants, connected to medium (or high) voltage electric networks, can have capacity of tens (or, in some cases, even a few hundred) of MW.

In countries with an operative day-ahead electricity market, large power plants based on renewable energies can act, as any other electricity producer, providing power generation sale offers to electricity markets. Obviously, producers corresponding to power plants based on variable renewable resources, such as wind or solar radiation, use forecasts of hourly energy generation to prepare energy sale offers. The use of these forecasts presents a risk: in electricity markets, when power producers are not following their schedule (that presented to the Market Operator), they are penalized with retributions lower than those established in markets for those hours with deviation between the energy actually produced and that presented in offers, so for a PV power producer, high quality forecasting systems are needed for reducing penalties in electricity markets, and for optimizing profits.

In the last decade, tens of short-term wind power forecasting models have been described in the international literature. Nevertheless, despite the fact that future contributions of PV plants to the global electricity consumption will be comparable to that corresponding to wind farms, short-term forecasting models for PV plants are in their early stages. Most of the published works corresponding to short-term forecasting models for PV plants are oriented to solar radiation predictions [6–9], while only a few works describe models aimed at directly forecasting the hourly power production in PV plants [10–17]. Most of these published models are based on artificial neural networks (ANNs). A hybrid approach with the combination of a data filtering technique based on wavelet transformation and ANNs is presented in [10] and used to obtain one-hour-ahead power output forecasts. Several forecasting techniques are evaluated and compared in [11] for predicting the power output of a PV plant with forecasting horizons of 1 and 2 h ahead; the best results are obtained with models based on ANNs optimized with Genetic Algorithm (GA). A model based on recurrent neural networks to forecast hourly insolation and temperature for the next 24 h is described in [12]; both forecasts are used to calculate the hourly power generation in the PV plant. Support vector machines are used in [13] and [14] to forecast directly the hourly power generation for the next 24 h. In [15] a multilayer perceptron ANN optimized with GAs is used to provide hourly power generation in a PV installation for the 24 h of the next day. In all these works describing forecasting models with horizons covering 24 h, some forecasted weather variables (such as global solar radiation, temperature, relative humidity or cloudiness, obtained from a NWP tool), are used as inputs in the forecasting model. Even these forecasted weather values are used in [16] to forecast the hourly power production for all PV plants in a local or regional scale. Genetic programming of evolution of fuzzy rules has been proposed in [17] to estimate the output of a PV plant, allowing the selection of the best forecasting model.

But in all the referenced works, the proposed forecasting models only provide the electric power point (spot) forecasts. They do not supply any additional information that enables the evaluation of the risk associated with the use of such forecasted values. Although several wind power forecasting models published recently deal with this evaluation, none have been applied to PV plants.

This paper presents a new short-term power forecasting model for PV plants, named HIstorical SImilar Mining (HISIMI) model, which provides the user with useful dimensions of electric power forecast. Thus, this forecasting model, based on transitions (in the past time) between different power intervals of electric generation in the PV plant, is able to achieve the point forecast, and also the uncertainty associated with that value. HISIMI model uses a database comprised of historical values of weather variables forecasted by a mesoscale NWP tool and the corresponding historical real power production in the PV plant. Spot (point) forecasts obtained with the HISIMI model have been compared with those obtained from a multilayer perceptron based model, and the "persistence" model, with the same input database: the HISIMI model has shown lower forecasting errors. Furthermore, this model also provides the uncertainty associated with the point forecast, increasing the value of the forecasting information. This uncertainty is provided by the HISIMI model in the form of the probability value that the PV power production is included in each one of the above mentioned power intervals.

The paper is structured as follows: Section 2 presents the structure of the proposed power forecasting model (HISIMI); Section 3 describes the methodology followed in the development of the HISIMI model, which structure is optimized with a genetic algorithm; Section 4 shows the results obtained with the HISIMI model in the forecast of PV power production (point or spot forecast) in a grid-connected PV plant; Section 5 analyses the additional and useful forecasting information supplied by this new model in the form of uncertainty representation; lastly, Section 6 presents the conclusions.

2. PV Power Forecasting Model

In this paper, the PV power forecasting model uses historical data collected from the PV plant under study; this model performs a "kind of search" of this historical database aimed at utilizing similar historical cases regarding electric power transitions in order to forecast the electric power generation. The proposed search mechanism is based on data mining techniques. Thus, the model was named HIstorical SImilar MIning (HISIMI).

The data, that comprises the model's historical database, are forecasted values for weather variables obtained with a mesoscale NWP tool and the corresponding electric power generation measures obtained from the PV plant using a Supervisory Control and Data Acquisition (SCADA) system. In order to extract information regarding electric power transitions between consecutive past time instants, the historical similar mining mechanism is applied to the cases in the historical database. Thus, good power forecast results require the use of a database with historical cases of good quality, with a high volume of reliable information, which also implies the need to find mechanisms to deal with such information, capable of searching the most relevant historical cases of power transitions to forecast the electric power corresponding to the current case. After obtaining the power transition information, an array called Probability Matrix (PM_{t+k}) is created, which contains power transition probabilities between the future instants $t+k-1$ and $t+k$, where t is the instant when the forecast is generated -present instant-; and k is the number of time steps of the forecast (forecasting horizon). Thus, the HISIMI model provides valuable forecast information, using probability values: prediction of uncertainties, associated with electric power forecasted values, and electric power forecasts (point or spot forecasts). The following subsections contain more detailed explanations regarding the HISIMI model.

2.1. Database of the Forecasting Model

The model uses a database that contains records with the historical values of variables corresponding to the PV plant. Thus, a record constitutes a historical case which contains values forecasted by a NWP tool, for several weather variables (two values for each variable, which correspond to instants $c-1$ and c); the value of the solar hour for both instants; and also the two corresponding real PV power production values (P_{c-1} and P_c) of the PV plant. The pairs of past instants $c-1$ and c are necessary to model power transitions. The index c ranges from 2 to present instant t and it is expressed in hours. At this point, notice that the size of the records depend on the number of forecasted weather variables and the number of the remaining variables related to the short-term forecast of power generation in the PV plant.

2.2. Mechanism Based on Data Mining (MDM)

This mechanism used to process electric power transitions information plays a key role in the success of the developed model. The purpose of the MDM is to examine historical cases, and to assign different weight values to such cases according to similarity between them and the current case. This allows selecting—via weights—only information concerning historical cases relevant to the current case.

The proposed MDM is based on a local Gaussian function (Figure 1) that is expressed in Equation (1):

$$g(v, \mu, \sigma) = \frac{1}{\sigma\sqrt{2\pi}} e^{-\left(\frac{(v-\mu)^2}{2\sigma^2}\right)} \quad v \in \Re \tag{1}$$

where v represents the prospection variable; μ represents the "mean or central" value assigned to the prospection variable; σ represents the "standard deviation" assigned to the prospection variable, that defines a width of the range of exploration; and g represents the weight function corresponding to a value of the prospection variable v.

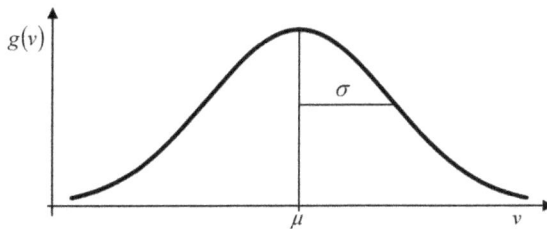

Figure 1. Graphical representation of local Gaussian function.

Note that, in Figure 1, the proposed local Gaussian function allows the selection of the central or mean value (μ) and the standard deviation value, σ. The selected central values for the Gaussian functions correspond to the current case values, and the standard deviation values can be chosen by an optimization process, as explained later. With these two parameters (μ and σ), which determine the Gaussian function, the weight values can be obtained according to the neighborhood or similarity between the historical cases and the current case. Thus, the MDM identifies historical cases with the highest similarity to the case for which it is intended to provide the electric power forecast, that is, the MDM determines higher weight values for a series of cases of electric power transitions that will be used to provide the forecast information.

Because an optimization process (to be explained later) can select the best inputs (variables from records) to be used by the model, assuming a HISIMI model that uses l variables, the associated value that defines the "similarity" with the current case, FH_c, is defined in Equation (2):

$$FH_c = \prod_{i=1}^{l} g_{i,c} \tag{2}$$

where $g_{i,c}$ represents the weight value associated with the local Gaussian function corresponding to the input variable i.

2.3. Power Intervals

The objective of the model is to achieve a representation of power transitions that occurred in the past with similar input variable values. Even with a reduced database, there could be a large set of different power transitions. Thus, transitions between specific electric power values are not considered, but transitions between intervals of electric power values (power intervals) of the PV plant are considered.

This aspect requires defining a set of power intervals that allows transforming the electric power continuous variable into a discrete variable. So, a total of n non-overlapped power intervals with the same width, covering all possible values for the electric power generation in the PV plant, are defined. Each interval is defined by its minimum and maximum power values (in kW), a and b, respectively. The average value of the interval m is defined in Equation (3), where a_m and b_m correspond to the minimum and maximum values of that interval:

$$I_m = \frac{1}{2}(a_m + b_m) \tag{3}$$

With the average values of all the power intervals, the average power values vector, $[AP]$, is created, as shown in Equation (4), where $I_1 < I_2 < \ldots < I_n$:

$$[AP] = \begin{bmatrix} I_1 \\ I_2 \\ \vdots \\ I_n \end{bmatrix} \tag{4}$$

This technique groups power values in a neighborhood that will be represented by the average power value of the corresponding interval.

2.4. Probability Matrix (PM)

We define two discrete random variables: X associated with the interval corresponding to electric power value at future instant $t+k-1$, and Y associated with the interval corresponding to electric power value at future instant $t+k$. Thus, the electric power interval $x \in X$ varies (and also the interval $y \in Y$) from 1 to n. Probabilities of transition of the power interval from instant $t+k-1$ to the next one, $t+k$, can be expressed by a square matrix with n rows and n columns applying the mechanism MDM. This matrix, named as the pseudo-probabilities matrix for hour $t+k$, PPM_{t+k}, is shown in Table 1. Each element in this matrix represents the pseudo-probability of a power transition from one power interval in instant $t+k-1$ (interval x, that is, row x in the matrix) to another power interval in instant $t+k$ (interval y, that is, column y in the matrix). The element $PPM_{t+k}(x, y)$, corresponding to row x and column y, is calculated using the sum of values FH_c from Equation (2) considering all cases in the database.

The "normalization" of values of the matrix of Table 1 to values between 0 and 1, leads to the matrix PM_{t+k}, which contains the bivariate distribution of power transitions, from one power interval x (associated with instant $t+k-1$) to another power interval y (associated with instant $t+k$). Note that results are obtained after applying the mechanism MDM, which defines the space of global events of power transitions.

The elements of the PM_{t+k} matrix can be associated with a joint probability distribution $f_{XY}(x, y)$ (Table 2), that satisfies Equation (5):

$$\begin{aligned} f_{XY}(x,y) &\geq 0 \\ \sum_x \sum_y f_{XY}(x,y) &= 1 \end{aligned} \tag{5}$$

where $f_{XY}(x, y)$ represents the probability that the interval of the electric power variable is x in a given instant, and y in the following one, *i.e.*, $P(X = x, Y = y)$.

Table 1. Representation of a pseudo-probabilities matrix for *n* power intervals.

Pseudo-probabilities		Power interval in *t+k*			
		1	2	. . .	*n*
	1	. . .	. . .	. . .	. . .
Power interval in t+k−1	2	. . .	. . .	. . .	. . .
	. . .	. . .	. . .	. . .	. . .
	n	. . .	. . .	. . .	. . .

Table 2. Power transition probability matrix.

$f_{XY}(x, y)$		*y*			
		1	2	. . .	*n*
	1	$f_{XY}(1, 1)$	$f_{XY}(1, 2)$	. . .	$f_{XY}(1, n)$
x	2	$f_{XY}(2, 1)$	$f_{XY}(2, 2)$	. . .	$f_{XY}(2, n)$
	. . .	. . .	. . .	. . .	. . .
	n	$f_{XY}(n, 1)$	$f_{XY}(n, 2)$	. . .	$f_{XY}(n, n)$

Then, two marginal probability functions f_{m1} and f_{m2} can be defined, for each transition in *t+k*, by Equations (6) and (7) respectively:

$$f_{m1}(x) = f_X(x) = P(X = x) = P(X = x, Y = 1) + \cdots + P(X = x, Y = n) = \sum_{R_x} f_{XY}(x, y) \quad (6)$$

where R_x denotes the set of all $f_{XY}(x,y)$ for which $X = x$:

$$f_{m2}(y) = f_Y(y) = P(Y = y) = P(X = 1, Y = y) + \cdots + P(X = n, Y = y) = \sum_{R_y} f_{XY}(x, y) \quad (7)$$

where R_y denotes the set of all $f_{XY}(x,y)$ for which $Y = y$.

Note that for each new forecast (for future time instant *t+k*), a probability matrix is created and therefore also a bivariate distribution.

2.5. Model Outputs

2.5.1. Uncertainty Prediction

Some of the main outputs of the HISIMI model are the probability values of power transitions for each future instant *t+k* (transition from instant *t+k−1* to the instant *t+k*). This information can be processed in order to obtain different types of useful predictions. In order to obtain predictions of forecast uncertainty, a new discrete probability distribution for each instant can be obtained as the product of two discrete distributions, f_{m1} and f_{m2}, obtained for two consecutive forecasting instant, *t+k−1* and *t+k*, denoted as $f_{m1;t+k-1}$ and $f_{m2;t+k}$.

We define an "uncertainty vector" for instant *t+k*, $[u]_{t+k}$, for the uncertainty prediction, as the product of the values of the marginal probability functions defined in Equations (6) and (7), as is given in Equation (8).

$$[u]_{t+k} = \begin{bmatrix} f_{m1;t+k-1}(1) \times f_{m2;t+k}(1) \\ f_{m1;t+k-1}(2) \times f_{m2;t+k}(2) \\ \vdots \\ f_{m1;t+k-1}(n) \times f_{m2;t+k}(n) \end{bmatrix} \quad (8)$$

Afterwards, the values of the vector $[u]_{t+k}$, defined in Equation (8), are normalized (to values between 0 and 1), leading to a new vector, $[u_n]_{t+k}$, in which each element of this new vector, corresponding to a power interval, is associated with the probability that the forecasted electric power value belongs to that interval; thus, this vector gives a measure of uncertainty associated with electric power forecasts.

2.5.2. Point Forecast

The point forecast PF_{t+k} (in kW) can be obtained by computing the expected power value for a future instant $t+k$ as seen in Equation (9):

$$PF_{t+k} = \sum_y u_{n,t+k}(y) \times AP(y) \tag{9}$$

where $AP(y)$ is the element of the vector $[AP]$ corresponding to the power interval y, and $u_{n,t+k}(y)$ is the element of the vector $[u_n]_{t+k}$ corresponding to the power interval y.

3. Optimization of the PV Power Forecasting Model

The proposed HISIMI model includes a set of parameters whose values can be optimized. These parameters correspond to the number of power intervals, and the standard deviation of the Gaussian functions used for the prospection variables. In order to optimize the values of these parameters, an optimization process ruled by a genetic algorithm (GA) [18] was developed. In this process, the selection of the best input (prospection) variables, among those available ones, was also included. All available input variables must be normalized in the range 0 to 1. This normalization allows using the same range in the standard deviation of all the prospection variables.

Binary encoding with 1 and 0 is used to store the value of these parameters in the chromosome that defines the solution stored in each individual. The structure chosen for the chromosome used by the GA is shown in Figure 2. The chromosome is composed of three or more genes, each one with a fixed size. First p bits, which compose the first gene, correspond to the inputs (prospection) variables used by the HISIMI model among those available (we suppose a total of p available variables in the historical database), that is, an "1" value in the bit bi_j means that the input j is used by the model as prospection variable, while a "0" value means that the input j is not used by the model. At least one of these first p bits must be activated (value "1") because the model represented by the individual needs one or more inputs (prospection variables). The second gene, with a 6 bits size, corresponds to the number of intervals minus two, expressed in binary, used by the model: a value "000000" means 2 intervals, while a value "111111" means 65 intervals. The third gene corresponds to the standard deviation of the Gaussian function for the first input (prospection) variable, σ_1 (the first variable selected as input); it is composed of 16 bits, and its value is equal to the binary number contained in the 16 bits plus one and divided by 32,768. So, the standard deviation can take values from 2^{-15} to 2 (remember that input variables are normalized). The fourth gene, if available, corresponds to the standard deviation of the Gaussian function of the second (prospection) variable selected as input, and so on for the following genes. In Figure 2, the standard deviation of the Gaussian function for the last input variable selected is σ_l, assuming that l inputs have been selected in the first gene. The maximum number of genes is equal to the number of available input variables plus the two first genes.

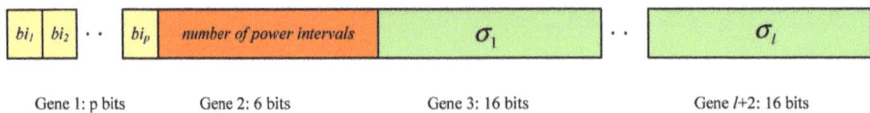

| bi_1 | bi_2 | $\cdot\,\cdot$ | bi_p | number of power intervals | σ_1 | $\cdot\,\cdot$ | σ_l |

| Gene 1: p bits | Gene 2: 6 bits | Gene 3: 16 bits | Gene *l*+2: 16 bits |

Figure 2. Structure of the chromosome used in the optimization of the HISIMI model.

As a first step in the optimization process, an initial population is created randomly: the value of each bit in each individual is randomly assigned. After creating the two first genes, the remaining genes that compose the individual are completed according to the number of variables selected (as inputs of the HISIMI) in the first gene.

The fitness function for the GA optimization was the inverse of the RMS error with the data set used in the construction of the HISIMI model. RMS (root mean square) error is defined in Equation (10), where N represents the number of data evaluated, P_i the real power generation value and $\hat{P}_i$ the value obtained (forecasted) with the model, and the index i covers all instants corresponding to the data set which error is evaluated:

$$RMS = \sqrt{\frac{1}{N}\sum_i (P_i - \hat{P}_i)^2} \qquad (10)$$

So, individuals that represented better forecasting models would achieve greater fitness values. In the creation of a new population, roulette wheel selection, elitism (only the best individual), two-point crossover and mutation were used. After a sufficient number of generations, the parameters of the best HISIMI model were obtained, that is, the optimization process selected the inputs used for the best model, among those available inputs, the number of power intervals for the electric power generated by the PV plant, and the values of the standard deviations of the Gaussian functions used in the MDM mechanism.

In order to prevent over-fit of the HISIMI model to the data used to build the model, we used 5-fold cross-validation [19]: The data set used to develop the model was divided into 5 subsets of approximately equal size. The evaluation of the model is carried out in five stages; in each stage, one subset is taken as the cross-validation data set, while the other four subsets are used to build the model. The RMS error of the studied model is the average of the RMS errors with the cross-validation data sets in the five stages.

4. Model Testing

The methodology described in the previous two sections was applied to develop a short-term forecasting model for a grid-connected PV plant. The plant, with ground mounted fixed panels and a capacity of 2.8 MWp, is located in Spain. The data available to develop the model include the hourly power generation in the PV plant for a whole year. The data show high intra-hour variability of the power output of the PV plant. Figure 3 illustrates the percentage of hours with power output variations of more than 10%, 20%, *etc.*, with respect to power rating of the PV plant, in a monthly basis, for the period between 09:00 to 14:00 (solar hour). In that figure, the vertical axis represents the percentage of hours, in which the absolute difference (variation) for the power output of the PV plant, from one hour to the following one, is greater than 10%, 20%, *etc.*, of the power rating. Notice that at least 30% of hours presents variability over 10% (280 kW) of power rating (2.8 MW) for all the months. Furthermore, for the data of the whole year, 43% of hours in the considered period presents variability over 10% of the power rating.

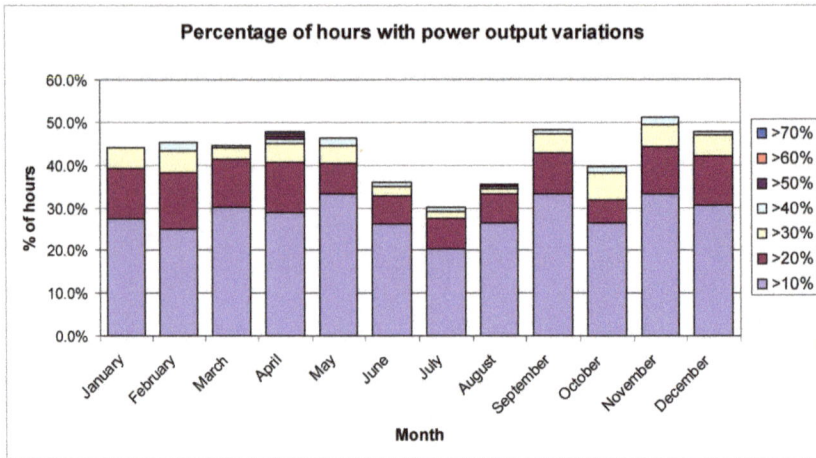

Figure 3. Percentage of hours with power output variations with respect to power rating of the PV plant.

The available data also include forecasted values for the hourly average surface shortwave radiation (v_1) and temperature (v_2) obtained with an NWP tool. This tool was the Weather Research and Forecasting (WRF) model [20], a mesoscale NWP model that can simulate atmospheric dynamics and provide numerical predictions for a wide set of weather variables in a selected geographic zone. The hourly average surface shortwave radiation and temperature values correspond to those forecasted (with the NWP tool) with the data assimilation (moment when real weather measures were supplied to the model to predict the future values) of the hour 00:00. The forecasted hourly average values include all the values for the next 24 h, making that forecasting horizons for the HISIMI model range from 1 to 24 h. Obviously, the maximum forecasting horizon with the HISIMI model coincides with that of the weather variables forecasted with the NWP tool (24 h in our case).

Because the production in a PV plant is very dependent on the solar hour, we included two variables in the historical database to represent it (named v_3 and v_4). These two variables are expressed in Equation (11), where h corresponds to the solar hour for the location of the PV plant for the corresponding instant:

$$\begin{cases} v_3 = \sin\left(2\pi\frac{h-12}{24}\right) \\ v_4 = \cos\left(2\pi\frac{h-12}{24}\right) \end{cases} \tag{11}$$

So, a record in the database contains ten values: two values for the forecasted hourly surface shortwave radiation, two values for the hourly average surface temperature, two values for variable v_3, two values for variable v_4, and two values for the hourly power production in the PV plant. The two values of each variable correspond to two consecutive instants, $c-1$ and c.

The database with the historical cases was divided into two sets: 80% of the records were used as the training set, while the remaining 20% of the records were used as the testing set. Only the training set was used to build the HISIMI forecasting model, while the testing set was only used for comparative purposes with other forecasting models. Figure 4 shows the variations in the power output of the PV plant during the diurnal period of 9:00–14:00 (solar hour) for both data sets of training and testing. The percentages of hours with variations over 10% of the power rating are quite similar for such data sets.

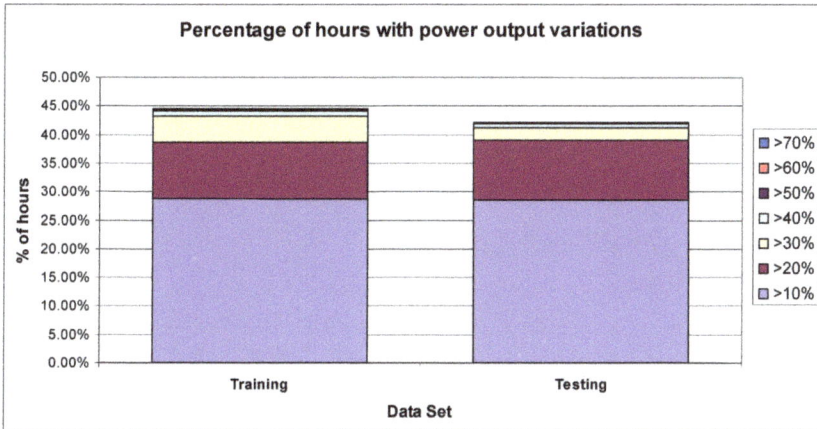

Figure 4. Percentage of hours with power output variations with respect to power rating of the PV plant, for the data sets of training and testing.

The structure of the HISIMI model was optimized with a genetic algorithm (described in Section 3). The population size was 50 individuals and the number of generations 50. The crossover rate was 90% and the mutation rate 2%. Elitism was applied copying the best individual from one generation to the following one. The fitness function was the inverse of the RMS error with the data of the training set using 5-fold cross-validation. The model obtained after the optimization process used the input variables v_1 (forecasted hourly average surface shortwave radiation), v_2 (forecasted hourly average surface temperature), and v_4. The number of power intervals was 9 and the range of any interval was 314 kW. The two extremes of the last interval (ninth) were centered on the maximum value of the power output of the PV plant for the training data set (2512 kW in our case) and the ones of first interval were centered on the minimum power output, *i.e.*, they were centered in 0 kW. Thus the first interval spanned from −157 to 157 kW, the second from 157 to 471 kW, and so on until the ninth interval from 2355 to 2669 kW. The standard deviations for the three used inputs of the HISIMI model were 0.314453125 (for v_1), 0.193359375 (for v_2) and 0.076171875 (for v_4). Figure 5 plots the RMS error with the data of the training set, using the 5-fold cross-validation, throughout the optimization process for the best individual in each generation and the average value of RMS error for all the individuals in each generation.

Energies **2013**, *6*, 2624–2643

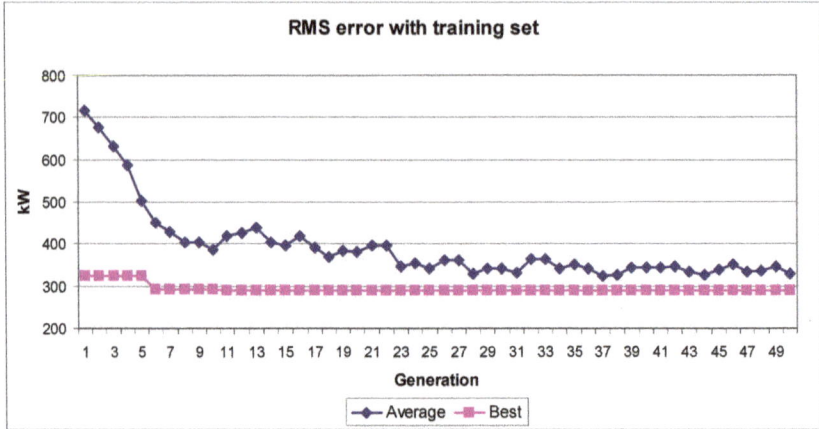

Figure 5. RMS error of the best individual and average RMS error of all the individuals in each generation.

Lastly, the obtained HISIMI model was applied to the testing set. The RMS error for all the data in the testing set (forecasting horizons from 1 to 24 h) was 283.89 kW, just 10.14% with respect to the total capacity of the PV plant.

In order to evaluate the HISIMI model performance, two other forecasting models were built for comparative purposes. The first one was a variation of the persistence model. The classical persistence model offers, as forecast for any horizon, the last known value, *i.e.*, the average power generation value in the last hour. In our case, we have modified the persistence model so that it offers the power generation in the PV plant in the previous day at the same hour as that corresponding to the forecasting horizon; this variation of the persistence model was used in [15] for comparative purposes. The second model was an artificial neural network based model: a multilayer perceptron neural network, MLP, with one hidden layer, which could use any of the available variables as inputs, and offered only one output: the hourly average power production in the PV plant. 75% of the cases in the training data set was used to train the network, while the remaining 25% of the cases was used as the cross-validation set. The structure of the MLP based model was also optimized with a genetic algorithm. The transfer function for the neurons in the hidden layer was the hyperbolic tangent and a linear hyperbolic tangent function was used for the output neuron. In the optimization process, the number of neurons in the hidden layer, the inputs used by the network, and the parameters of the back-propagation training algorithm (learning factor and momentum) were selected. The population size was 50 individuals and 50 generations were completed. The final MLP neural network obtained after the optimization process had 15 neurons in the hidden layer. Table 3 summarizes the main parameters of the optimization of the HISIMI and MLP models.

Table 3. Main parameters of the optimization of the HISIMI and MLP models.

Model	HISIMI	MLP
Population size	50	50
Number of generations	50	50
Crossover rate	90%	90%
Mutation rate	2%	1%
Inputs selected	v_1, v_2, v_4	v_1, v_3, v_4
Power Intervals	9	-
Neurons in hidden layer	-	15

Once these comparative models were built, they were applied to forecast the PV plant's hourly power generation for the data of the testing data set. RMS errors were 286.11 kW for the MLP based model, and 445.48 kW for the persistence model. Therefore, the optimized HISIMI model obtained better results than the other two models (since the HISIMI model achieved the aforementioned 283.89 kW). The improvement in RMS error with respect to the results obtained with another model is calculated by Equation (12), where $RMS_{reference}$ corresponds to the RMS error obtained with the model used as reference, and RMS_{model} corresponds to the RMS error of the compared model. The RMS forecasting error for the HISIMI model was 0.8% better than that obtained with the MLP model, and 36.3% better than that obtained with the persistence model. Table 4 summarizes the forecasting results obtained with the three models:

$$\text{Improvement (\%)} = \frac{RMS_{reference} - RMS_{model}}{RMS_{reference}} \cdot 100 \qquad (12)$$

Table 4. Summary of forecasting results.

Forecasting Results	HISIMI	MLP	Persistence
RMS error (kW)	283.89	286.11	445.48
Normalized RMS (%)	10.14	10.22	15.91
Improvement with respect to Persistence (%)	36.3	35.8	-
Improvement with respect to MLP (%)	0.8	-	-

Figure 6 shows real power production and the corresponding hourly PV power spot forecast obtained with the HISIMI and MLP models for three consecutive days (in the testing data set), which are cloudy and rainy days. Power forecasts, from the studied models, were carried out in the first hour in the morning, covering all hours of the day. Figure 7 represents the scatters plots of forecasted values *versus* actual values of power output for HISIMI and MLP models.

Figure 8 shows the histograms of the absolute forecasting errors, for both models (HISIMI and MLP), for all diurnal hours in the testing data set. Absolute errors are expressed, in the horizontal axis of Figure 8, in percentage with respect to the power rating of the PV plant. The vertical axis represents the percentage of diurnal hours in the testing data set. Absolute errors for both models are quite similar, although the HISIMI model presents more hours with lower errors. For example, in 37.65% of the hours, the absolute forecasting error of the HISIMI model remains under 2.5% of the power rating, while for the MLP model this percentage of hours is 36%.

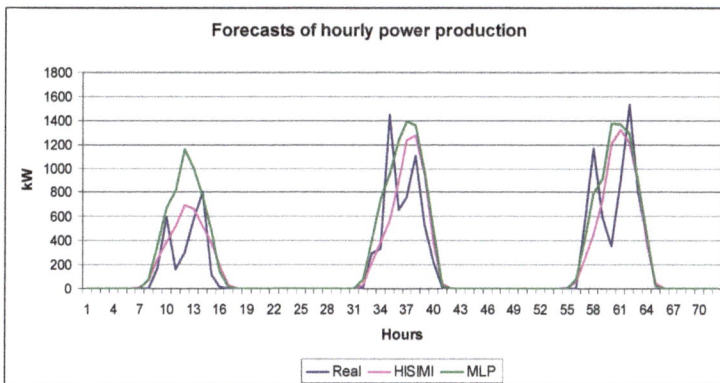

Figure 6. Forecasts of the hourly power production for three cloudy and rainy days in the testing set.

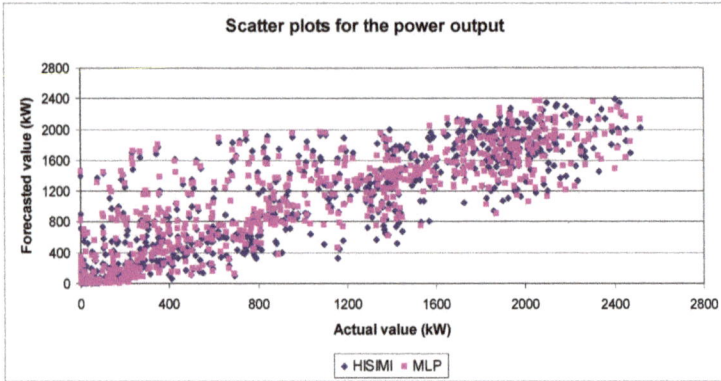

Figure 7. Scatter plots of forecasted values *versus* actual values of power output for HISIMI and MLP models.

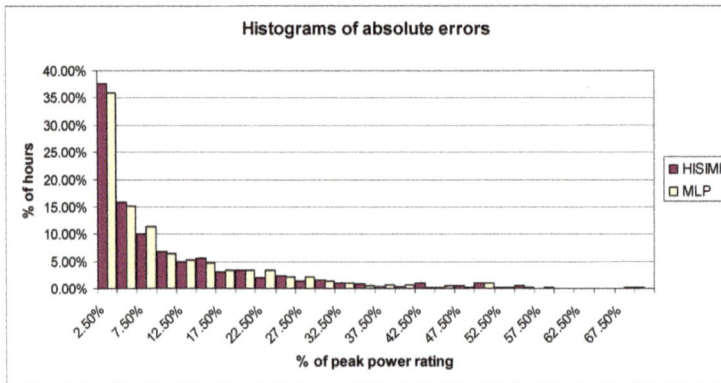

Figure 8. Histograms of absolute errors for the HISIMI and MLP models in the testing data set.

Thus, spot forecasting results obtained by HISIMI model are better than those obtained with the other two models (MLP model and persistence model); furthermore, a key advantage of the HISIMI model is the uncertainty prediction (probabilities) associated with the numerical point (spot) forecasting value (analyzed in the following Section 5). The two other forecasting models are not capable of providing such useful forecasting information (probabilities).

5. Analysis of Information Provided by the PV Power Forecasting Model

Most of forecasting models described in the literature only provide spot power forecasts (point forecasts), while the HISIMI model provides more complete information based on power transitions for each forecasting time: spot forecasts and uncertainty predictions are computed using discrete probability functions.

In our case of a real-life grid-connected photovoltaic plant, the daily PV power production forecast was carried out with the data of 00:00, with forecasting steps of one hour. The uncertainty associated with each point forecast can easily be calculated using the primary output of the HISIMI model, that is, the discrete probability distribution associated with the electric power transition for each step in the forecasting horizon, *i.e.*, for each one of the hour periods from 00:00 to 23:00.

Figure 9a plots the real hourly PV power production values as well as the spot forecasted values of electric power from the HISIMI model, for a sunny day belonging to the testing data set. The vertical axis shows limits of the power intervals for the HISIMI model: the first interval corresponds to values between −157 and 157 kW (in Figure 9a only the positive half of the interval is represented); the second interval corresponds to power output values between 157 and 451 kW; and so on. Figure 9b gives the probability distributions, corresponding to the uncertainty of the point forecast, for central hours of the day (from 9:00 to 14:00). The horizontal axis of Figure 9b shows the power intervals, while the corresponding probability values are represented in the vertical axis. Thus, for example, the hour 9 (period between 9:00 and 10:00) presents two intervals with significant probabilities: the sixth and the seventh. Notice that for a solar hour containing only two consecutive power intervals with significant probabilities (above the value 0.1), the uncertainty is relatively low (with respect to solar hours containing three o more power intervals with significant probabilities), because the point forecast should correspond mainly to a weighted average value of the powers represented by both intervals. For the day represented in Figure 9a, the uncertainty about the spot forecasts of electric power is very low, because there are few power intervals with significant probabilities in each hour (only one or two consecutive power intervals).

Figure 10a plots the real hourly PV power production values and the forecasted ones of the HISIMI model for a partly cloudy day belonging to the testing data set. In this case, the forecasts differ from the actual values of power output, especially in hours from 7:00 to 10:00, and in hours from 12:00 to 15:00. Figure 10b shows the uncertainty associated with the spot forecasts for those hours: only for the hour from 12:00 to 13:00 there are two power intervals with significant probabilities. The other five hours present at least three power intervals with significant probabilities.

Figure 11a plots the values of the real hourly PV power production and the forecasted ones of the HISIMI model for a rainy day (most of the hours with rain) belonging to the testing data set. Spot forecasts present significant errors. Figure 11b shows the uncertainty associated with the spot forecasts for the hours between 9:00 and 14:00. All hours in the represented period have at least three power intervals with significant probability. For example, there are five power intervals with probability over 10% for hour 9:00; this also occurs for the four selected hours between 9:00 and 13:00; and four power intervals can be identified in the selected last hour (from 13:00 to 14:00). The uncertainty (associated with the spot forecasts) provided by the HISIMI model is notably high for the day represented in Figure 11a.

Figure 12a shows the values of forecasted and real hourly power output for a cloudy and rainy day (cloudy day with showers). In that figure the spot forecasts obtained with HISIMI model are represented, as well as the spot forecasts obtained with the MLP model. Both models provide quite similar spot forecasts, with significant errors with respect to the actual power output value. Figure 12b represents the uncertainty prediction for six central hours of that day, obtained with HISIMI model. There are at least three power intervals with significant probability for six hours represented in the figure, denoting a high uncertainty. So, although the HISIMI model and the MLP model provide very similar spot forecasts, the HISIMI also provides information to help in the evaluation of the risk assumed using those values of spot forecasts.

Analyzing the information produced by HISIMI model we highlight the differences between the results provided by our model and those offered by common short-term power point (spot) forecasting systems. With the model described in this paper, the user has access to more comprehensive information, including that related to predictions of uncertainty. The modeling of the uncertainty in the spot forecast, in the form of probability distributions, provide the possibility of analyzing the associated risk when the spot forecasted values are used to prepare energy sale offers for electricity markets.

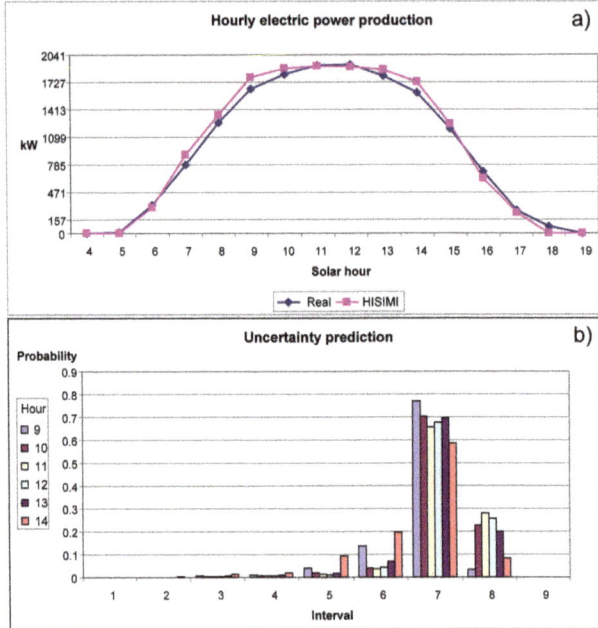

Figure 9. Forecasted hourly power production (**a**) and uncertainty prediction for six central hours; (**b**) (from 9:00 to 14:00) on a sunny day.

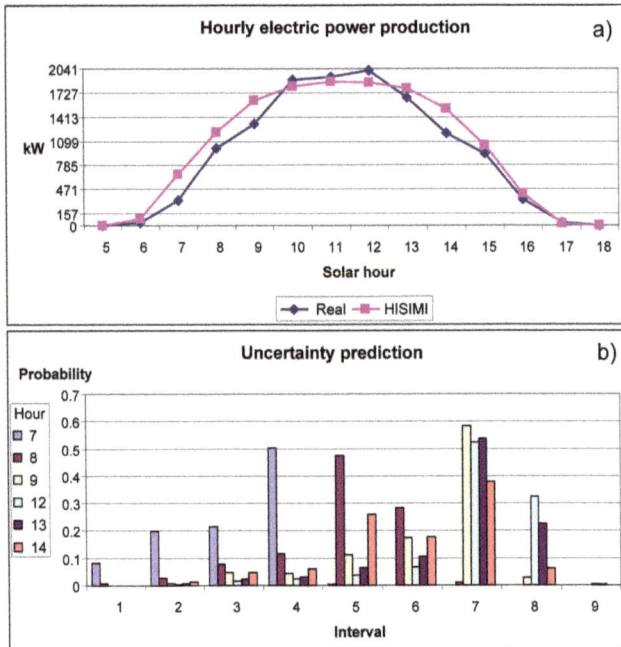

Figure 10. Forecasted production of hourly power (**a**) and uncertainty prediction for six hours; (**b**) on a partly cloudy day.

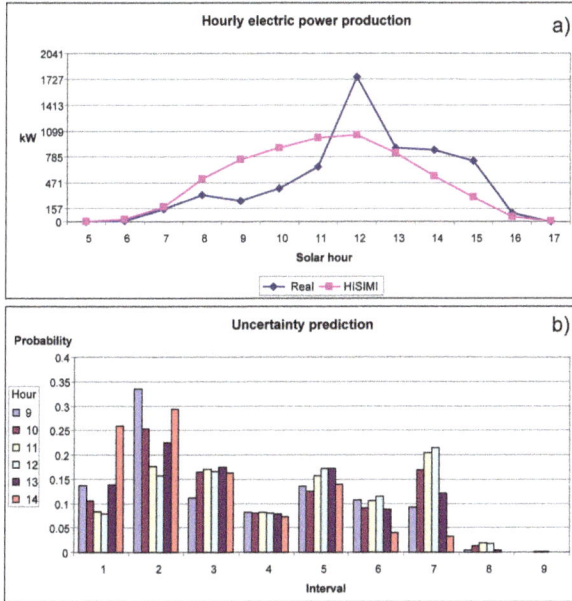

Figure 11. Forecasted hourly power production (**a**) and uncertainty prediction for six central hours; (**b**) (from 9:00 to 14:00) on a rainy day.

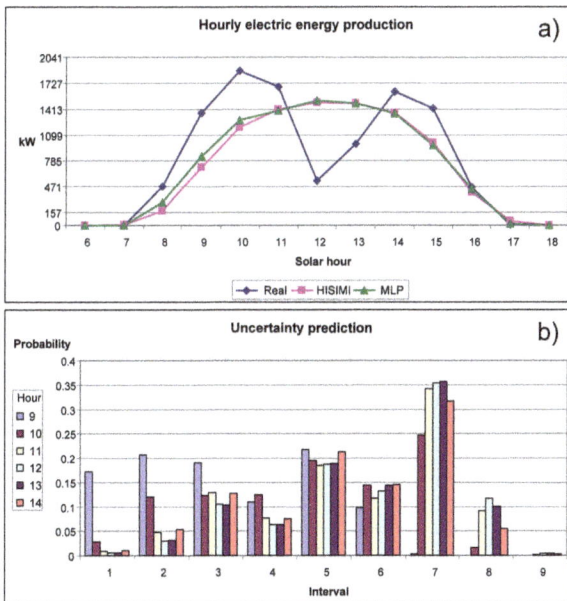

Figure 12. Forecasted hourly power production (**a**) and uncertainty prediction for six central hours; (**b**) (from 9:00 to 14:00) on a cloudy and rainy day.

6. Conclusions

A new short-term power forecasting model for PV plants, HISIMI, has been described in the paper. This model presents an innovative set of characteristics, which are:

- The HISIMI model allows a stochastic modeling based on similarity between input values in a database of historical cases. This similarity focuses mainly on variables forecasted by NWP tools.
- A database with a significant number of historical cases is used to model stochastic forecast, by creating discrete probability distribution functions.
- A genetic algorithm optimizes the structure of the HISIMI model, allowing the selection of the best inputs (variables) to be used by the model as well as the optimal values of basic parameters that define the model.
- The stochastic modeling of electric power transitions allows for the estimation of uncertainties in point (spot) forecasts of electric power. Uncertainty results obtained from HISIMI provide the probability distributions associated with the spot values.
- Spot forecasts are calculated using such discrete probability distributions.

Another useful characteristic of the HISIMI model is its ability to easily update the database used by this model. As soon as new past values are available, they can be included in the database. If window techniques are used, *i.e.*, the database is limited to the last period defined by a time window (for example, last six months), even the HISIMI model could be easily be adapted to a new NWP tool that would provide the prospection variables.

The forecasting model of this paper has been tested using real-life data which show high intra-hour variability of the power output of a photovoltaic plant. This model has improved spot forecasting results with respect to the ones from the persistence model and the MLP model (non-stochastic models).

The new model presented in this paper overcomes common short-term forecasting models: besides the value for spot forecasts, provided by conventional forecasting models, the proposed model achieves uncertainty values of the spot forecasts. This new forecasting result (*i.e.*, uncertainty) allows for integration of the HISIMI model into applications where there is a risk associated with forecasting errors. This risk is evident in electricity markets where forecasting errors, and the consequent deviations between values of real power generation and offered (forecasted) power, can lead to economic penalties: in the case of these penalties, the information regarding the uncertainty is very useful to advance an expected value of penalty, so the market agent can plan to cover the risk.

Acknowledgments: The authors would like to thank the "Ministerio de Ciencia e Innovación" of the Spanish Government for supporting this research under the Project ENE2009-14582-C02-02.

References

1. International Energy Agency. *Technology Roadmap: Wind Energy*; IEA Publications: Paris, France, 2010; Available online: http://www.iea.org/publications/freepublications/publication/Wind_Roadmap-1.pdf (accessed on 12 December 2012).
2. International Energy Agency. *Technology Roadmap: Solar Photovoltaic Energy*; IEA Publications: Paris, France, 2010; Available online: http://www.iea.org/publications/freepublications/publication/pv_roadmap-1.pdf (accessed on 12 December 2012).
3. Parsons, B.; Milligan, M.; Zavadil, B.; Brooks, D.; Kirby, B.; Dragoon, K.; Caldwell, J. Grid impacts of wind power: A summary of recent studies in the United States. *Wind Energy* **2005**, *7*, 87–108. [CrossRef]
4. Ortega-Vazquez, M.A.; Kirschen, D.S. Assessing the impact of wind power generation on operating costs. *IEEE T. Smart Grids* **2010**, *1*, 295–301. [CrossRef]
5. Angarita-Márquez, J.L.; Hernandez-Aramburo, C.A.; Usaola-Garcia, J. Analysis of a wind farm's revenue in the British and Spanish markets. *Energy Policy* **2007**, *35*, 5051–5059. [CrossRef]
6. Hocaoglu, F.O.; Gerek, O.N.; Kurban, M. Hourly solar radiation forecasting using optimal coefficient 2-D linear filters and feed-forward neural networks. *Solar Energy* **2008**, *82*, 714–726. [CrossRef]

Energies **2013**, *6*, 2624–2643

7. Mellit, A.; Massi Pavan, A. A 24-h forecast of solar irradiance using artificial neural network: Application for performance prediction of a grid-connected PV plant at Trieste, Italy. *Sol. Energy* **2010**, *84*, 807–821. [CrossRef]

8. Kaplanis, S.; Kaplani, E. Stochastic prediction of hourly global solar radiation for Patra, Greece. *Appl. Energy* **2010**, *87*, 3748–3758. [CrossRef]

9. Wang, F.; Mi, Z.; Su, S.; Zhao, H. Short-term solar irradiance forecasting model based on artificial neural network using statistical feature parameters. *Energies* **2012**, *5*, 1355–1370. [CrossRef]

10. Mandal, P.; Madhira, S.T.S.; Haque, A.U.I.; Meng, J.; Pineda, R.L. Forecasting power output of solar photovoltaic system using wavelet transform and artificial intelligence techniques. *Procedia Comput. Sci.* **2012**, *12*, 332–337. [CrossRef]

11. Pedro, H.T.C.; Coimbra, C.F.M. Assessment of forecasting techniques for solar power production with no exogenous inputs. *Solar Energy* **2012**, *86*, 2017–2028. [CrossRef]

12. Yona, A.; Senjyu, T.; Saber, A.Y.; Funabashi, T.; Sekine, H.; Kim, C.-H. Application of Neural Network to One-day-ahead 24 hours Generating Power Forecasting for Photovoltaic System. In Proceedings of the International Conference on Intelligent Systems Applications to Power Systems, Kaohsiung, Taiwan, 5–8 November 2007; pp. 442–447.

13. Da Silva Fonseca, J.G.; Oozeki, T.; Takashima, T.; Koshimizu, G.; Uchida, Y.; Ogimoto, K. Use of support vector regression and numerically predicted cloudiness to forecast power output of a photovoltaic power plant in Kitakyushu, Japan. *Prog. Photovolt. Res. Appl.* **2012**, *20*, 874–882. [CrossRef]

14. Shi, J.; Lee, W.J.; Liu, Y.; Yang, Y.; Wang, P. Forecasting power output of photovoltaic systems based on weather classification and support vector machines. *IEEE Trans. Ind. Appl.* **2012**, *48*, 1064–1069. [CrossRef]

15. Fernandez-Jimenez, L.A.; Muñoz-Jimenez, A.; Falces, A.; Mendoza-Villena, M.; Garcia-Garrido, E.; Lara-Santillan, P.M.; Zorzano-Alba, E.; Zorzano-Santamaria, P.J. Short-term power forecasting system for photovoltaic plants. *Renew. Energy* **2012**, *44*, 311–317. [CrossRef]

16. Lorenz, E.; Heinemann, D.; Kurz, C. Local and regional photovoltaic power prediction for large scale grid integration: Assessment of a new algorithm for snow detection. *Prog. Photovolt. Res. Appl.* **2012**, *20*, 760–769. [CrossRef]

17. Krömer, P.; Prokop, L.; Snášel, V.; Mišák, S.; Platoš, J.; Abraham, A. Evolutionary Prediction of Photovoltaic Power Plant Energy Production. In Proceedings of International Conference on Genetic and Evolutionary Computation, Philadelphia, PA, USA, 7–11 July 2012; pp. 35–42.

18. Sivanandam, S.N.; Deepa, S.N. *Introduction to Genetic Algorithms*; Springer-Verlag: Berlin, Germany, 2008.

19. Arlot, S. A survey of cross-validation procedures for model selection. *Statist. Surv.* **2010**, *4*, 40–79. [CrossRef]

20. Janjic, Z.; Black, T.; Pyle, M.; Rogers, E.; Chuang, H.-Y.; DiMego, G. High Resolution Applications of the WRF NMM. In Extended Abstract of 21st Conference on Weather Analysis and Forecasting/17th Conference on Numerical Weather Prediction, Washington, DC, USA, 31 July–5 August 2005.

energies

MDPI

Article

Hybrid Predictive Models for Accurate Forecasting in PV Systems

Emanuele Ogliari, Francesco Grimaccia *, Sonia Leva and Marco Mussetta *

Department of Energy, Polytechnic University of Milan, Via La Masa 34, I-20156 Milano, Italy;
emanuele.ogliari@mail.polimi.it (E.O.); sonia.leva@polimi.it (S.L.)

* Authors to whom correspondence should be addressed; francesco.grimaccia@polimi.it (F.G.);
marco.mussetta@polimi.it (M.M.); Tel.: +39-02-2399-4407 (M.M.); Fax: +39-02-2399-8566 (M.M.).

Received: 9 January 2013; in revised form: 8 February 2013; Accepted: 26 February 2013; Published: 3 April 2013

Abstract: The accurate forecasting of energy production from renewable sources represents an important topic also looking at different national authorities that are starting to stimulate a greater responsibility towards plants using non-programmable renewables. In this paper the authors use advanced hybrid evolutionary techniques of computational intelligence applied to photovoltaic systems forecasting, analyzing the predictions obtained by comparing different definitions of the forecasting error.

Keywords: hybrid techniques; PV forecasting; artificial Intelligence; neural networks

---

1. Introduction

As a consequence of the high increase in the installed capacity of grid-connected PV plants in recent years, it is quite difficult to plan the growing amount of energy from renewable sources fed into the grid, which is up to now non-programmable. This factor cannot be ignored any longer in the management and control of the load in the transmission network and distribution.

Moreover, considering that the distribution networks are now becoming from passive to active and that production or consumption plants [1] are gradually becoming more and more important actors in the management of the global electrical system, it is necessary to wonder what is the reliability and accuracy level of the forecasting systems, with particular reference to wind and solar plants [2].

In this perspective, it is easy to understand how the reliability factor of the forecast becomes a key issue in the set of rules for the identification of incentives or penalty mechanisms, in particular in finding the best mix between programmable and non-programmable sources (as defined in [3]).

In this case, a different definition of prediction error potentially triggers a significant impact in the economy for the share of the daily energy actually produced every day in comparison with the relevant declared forecasting.

For the system management in the areas of regulation, as regards the operators of transmission networks, in addition to real-time and accurate detection of the power fed into the grid, a precise prediction of energy supply in the short and medium term is also of the utmost importance.

In recent years, the national European regulatory authorities (particularly in Italy [4]) began to define a number of law provisions aimed at improving the prediction of the input power from renewable sources that cannot be planned, increasing gradually producers' responsibility. In addition, several transmission networks managers will rely on the operators in the field of monitoring of photovoltaic systems for the prediction of energy generation from solar energy [5].

In this context the need for a one-day ahead forecasting of the energy production on an hourly basis, by means of soft computer techniques starting from weather forecast provided by meteorological service, can play a fundamental role and becomes extremely useful for optimal management of the energy system. For example the same problem with different approach has been studied in [6] where a MPC-based (Model Predictive Control) strategy is performed in real-time with accurate short-term PV power predictions.

Usually the complex nature of many practical problems involves an effective use of Artificial Neural Networks (ANNs) to solve them. ANNs are useful tools when it is necessary to understand the complex and nonlinear relationships among data, without any previous assumption concerning the nature of these correlations. The training is one of the most critical phase. In this step the weights of the neural connections have to be properly set in order to have an appropriate simulation of the performance of a PV plant. Recently, even in other application fields like for example traffic flows, hybrid evolutionary algorithms have been applied to obtain more appropriate parameter combination to achieve more accurate forecasting [7]. In this paper the parameters of a neural network is optimized in order to reach a good and accurate output using a different kind of hybrid technique.

The ANN learning process should result in finding the weights configuration associated to the minimum output error, namely the optimized weights configuration. Usually problems are associated to an objective function to be optimized. Thus this function, called also *"fitness"*, cost or energy function, provides the interface between the physical problems and the optimization algorithm itself. The huge number of variables is the first difficulty when dealing with one of these optimizing issues. Secondly, there are lots of configurations with different values of the objective function that are quite similar each other and very close to the global optimum case, even if these configurations are sub-optimal. Generally finding a solution in an optimization process means to reach a balance among different and often conflicting goals; as a consequence such a search could be extremely difficult.

Among the various renewable energy sources, this study refers specifically to photovoltaic plants, without precluding the applicability of the proposed methods to other energy sources.

In this context the paper introduce a specific hybrid evolutionary algorithm to artificial neural networks in order to speed up the convergence when applied to ANN training phase and reduce the overall error in PV plant production forecasting applications. ANN and its training with combination of other computational intelligence (CI) techniques are nowadays very well established, nevertheless the paper does not aim to present a pure theoretical contribution, but introduces a novel application in PV power forecasting to be potentially used by power plant management operators and institutions. The whole forecasting flow is shown in Figure 1. The next sections describe in detail implementation of such technique showing the advantage in integrating evolutionary algorithms (EAs) in ANN models.

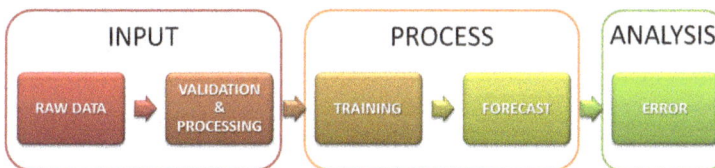

Figure 1. Forecasting flow process.

2. Hybrid Evolutionary Techniques Combined with ANN

Error Back Propagation (EBP) algorithm is a well-known analytical algorithm used for neural networks training. In literature, there are several forms of back-propagation, all of them requiring different levels of computational efforts; the conventional back-propagation method is, however, the one based on the gradient descent algorithm. The strong dependence upon the starting hypothesis that severely affects the result is one of the drawbacks of this method. A bad choice of the starting point may result in the possibility to get stuck in a local minimum and consequently to find a solution that is not the best one. Besides, most of the typical requiring optimization problems often have non-differentiable or/and discontinuous regions in the solution domain therefore some difficulties interfere in the application of these traditional methods based on derivatives calculations. These aspects are often overcome by evolutionary methods. The most effective evolutionary algorithm developed until now is Genetic Algorithm (GA), which is now quite familiar to the engineering community and widely used ([8,9] and references therein). Genetic algorithms are very efficient at exploring the entire search space, but are relatively poor in finding the precise local optimal solution in the convergence region. Some additional operators can be introduced for GA in order to get a better predictive power of ANNs selecting an optimal combination of input variables. Moreover, in recent years also the Particle Swarm Optimization (PSO) algorithm is gaining increasing attention for the integration in the training phase of ANNs [10,11].

Recently hybrid evolutionary techniques have been developed in order to combine the best properties of classical GA and PSO to overcome the problem of premature convergence. Some comparisons of the performances of them [12] emphasize the reliability and convergence speed of both methods, but still keep them separate. These procedures show a marked application driven characteristic for any respective technique: PSO seems to have faster convergence in the first runs, but often it is outperformed by GA for long simulations, when the last one finds better solutions. Some attempts to exploit the qualities of the two algorithms have been done in the last ten years with a kind of integration of the two strategies [13], but the authors aimed to reach a stronger co-operation of the two techniques stressing its hybrid nature and maintaining the GA and PSO integration for the entire run of the algorithm. Thus in the last years the authors have developed an innovative hybrid strategy called GSO, Genetical Swarm Optimization, which proved to improve traditional evolutionary mechanisms for a wide range of applications by means of an effective combination of natural selection and knowledge sharing. In particular, in [14], some comparisons of GSO and classical methods performances were presented, emphasizing the reliability and convergence speed of the first one and applying it to different case studies.

The basic concepts of GSO have been presented in [15]: in every iteration, the population is randomly divided into two parts that are evolved with GA and PSO techniques respectively. Then the fitness of the newly generated individuals is evaluated and they are recombined in the updated population, which is again divided into two parts in the next iteration for the next run of genetic or particle swarm operators. The population update concept can be easily understood thinking that a part of the individuals is substituted by new generated ones by means of GA, while the remaining are the same of the previous generation but moved on the solution space by PSO. The driving parameter of GSO algorithm is the so called hybridization coefficient (hc); it expresses the percentage of population that in each iteration is evolved with GA with respect of PSO technique. GSO has been tested on problem of different dimensions: while for a small number of unknowns GSO performance is similar to GA and PSO ones, if the size of the problem increases, GSO behavior improves and outperforms GA and PSO during iterations. Moreover, the best hc value found in that preliminary study does not depend on the dimension of the problem, as it has been reported also in [14]. Furthermore, the obtained best hc value between 0.2 and 0.3 means that for a big-sized problem, the basic PSO can be strongly improved by adding a small percentage of genetic operators on the population. In further studies a convenient value was found to be in the same range for several fitness functions, but the authors extended the class of GSO algorithms by considering several variation rules for hc, in order to explore different hybridization strategies for the GSO algorithm and to compare new approaches with others already present in literature. The full set of hybridization rules considered by the authors is also reported in [16].

In [15] the authors introduced new rules for varying the hc value during the run, to combine more efficiently the properties of GA and PSO, in order to have a general procedure. In fact for engineering optimization problems the best mix of GA and PSO operators cannot be known a priori. In particular there are situations where a fixed hc is the right choice, and others where a variable $hc(k)$ during the run is better. This means that also the "amount" of hybridization plays a role in affecting the performances of this procedure. Therefore the authors chose to let the procedure adjust the $hc(k)$ value by itself during the iterations, according to a predefined set of rules defining two different approached defined as dynamical and self-adapting, where the rule implemented comes in part from the very simple and reliable swarm techniques.

The overall results reported by the authors in cited papers show that, although the static GSO is generally the faster and more robust strategy in order to optimize multi-modal functions, a self-adaptive approach is a suitable and reliable solution especially when the proper *hc* value is not known for a specific problem. The overall results reported by the authors in cited papers show that GSO is a good candidate to be used in classical neural networks to replace training procedure as for example the common EBP (Error Back Propagation).

In this work a dynamic GSO was combined with a classical EBP in order to improve the speed of convergence of the neural network training phase and, at the same time, to improve the performance of the predictive system. In [17] the authors started to apply hybrid evolutionary learning algorithm to increase the accuracy of the daily forecast finding the best neural weights configuration. Here a similar mixed approach is used to optimize the neural weights in a more complex predictive model where the one-day ahead production estimation is performed on a hourly base [18].

Before showing how such a technique has been applied to a specific real case study, in the next section we will discuss some error definitions in order to identify the most appropriate formula that better describes the gap between declared and really produced energy in the context of future incentives or penalty mechanisms. (e.g., according to Italian regulation Authority [4]).

3. Error Definitions

The application of the technique described in the previous paragraph to the problem of PV production forecasting requires a proper and shared definition of the error estimation with the aim to assess the amount of the daily produced and declared energy. In order to correctly define the accuracy of the prediction and the relative error, it is necessary to analyze different definitions of error. The starting point reference is the hourly error e_h, defined as the difference between the average power produced in the hour $P_{m,h}$ and the given prediction $P_{p,h}$ [4] provided by the neural model:

$$e_h = P_{m,h} - P_{p,h} \ [\text{Wh}] \tag{1}$$

From this basic definition, other definitions can be introduced:

- **Absolute hourly error** $e_{h,abs}$, which is the absolute value of the previous definition (e_h can give both positive and negative values):

$$e_{h,abs} = |e_h| \ [\text{Wh}] \tag{2}$$

- **Daily error** e_d, given by the summation extended to 24 hours e_h time error:

$$e_d = \sum_h e_h \ [\text{Wh}] \tag{3}$$

- **Daily absolute error** $e_{d,abs}$, given by the summation extended to 24 hours of the Absolute Hourly Error:

$$e_{d,abs} = \sum_h |e_h| \ [\text{Wh}] \tag{4}$$

- **Time error percentage based on the rated power** of the photovoltaic (P_r):

$$e_{\%,r} = \frac{|e_h|}{P_r} \ [\%] \tag{5}$$

229

- **Time error percentage based on the hourly output expected power** $(P_{p,h})$:

$$e_{\%,p} = \frac{|e_h|}{P_{p,h}} \ [\%] \tag{6}$$

Following Italian regulation authority, the penalties concerning the transitional period of the year 2013 will be calculated on its basis.

- **Daily time error percentage**, based on the hourly output expected power $(P_{p,h})$:

$$e_{\%,d,p} = \sum_h \frac{|e_h|}{P_{p,h}} \ [\%] \tag{7}$$

Moreover some considerations on the accuracy of measurements related to the plant available instrumentation affecting the input raw datasets and dispersion evaluation on these data from the expected values should be considered in a validation phase before starting the ANN training process itself.

4. Case Study

After a trial campaign, the network architecture that has provided better results presents two hidden layers. In particular, the number of neurons in the input layer is 7, as shown in Figure 2, which describes the meteorological parameters provided from the weather forecast service.

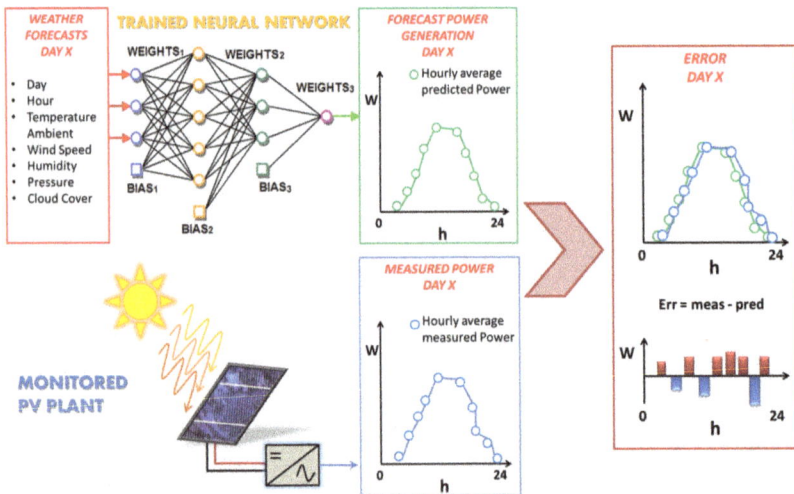

Figure 2. Simplified view of the implemented feed-forward ANN with details on input, output, and hidden layers.

This network structure is less complex than the one proposed by another comparative study on power forecasting methods in PV plants [18], where the authors used neural networks with hidden layers and a number of neurons in a higher range (from 11 to 15); in our work in fact we use two hidden layers with, respectively, 9 and 7 neurons, while the output layer is composed by 1 neuron. The authors, in fact, performed preliminary simulations for different photovoltaic power plants in order to test their method at various scale, with different productive capacities, to make this procedure more general [19].

Different time horizons can be considered in power forecasting: short term (day and several days), medium term (week and several weeks), long term (year and several years), and forecasts for different lead time can be used for different aims. Here, we chose a very short term time base for this specific application, giving one-day ahead forecasting on an hourly basis in accordance with [20], since this is the typical resolution requested by power plant operators. In this work the neural network optimized by GSO exhibit good predictive performances in all the operative conditions, in a complete sunny day, a partly cloudy one and even a plant maintenance day. Here a combination of GSO and EBP was used for the hybrid learning process of the artificial neural network, as described in [19], and weights values of the ANN were changed to reach the minimum error in the network output in a faster way compared with standard EBP alone. In literature other evolutionary procedures as standard PSO and GA were compared with the classical EBP in order to perform a comparison in terms of convergence rate and final obtained result (e.g., [21]), but GSO has already proven to outperform both GA and PSO [14].

In particular, here the GSO-based training phase is first conducted for 9000 iterations (with a population of 50 individuals) over a period of one year, to process a global search of weights values; the EBP training is then used for additional 5000 iterations to refine the optimal weights configuration. To show the effectiveness of this hybrid approach, results are then compared with those obtained by running a standard EBP learning for a comparable amount of computational time, *i.e.*, for 500,000 iterations.

5. Data Validation and Training

The neural network has to be calibrated with previously collected data on the energy production, along with the weather forecasts, for a sufficiently long time (a full calendar year would be an appropriate reference). On the basis of the actual measurements, the error is calculated after applying different criteria for defining it as discussed in the next section. The previously described method has been applied to the production forecast of a PV test plant managed by the Department of Energy of Politecnico di Milano.

The considered input parameters are the following physical quantities provided by the weather forecast service:

- environmental temperature (°C);
- atmospheric pressure (hPa);
- wind speed (m/s);
- humidity (%);
- cloud coverage (%).

In particular, since we cannot have a forecast of the temperature for each specific module in a plant, we take into account factors that are correlated with that value, *i.e.*, the environmental temperature, wind speed and humidity.

The average hourly power ($P_{p,h}$) forecast in the "h" hour is then calculated as the output value for the next day.

This value is compared with the following meteorological and physical quantities:

- $P_{p,h}$ is the average hourly power produced by the PV plant in the "h" hour (W);
- GHI is the Global Irradiance on the horizontal plan (W/m^2);
- GTI is the Global Irradiance on the tilted plan (W/m^2) (as defined in [22]);
- $cvg\%$ is the percentage of the cloud coverage (%) (as defined in [23]).

The irradiance data are compared with the theoretical values on the tilted plan *Irr,Th.Tilt* assessed by a deterministic algorithm on the basis of the geographical coordinates with the aim to validate the forecast data. The input data have been validated in order to verify their true significance or to provide the proper training to the network. For example all irradiation samples by night were omitted to exclude a high rate of forecasting elements that could highly affect the resulting error (irradiance forecasts during the night are zero). All the missing data have been excluded not contributing to the forecast. Starting from the comparison between the actual produced power and the predicted one, it is calculated for each day:

- the hourly error e_h;
- the error percentage, based on the rated power $e_{\%,r}$;
- the error percentage on the hourly power forecast $e_{\%,p}$;
- the daily error percentage based on the hourly power forecast $e_{\%,d,p}$.

6. Results and Discussion

The described forecasting technique has been applied over a one year production period. Two days are displayed hereunder, the first one showing good weather conditions (Figure 3), the second day showing bad ones (Figure 4). Table 1 reports detailed energy production and error calculation for the two sample days.

Figures 5 and 6 respectively report the comparison between the forecast results and errors obtained by the hybrid training (GSO + EBP) and the standard EBP alone, showing an improvement of performances using the hybrid approach for the same computational time. In particular, these results are summarized in Table 2, where the two training approaches are compared considering some of the error definitions previously introduced in Section 3.

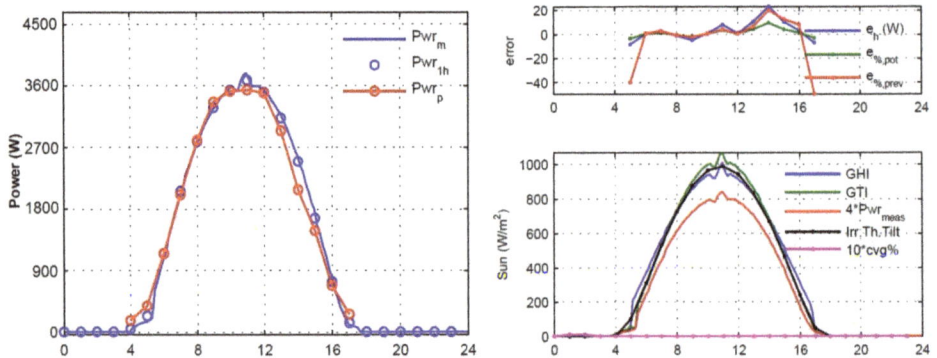

Figure 3. Daily detail of the curves of radiation (theoretical, forecast and actual) with the calculation of the error values during a clear sky day.

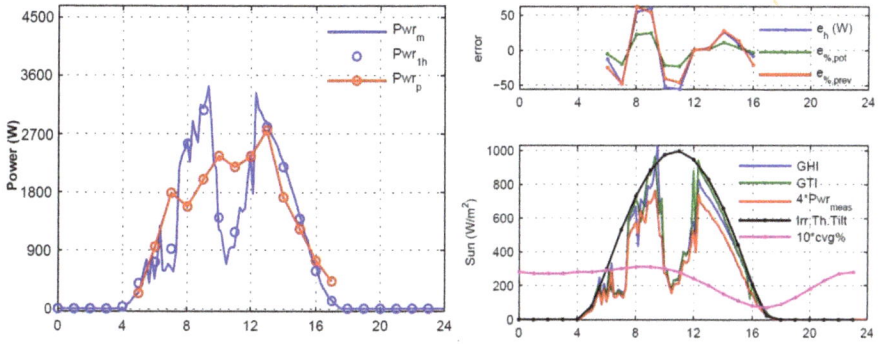

Figure 4. Daily detail of the curves of radiation (theoretical, forecast and actual) with the calculation of the error values during day with bad weather conditions.

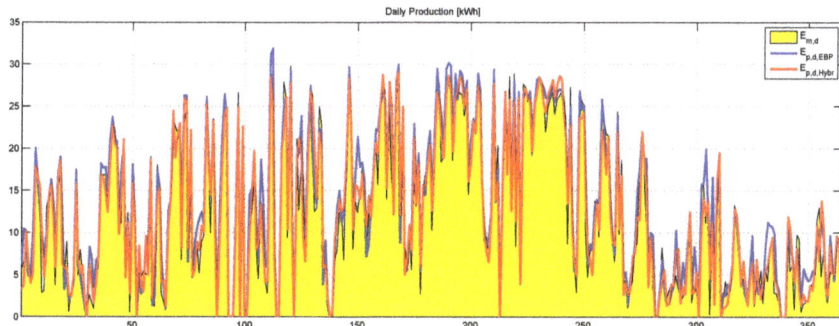

Figure 5. Daily produced and predicted energy comparison.

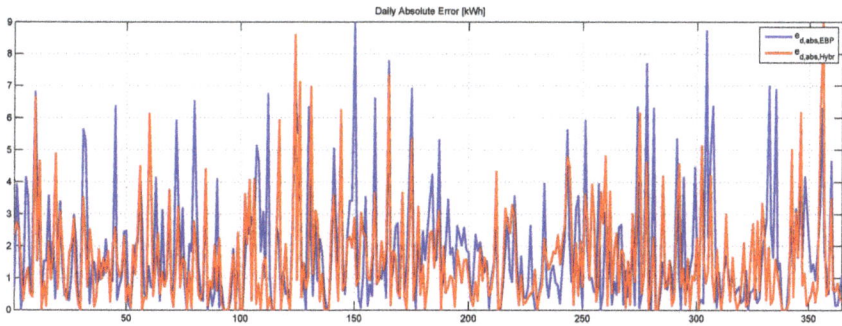

Figure 6. Daily absolute error $e_{d,abs}$ comparison.

Table 1. Production data and error calculation for two-day examples.

Physical quantities	Clear sky day	Cloudy day
Daily Production	28.37 kWh	19.16 kWh
Daily Forecast	27.68 kWh	19.65 kWh
Daily Error e_d	684 kWh	−491 kWh
Daily Error percentage on the base of the daily production	2.41%	−2.56%
Daily Error absolute $e_{d,abs}$	2.29 kWh	5.96 kWh
Daily Error percentage on the base of hourly forecast power $e_{\%,d,p}$	8.08%	30.31%

Table 2. Main production data and error calculation with reference to the entire period.

Forecast error definition		EBP	Hybrid (GSO + EBP)	
Nr. Samples		2879	2879	h
Average Hourly Abs. Error	$avg\,(e_{h,abs})$	0.338	0.317	kW
Yearly Total:				
Overall Production	$E_m = \int_{year} P_{m,h}$	4512.3	4512.3	kWh
Overall Forecast	$E_p = \int_{year} P_{p,h}$	4852.6	4562.4	kWh
Overall Yearly Error	$E_m - E_p$	−340.3	−50.12	kWh
Error abs. tot	$\sum e_{d,abs}$	973.2	912.98	kWh
Error abs.% tot	$\frac{\sum e_{d,abs}}{E_p} \cdot 100$	20.06%	20.01%	
Daily Averages:				
Average Daily Production	$E_m / 365$	12.36	12.36	kWh
Average Daily Forecasting	$E_p / 365$	13.29	12.50	kWh
Average Daily Error	$(E_m - E_p)/365$	−0.932	−0.137	kWh
Average Daily Abs. Error	$avg\,(e_{d,abs})$	2.666	2.501	kWh

The developed model allows an accurate prediction generally for clear days as well as for permanently covered days. However, as shown in Figure 4, there was a lower precision in the days characterized by a strong and rapid variability. In fact, since data provided by the weather service are on hourly basis, we lose information that is particularly important in days with highly variable cloudy conditions. In particular, this problem could be also due to the slowness of $cvg\%$ forecast with respect to the actual fluctuations of the other meteorological parameters: probably this index is averaged over several hours and, therefore, it is unable to represent accurately the real variability of the actual cloudy coverage.

Additional considerations can also be made with regard to the formulation of the forecasting error. For example the error percentage on the hourly power forecast is always high in correspondence of values of low solar radiation. This event occurs both during cloudy days and during sunrise and sunset of any day. As in these periods of time the power generation is relatively small, the forecasting error should not be counted in the same way as the forecasting error calculated during the hours with high level of solar radiation. In these conditions, the authors suggest a threshold of solar radiation under which the data have not to be considered. Besides, in this error definition, the measured power data should be adopted instead of the forecast power ones.

7. Conclusions

Due to the increase of renewable energy penetration in the electric grid, it is quite important to estimate the amount of energy from such non-programmable sources.

In this paper a novel hybrid evolutionary approach is used for training artificial neural network in order to achieve more accurate forecasting of photovoltaic systems based on weather forecast as input data. Moreover, analyzing all test results obtained by comparing different definitions suggested for the forecasting error, a sensible reduction of the error itself can be achieved by increasing the time range of observation and of course the quality and resolution of the data provided from the local weather forecast service.

The developed model allows both better predictions and potential novel applications in PV power plant management operations.

References

1. Brenna, M.; Dolara, A.; Foiadelli, F.; Lazaroiu, G.C.; Leva, S. Transient analysis of large scale PV systems with floating DC section. *Energies* **2012**, *10*, 3736–3752. [CrossRef]
2. Zhang, Q.; Lai, K.K.; Niu, D.; Wang, Q.; Zhang, X. A fuzzy group forecasting model based on least squares support vector machine (LS-SVM) for short-term wind power. *Energies* **2012**, *9*, 3329–3346. [CrossRef]
3. EnergyLab Foundation. Renewable Energies: State of the Art–Executive Summary. Available online: http://www.energylabfoundation.org (accessed on 26 February 2013).
4. Autorità per l'energia Elettrica e il Gas. *Revisione del Servizio di Dispacciamento DellEnergia Elettrica per le Unità di Produzione di Energia Elettrica Alimentate da Fonti Rinnovabili non Programmabili (in Italian);* DCO 281/2012/R/efr; The Regulatory Authority for Electricity and Gas: Milano, Italy, 2012.
5. Meteocontrol—Energy & Weather Services. Available online: http://www.meteocontrol.de/wir-ueber-uns/aktuelles/detailansicht/article/neuer-massstab-fuer-solarstromprognosen/ (accessed on 24 October 2012).
6. Pérez, E.; Beltran, H.; Aparicio, N.; Rodriguez, P. Predictive power control for PV plants with energy storage. *IEEE Trans. Sustain. Energy* **2012**, *99*, 1–9. [CrossRef]
7. Hung, W.-M.; Hong, W.-C.; Chen, T.-B. Application of Hybrid Genetic Algorithm and Simulated Annealing in a SVR Traffic Flow Forecasting Model. In Proceedings of the 2009 IEEE Congress on Evolutionary Computation (CEC 2009), Trondheim, Norway, 18–21 May, 2009; pp. 728–735.
8. Rahmat-Samii, Y.; Michielssen, E. *Electromagnetic Optimization by Genetic Algorithms*; Wiley: New York, NY, USA, 1999.
9. Papakostas, G.A.; Boutalis, Y.S.; Samartzidis, S.; Karras, D.A.; Mertzios, B.G. Two-stage hybrid tuning algorithm for training neural networks in image vision applications. *Int. J. Signal Imaging Syst. Eng.* **2008**, *1*, 58–67. [CrossRef]
10. Grimaldi, E.; Grimaccia, F.; Mussetta, M.; Zich, R.E. PSO as an Effective Learning Algorithm for Neural Network Applications. In Proceedings of the 3rd International Conference on Computational Electromagnetics and Its Applications ICCEA, Beijing, China, 1–4 November 2004; pp. 557–560.
11. Niu, Q.; Zhou, Z.; Zhang, H.-Y.; Deng, J. An improved quantum-behaved particle swarm optimization method for economic dispatch problems with multiple fuel options and valve-points effects. *Energies* **2012**, *9*, 3655–3673. [CrossRef]
12. Boeringer, D.W.; Werner, D.H. Particle swarm optimization *versus* genetic algorithms for phased array synthesis. *IEEE Trans. Antennas Propag.* **2004**, *52*, 771–779. [CrossRef]
13. Juang, C.-F. A hybrid of genetic algorithm and particle swarm optimization for recurrent network design. *IEEE Trans. Syst. Man. Cybern.-Part B Cybern.* **2004**, *34*, 997–1006. [CrossRef]
14. Gandelli, A.; Grimaccia, F.; Mussetta, M.; Pirinoli, P.; Zich, R.E. Development and Validation of Different Hybridization Strategies between GA and PSO. In Proceedings of the 2007 IEEE Congress on Evolutionary Computation, Singapore, 25–28 September 2007; pp. 2782–2787.
15. Grimaccia, F.; Mussetta, M.; Zich, R.E. Genetical swarm optimization: Self-adaptive hybrid evolutionary algorithm for electromagnetics. *IEEE Trans. Antennas Propag.* **2007**, *55*, 781–785. [CrossRef]
16. Gandelli, A.; Grimaccia, F.; Mussetta, M.; Pirinoli, P.; Zich, R.E. Genetical swarm optimization: An evolutionary algorithm for antenna design. *J. Autom.* **2006**, *47*, 105–112.
17. Caputo, D.; Grimaccia, F.; Mussetta, M.; Zich, R.E. Photovoltaic Plants Predictive Model by Means of ANN Trained by a Hybrid Evolutionary Algorithm. In Proceedings of the 2010 International Joint Conference on Neural Networks (IJCNN), Barcelona, Spain, 18–23 July 2010.
18. Huang, Y.; Lu, J.; Liu, C.; Xu, X.; Wang, W.; Zhou, X. Comparative Study of Power Forecasting Methods for PV Stations. In Proceedings of the International Conference on Power System Technology, Hangzhou, China, 24–28 October 2010; pp. 1–6.

19. Simonov, M.; Mussetta, M.; Grimaccia, F.; Leva, S.; Zich, R.E. Artificial intelligence forecast of PV plant production for integration in smart energy systems. *Int. Rev. Electr. Eng.* **2012**, *7*, 3454–3460.

20. Wang, F.; Mi, Z.; Su, S.; Zhang, C. A Practical Model for Single-Step Power Prediction of Grid-Connected PV Plant Using Artificial Neural Network. In Proceedings of the IEEE PES Innovative Smart Grid Technologies Asia (ISGT), Perth, Australia, 13–16 November 2011; 2011; pp. 1–4.

21. Gudise, V.G.; Venayagamoorthy, G.K. Comparison of Particle Swarm Optimization and Backpropagation as Training Algorithms for Neural Networks. In Proceedings of the 2003 IEEE Swarm Intelligence Symposium, SIS '03, Indianapolis, IN, USA, 24–26 April 2003; pp. 110–117.

22. Dolara, A.; Grimaccia, F.; Leva, S.; Mussetta, M.; Faranda, R.; Gualdoni, M. Performance analysis of a single-axis tracking PV system. *IEEE J. Photovolt.* **2012**, *2*, 524–531. [CrossRef]

23. Arking, A.; Jeffrey, D.C. Retrieval of cloud cover parameters from multispectral satellite images. *J. Clim. Appl. Meteor.* **1985**, *24*, 322–333. [CrossRef]

MDPI

St. Alban-Anlage 66

4052 Basel

Switzerland

Tel. +41 61 683 77 34

Fax +41 61 302 89 18

www.mdpi.com

Energies Editorial Office

E-mail: energies@mdpi.com

www.mdpi.com/journal/energies

www.ingramcontent.com/pod-product-compliance
Lightning Source LLC
Chambersburg PA
CBHW051728210326
41597CB00032B/5647